PLANNING FOR FOREST RESOURCES AND BIO-DIVERSITY MANAGEMENT

About the author

Kailash Chandra Bebarta, a member of Indian Forest Service, is working as a practicing forester in Forest Department of Chattisgarh. He has post-graduate diploma in forestry (1983) and business administration (1993). He acquired specialized interests in natural resource management from various different assignments in forestry organizations. He has published many papers and a book entitled "Teak: Ecology, Silviculture, Management and Profitability". He has held many key forestry positions in forest department and Forest Development Corporation of Madhya Pradesh. His last substantive assignment being Working Plan Officer for Rajnandgaon forest division (Chattisgarh) provided the necessary impetus, insights and skills to evolve sustainable forest management model for the said division. His current area of research is sustainable forest management at macro- and micro levels.

PLANNING FOR FOREST RESOURCES AND BIO-DIVERSITY MANAGEMENT

Principles, Organization and Methodology

KAILASH CHANDRA BEBARTA I.F.S.

CONCEPT PUBLISHING COMPANY, NEW DELHI-110059

ISBN 81-7022-953-7

First Published 2002

Published and Printed by
Ashok Kumar Mittal
Concept Publishing Company
A/15-16, Commercial Block, Mohan Garden,
NEW DELHI - 110059 (India)

Phones : 5648039, 5649024
Fax : 091-11-5648053
Email : publishing@conceptpub.com

IN MEMORY

OF

MY FATHER

Preface

Forest working plan has been the basis of forest management practices since the last leg of twentieth centuary when India heralded into an age of scientific management in forestry. Over the years, India's forestry has travelled a long distance through the rich tradition of managing the forest resources according to the working plan. The National Forest Policy, 1988 has reiterated the need of having a working plan for management of government forests. Again, the importance of management plan in the field of forestry has been re-emphasized by few epoch making decisions of honorable Supreme Court of India.

There is no comprehensive book dwelling on fundamental principles and methodology of preparing a forest management plan. Generally, the knowledge and expertise about preparing a plan is passed on to newer generation of foresters by the older ones. This system of linkages runs the risks of breakages at any point of time. Moreover, in the absence of systematic guidelines and continued practices, there is every possibility that the knowledge and skill may be diluted with the passage of time. Apart from this, over the years, certain fundamental principles of forest management have undergone radical change. New concepts and methodology have emerged in the horizon of forest management practices. Forestry principles and practices are being redefined on the basis of emerging ethos of sustainability. Therefore, considered in these perspectives, the author felt the need of writing a book elucidating principles, rationale and methodologies that go into the making of a working plan for a forest division.

Preparing working plan for a forest division necessitates a coordinated application of principles of forestry, management and social sciences. It is a complex job, which needs synchronization and optimization of management objectives. Therefore, formulating a suitable management plan for a forest division has remained a challenge in the history of scientific forestry. Again, in the absence of any comprehensive book on the subject, the task becomes more arduous and intractable. Therefore, the author attempted to provide a book encompassing all relevant and substantive materials required to prepare a forest management plan. This will also go a long way in demystifying the process of working plan preparation among the young foresters who, after a couple of years of service, are required to draw a management plan for a forest division. In other words, the book aims at preparing young foresters for the exciting and challenging job of managing the forests sustainably in the coming millennium.

Principles and methodology about preparing forest working plan have been organized in five parts. The first part introduces the readers to the forest resource base of India and its history of past management. The second part covers the basic principles of forest planning and introduces the readers into features of different management plans. The third part discusses about the principles and methodology regarding different field works carried out in the course of planning process. The fourth part focuses on principles of data collection, classification and presentation. It also covers the organization of working plan reports at different stages and elements of technical writing. The fifth part deals with new tools of planning which are currently used in modern practices of forest management.

Many senior and junior colleagues have contributed their ideas and suggestions to the author during the course of writing this book. Besides, the association with the formulation of working plan of Rajnandgaon forest division of Chattisgarh, provided the author an immense opportunity to sharpen the knowledge and skills about forest management systems and practices. Consequently, the continued process of association and involvement in the making of working plan and the urge to document all relevant principles and methodology took the shape of this comprehensive book.

KAILASH CHANDRA BEBARTA

Raipur
Chattisgarh

Contents

PART - I

INTRODUCTION TO FOREST RESOURCES AND THEIR SILVICULTURAL MANAGEMENT

1
The Forest Resource Base

Introduction

Land is the basic life support system. It supports all forms of plant and animal lives. The important land-uses are: (i) agriculture, (ii) forest, and (iii) grassland pasture. Most of the flat and gently sloping lands are under agricultural practices. A large area is still under natural vegetation in different parts of the world. It consists of trees, shrubs, herbs and grasses. The area occupied predominantly by trees and other woody vegetation is called forest. The object of managing these areas is primarily to protect the environment and derive tangible benefits like wood and minor forest produces. Areas, which do not support trees of higher density and suitable for grazing are called grasslands or pasture lands.

Floristic Regions of India

Vegetation of India has wide physical as well as ecological diversity. The recognized floristic regions of country are as follows:

1. Western Himalayas.
2. Eastern Himalayas.
3. Indus Plain.
4. Gangetic Plain.
5. West Coast (Malabar).
6. Decan Plateau.
7. North East India (Assam).
8. Andaman and Nicobar Islands.

The western Himalayan region extends from Kashmir to Kumaon. Its vegetation resembles that of the Temperate Zone in Europe and America. The western Himalayas has coniferous and few broad-leaved species. In the lower up, forests of Chir pine, other conifers and broad leaved temperate trees are found. In the higher region, forests of deodar, blue pine, spruce and silver fir occur. The alpine zone extends from the upper limit of the temperate zone of about 4750 meters or even higher. The characteristic trees of this zone are high level silver fir, silver birch and junipers.

The Eastern Himalayas extends from Sikkim eastwards and embraces Darjeeling, Kurseong and the adjacent tract. The temperate zone has forests of *oaks*, *lurels, maples*, rhododendrons, *alder* and *birch*. Many *conifers*, *junipers* and *dwarf* willows also occur.

The Indus plain comprises part of Punjab, Rajasthan and Gujarat. They constitute the dry and hot area and have vegetation like Acacia, Prosopis and Anogeissus. Gangetic plain covers Uttar Pradesh, Bihar, Orissa and West Bengal. In most of the regions northward of Arravalli ranges agriculture is the major land use. However, this region consists of Sal as the dominant species in the forest vegetation. The major associates of Sal are Terminalia and Euginia.

The Central region constitutes the Vindhyan hills ranging an elevation from 500 metres to 650 metres. Teak is the predominant species of this region. Main associates of teak are Terminalia, Anogeissus and Lagerstroemia.

West Coast or the Mallabar region extends from Cape Comorin to the southern limits of Gujarat. It covers excessively humid belt of mountain country parallel to the West Coast of peninsula. The distinctive feature of this region is its heavy rainfall. This sustains three distinct types of forests, *viz.*, moist evergreen, moist deciduous and subtropical broad-leaved hill forests. The evergreen forests occur in some parts of Karnataka and Kerala. It is composed of *Cedrela toona, Dipterocarpus indica, Dyoxylum malabaricum, Artocarpus hirsuta, Mesua ferrea* and *Hopea parviflora*. The moist deciduous forests comprise *Tectona grandis, Terminalia tomentosa, Terminalia peniculata, Dalbergia latifolia, Lagerstromia lanceolata* and *Xylia xylocarpa*. It occurs in Kerala, Tamil Nadu, Karnataka and Maharashtra. The sub-tropical broad-leaved hill forests consist of *Eurya Japonica, Gordenia, Obtusa* and *Michelia nilegarica*. These occur in Palni, Annamalai and Nilgiri districts of Tamil Nadu.

The Deccan region constitutes a unique geographic entity. It encompasses the major parts of peninsular India. Teak is the dominant species of this region. The common associates of teak are

Hardwickia, *Anogeisus*, *Boswelia* and *Dyospyros*. The major teak forests of India are found in this region. The Deccan Peninsula is rich in endemic flora with a varied number of plants which are not found elsewhere.

North-East region or Assam region comprises hills and valleys, which are laden with evergreen forests, occasional thick clumps of bamboo and tall grasses. On account of heavy rainfall, the vegetation is rich and has luxuriant growth. This region comprises of *Artocarpus chapalasa, Michelia champaca, Ficus elastica, Mesua ferrea*, and *Sterculia alata*.

Andaman and Nicobar Islands constitute a different region on account of its geographic position. It receives heavy monsoon rainfall. It has rich and varied vegetation ranging from evergreen in the center to mangrove in the coastal areas. *Dipterocarpus, Terminalia* and *Lagerostroemia* are the dominant species found in these forests. India is abound with almost all kinds of diversity of vegetation. A large variety of vegetation forms are found in the country. Out of about 2 lakh known species of plants in the world, about 20,000 occur in India, which indicates the richness of flora in the country. Maximum part of the Indian subcontinent falls in the biotic region of tropical deciduous forest and tropical scrub forest. Tropical rain forests occupy a narrow belt along the West Coast, north-east India and in Andaman and Nicobar Islands. The xerophytic form of vegetation is found in Rajasthan and adjoining areas. Sub-tropical, temperate and alpine forms of vegetation are found in Himalayas and other hill ranges.

Forest — The Definition

The ward "Forest" is derived from Latin word '*foris*' meaning outside the village boundary or away from inhabited land. Forest is a large tract covered with trees and undergrowth sometime mixed with pastures. Generally, forest is referred to a place occupied by different kinds of trees, shrubs, herbs and grasses maintained for the production of wood and non-wood products. A forest is a large uncultivated tract of land covered with trees and underwood, woody growth and pasture. Technically, a forest is an area set aside for the production of timber and other forest produce, or maintained under woody vegetation for certain indirect benefit, which it provides, *e.g.* climate or protective (Dwivedi, 1980). Ecologically, it is defined as a plant community, predominantly of and other woody vegetation, usually with a closed canopy. Legally, it is an area of land proclaimed to be a forest under forest laws. FAO classified all such lands into forest which bear vegetative association dominated by trees of different sizes, exploited or not, capable of producing wood or other forest produce, exerting an influence on the climate or water regime or providing shelter for livestock and vagital for wildlife.

From the above definitions, it is clear that normally a forest has five components *viz*., (i) it is an uncultivated land, (ii) it is occupied by different kinds of natural vegetation predominantly by trees; or it is proposed to establish trees and other woody vegetation, (iii) the trees forming a closed or a partially closed canopy, (iv) the trees and other forest vegetation are managed for obtaining wood or other forest produces and/or benefits, and (v) the area presents a rich bio-diversity and provides shelter to wildlife, birds and other fauna.

Classification of Forests

Forestry is defined as the theory and practice of all that constitutes the creation, conservation and scientific management of forest and the utilization of their resources (Anon., 1966). It includes all thinking and all actions pertaining to creation, conservation, protection and management of forests, including harvesting, marketing and utilization of all forest products and services. It includes not only management of existing forests but also the creation of new forests. Forestry practices are evolved on the basis of objectives enunciated in National or State forest policy.

Forests can be classified on the basis of: (i) age, (ii) method of regeneration, (iii) composition, (iv) object of management, (v) ownership and (vi) growing stock. On the basis of age, forests are classified into (a) even-aged forest, and (b) uneven-aged forests. Even-aged forests are also called as regular forests. These forests consists of trees of approximately the same age. True even-aged forests can only be man-made forests. The forest is called uneven-aged or irregular when trees vary widely in age.

On the basis of regeneration, forests are categorized into: (a) high forests, when regeneration is obtained from seed, (b) coppice forests, when regeneration is obtained through coppice or some other vegetative parts of the tree, and (c) man-made forests when regeneration is obtained by artificial means. When the regeneration is obtained naturally, *viz*. by seeds or coppice, the forests are called natural forests and when it is obtained artificially, the forests are called man-made forests or plantations.

On the basis of composition, forests are classified into: (i) pure forests, and (ii) mixed forests. Pure forests are composed almost entirely of one species, usually to the extent of not less than 50 per cent. Mixed forests are defined as forest composed of trees of two or more species intermingled in the same canopy. The species compositon the mixture is usually distinguished on the basis of economic importance . Such composition consists of: (i) principal, (ii) accessory, and (iii) auxiliary species. Principal species are the species in the mixed forest, first in order of importance either by frequency, volume or silvicultural value. Accessory species are useful but of less value than the principal species, which assist in the growth of the latter and influence to a smaller degree the method of the treatment. Auxiliary species are the species of inferior quality or size, and have relatively little silvicultural value or importance. The accessory and auxiliary species are also called secondary species. On the basis of management, forests are classified into: (a) protection forests, (b) production forests, and (c) social forests. On the basis of ownership, forests are classified into (a) Government forests, (b) Private forests, and (c) forests owned by Panchayat, and other public bodies. Based on legal status, government forests are further classified into (a) reserve forests, (b) protected forest and (c) village forests. A reserved forest is a forest with complete protection, constituted according to the provisions of Chapter 2 of the Indian Forest Act, 1927. A protected forest is the forest area subjected to limited degree of protection constituted under the provision of Chapter 4 of the Indian Forest Act, 1927. A village forest is a forest assigned to a village community under the provisions of Chapter 3 of Indian Forest Act, 1927.

On the basis of growing stock a forest is classified into: (a) normal forest, and (b) abnormal forest. A normal forest is an ideal with regard to growing stock, age-class distribution and increment. Apart from this, the annual or periodic removal of timber from such forests equals to the increment and can be continued indefinitely without endangering further yields. Such forests serve as a standard for comparison and are rarely found in nature. Abnormal forest is one which is not normal, i.e. growing stock, age-class distribution of stems, increment, etc., are either in excess or more usually in deficit, than the normal forest.

National Commission on Agriculture (1976) proposed a different functional classification of forests. They are: (i) protection forests, (ii) production forests, and (iii) social forests. The forests managed for promoting protection function is called protection forest. In the management of such forest, the main object is to protect the site from soil erosion and land degradation. The areas where manipulation of the forest cover is not desirable are classified as protection forests. The forests located on higher hill slopes, national parks and sanctuaries, preservation plots, biosphere or nature reserves and wilderness areas are generally categorized as protection forests. The practice of forestry with a view to conserve flora, fauna, soil and water, to reduce incidence of floods and droughts, and to ameliorate climatic conditions, etc. is called protection forestry.

The ever-increasing demand of timber, fuel wood and various other forest products has to be met from production forests and the practice of forestry with object of producing maximum quantity of timber, fuel-wood and other forest produce is called production forestry. The production forestry is also called as commercial forestry. Commercial forestry aims to get maximum production of timber, fuelwood and other forest products in consonance with business principles. In production forestry, there is a greater concern for the production and economic returns. Here, the object is to meet the national requirements of timber and other forest products.

Social forestry is the practice, which aims at meeting the requirements of rural and urban population. The object of social forestry is to meet the basic needs of community and to improve the environmental conditions of living through: (i) meeting the fuelwood, fodder and small timber requirements, (ii) protection of agricultural fields against wind, (iii) meeting the environmental and recreational needs, and (iv) maximizing production and increasing farm returns. Farm forestry is a form of social forestry. It primarily aims at maximizing agricultural crop production along with on-farm tree products. More recently, there has been emphasis on dynamic and integrated land use planning. Efforts are on to maximize production on farmlands under farm forestry or agroforestry. Agroforestry has been defined as a sustainable land management system, which increases the yield from the land, combines the production of crop and forest plants and/or animals simultaneously or sequentially on the same unit of land and applies to the management practices that are compatible with the cultural practices of the local population.

Extent of Forest and Forest Types

The total geographical area of the country constitutes 328.8 million hectare. Out of it, approximately 63 million hectare land is having forest cover. However, as per statistics of various state forest departments, the total area of forests is 75.1 million hectares. This is about 19.27 per cent of the geographical area of the country. Forests in India are generally under the government ownership. The extent of forest area of various categories in various States/Union Territories is indicated in Table 1.1. It appears from perusal of the

Table 1.1 : Forest area in States/Union Territories as on 1988

(thousand ha)

	State/U.P.	Geographical area	Forest area			Total forest area
			Reserved	Protected	Unclassed	
1	2	3	4	5	6	7
1.	Andhra pradesh	275,068	50,479	12,365	970	63,814
2.	Assam	78,438	18,242	3,934	8,532	30,708
3.	Bihar	1,73,877	5,051	24,168	7	29,226
4.	Gujarat	196,024	13,819	997	4,577	19,393
5.	Haryana	44,212	247	1,104	322	1,673
6.	Himachal Pradesh	55,673	1,896	31,473	2,038	35,407
7.	Jammu&Kashmir	222,235	20,182	—	—	20,182
8.	Karnataka	191,791	28,611	3,932	6181	38,724
9.	Kerala	38,863	11,038	183	—	11,221
10.	Madhya Pradesh	443,446	82,700	66,678	5119	154,497
11.	Maharahstra	307,690	48,373	9,350	6,119	63,842
12.	Manipur	22,327	1,463	4,171	9,520	15,154
13.	Meghalaya	22,429	981	12	8,503	9,496
14.	Nagaland	16,579	86	507	8,036	8,629
15.	Orissa	155,707	27,087	30,080	17	57,184
16.	Punjab	50,362	44	1,107	1,750	2,901
17.	Rajasthan	342,239	11,585	16,837	3,278	31,700
18.	Sikkim	7,096	2,261	285	104	2,650
19.	Tamil Nadu	130,058	19,486	2,528	614	22,628
20.	Tripura	10,486	3,588	509	2,196	6,293
21.	Uttar Pradesh	2,94,411	36,425	1,499	13,739	51,663
22.	West Bengal	88,752	7,054	3,772	1,053	11,879

Table 1.1 *(contd.)*

1	2	3	4	5	6	7
23.	Anadaman and Nicobar Islands	8,649	2,292	4,242	—	7,171
24.	Arunchal Pradesh	83,743	15,321	8	36,211	51,540
25.	Dadra Nagar Haveli	491	203	—	—	203
26.	Goa, Daman & Diu	3814	165	—	1,259	1,424
27.	Mizoram	21,081	7,127	3,568	5,240	15,935
28.	Chandigarh	114	31	—	—	31
29.	Lakshdweep	32	—	—	—	—
30.	Pondichery	439	—	—	—	—
31.	Delhi	1483	42	—	—	42
	Total	3,287,263	4,16,516	2,23,309	125,385	7,65,210

Source:- FSI (1997)

data given in the said table that forests in India are not evenly distributed. The northern plains are completely devoid of forests. Only hill ranges such as Himalayas, Western Ghats, Satpuras, Vindhyas, Aravallis, etc. possess some area covered with forests. Forests of India have been classified into 16 major forest types from the point of ecology. The area under different types of forests in India is given in Table 1.2 (Champion and Seth, 1968).

Table 1.2 : Area under different Forest Types in India
(area in million hectare)

S.No	Forest Types	Area	Percentage of the total
1.	Tropical Wet Evergreen Forests	4.5	5.8
2.	Tropical Semi-Evergreen Forests	1.9	2.5
3.	Tropical Moist Deciduous Forests	23.3	30.3
4.	Littoral and Swamp Forests	0.7	0.9
5.	Tropical Dry Deciduous Forests	29.4	38.2
6.	Tropical Thorn Forests	5.2	6.7
7.	Tropical Dry Evergreen Forests	0.1	0.1
8.	Sub-tropical Broad-leaved Hill Forests	0.3	0.4
9.	Sub-tropical Pine Forests	3.7	5.0
10.	Sub-tropical Dry Evergreen Forests	0.2	0.2
11.	Montane Wet Temperate Forests	1.6	2.0
12.	Himalayan Moist Temperate Forests	2.6	3.4
13.	Himalayan Dry temperate Forests	0.2	0.2
14.	Sub-Alpine Forests		
15.	Moist-Alpine Scrub }	3.3	4.3
16.	Dry Alpine Scrub }		

Source: The State Of Forest Report: (1995, FSI).

Tropical Wet Evergreen Forests

These are tall, dense and multilayered forests generally found in regions with annual rainfall over 2500 mm. The total area under these forests is 4.5 million ha distributed mainly in Western Ghats, Upper Assam, Arunachal Pradesh, Andaman and Nicobar islands. The flora in these forests have Malayan affinity and consists of valuable species such as *Dipterocarpus macrocarpus*, *Shorea assamica, Mesua ferrea* and several other species. There is a large multiplicity of species in various layers. Bamboos and canes also occur in specific locations. Fern and epiphytes are also common in these forests.

Tropical Semi-evergreen Forests

These forests occur adjoining to tropical wet evergreen forests and form a transition between evergreen forest and moist deciduous forest. They are found locally in Western Ghats, Assam, Arunachal Pradesh, parts of Orissa and Andamàn and Nicobar Islands. The total area under this forest type is 1.9 million ha. The evergreen species dominant in these forests are *Dipterocarpus* spp., *Shorea·assamica, Mesua ferrea* etc. which are found mixed with deciduous species such as *Terminalia* spp., *Bombax cieba, Mangi-fera indica, Michelia champaca, Hopea parviflora*, etc. The growth of forest is relatively poorer than that of wet evergreen forest. The canopies and species are fewer than that of evergreen forest. Bamboos, canes, ferns, epiphytes etc. are also abundant.

Tropical Moist Deciduous Forests

These forests occur over an area of 23.3 million ha distributed mainly in Western Ghats, Assam, Arunachal Pradesh, Andhra Pradesh, Mizoram, Bihar, West Bengal, Orissa, Madhya Pradesh, Maharashtra and Uttar Pradesh. This forest type occurs in a shape of one strip along the foothills of Himalayas, another along the east side of Western and in a large area in Chhota Nagpur and north-east hills. These forests are common areas where rainfall ranges from 1500 to 2000 mm with a dry season of 4 to 6 months. The most important forest types are those consisting of sal (*Shorea robusta*) and teak (*Tectona grandis*). The teak forests are characteristics of southern form whereas, sal forest form the greater proportion in the northern form. Moist teak forests occur in Kerala, Karnataka, Andhra Pradesh, Maharashtra and Madhya Pradesh. Sal forests occur in Madhya Pradesh, Orissa, Bihar, Uttar Pradesh and West Bengal. Other species which occur mixed in both types are from genera *Terminalia, Albizia, Lagerstroemia, Dalbergia, Bauhinia, Bombax, Anogeissus,* etc. Bamboos are quite common. *Bambusa arundinacea* and *Dendrocalamus hamiltonii* are the most common bamboo in these types. These forests are usually 2 to 3 layered with lesser number of species than semi-evergreen forests. In forests of *terai* and eastern plains, alluvial savanna with scattered trees of *Bombax* and *Albizia* are usually found.

Littoral and Swamp Forests

These forests consist of evergreen species of varying density and height usually associated with wetness. These forests occupy an area of 0.7 million ha along the coast. These forests are mostly in developmental stage and are *seral* in nature. They occur throughout the country wherever wet and *waterlogging* conditions develop. The littoral and tidal forests also occur along the coast, the latter being especially associated with deltas of large rivers. Swamp forests occur in north-east India along the major river system. Mangrove forests consist of *Rhizophora, Brugutera*, *Ceriops* and *Heriteira* spp. The tidal and swamp forests are dominated by several evergreen and semi-evergreen species, while species like *Baringtoiria* spp., *Syzygium cumini, Dillenia* spp., etc. occur in seasonal swamps. Fresh water swamps contain species such as *Terminalia arjuna, Lagerstroemia speceosa, Trewia nudiflora*, *Myristica* spp., etc.

Tropical Dry Deciduous Forests

These forests occur from the north to the south of the country *i.e.* from Cape Comorin to the foothills of Himalayas. It appears in irregular wide strips in areas having rainfall between 750 mm to 1250 mm. These forests are concentrated in Rajasthan, Madhya Pradesh, Maharashtra, Orissa, Uttar Pradesh, Karnataka, Andhra Pradesh and Tamil Nadu. The total area under these forests is approximately 29.4 million hectare. These forests consist of trees having height about 20 mt to 25 mt with a light canopy consisting of deciduous trees. Dry teak and dry *sal* forests are commercially valuable forests and predominant in the southern and northern regions respectively. In some areas, both these species are absent and the area is occupied by the mixture of *Anogeissus pendula, Boswellia serrata, Hardwickia binata, Acacia nilotica, Madhuca indica, Butea monosperma,* etc. *Acacia catechu* and *Dalbergia sissoo* conspicuously appear on newly formed soils.

Tropical Thorn Forests

These forests occupy a large strip in the basin region of southern Punjab, Haryana, northern Gujarat and almost entire Rajasthan where rainfall ranges between 250 to 750 mm. Such forests are also found over a large areas in upper *Gangetic* plain and Deccan plateau. The total area under this forest type is about 5.2 million ha. These forests are open, consisting of trees of small height generally belonging to thorny leguminous species. The characteristic species include *Prosopis cineraria, Acacia leucocephalae, A. nilotica, Ziziphus* spp., *Salvadora* spp., etc. *Acacia tortilis* and *Prosopis juliflora* have been widely planted in this region. In south India, important species are *Acacia chundra* and *A. planifrons*. The forests which are highly degraded due to severe anthropogenic pressure occur in form of scrub forests in most of the areas.

Tropical Dry Evergreen Forests

These forests are found in relatively small area extending about 0.1 million hectare on the Carnatic coast. Such forests receive little or no summer rainfall. The forests consist of low but often dense forest with broad leaved evergreen trees in which thorny species predominate. The characteristic species are *Memecylonedule*, *Maba buxifolia*, etc.

Sub-tropical Broad-leaved Hill Forests

These forests occur in the lower slopes of the Himalayas in Bengal and Assam and on other hill ranges such as, *Khasi, Nilgiri* (Tamil Nadu), Mahabaleshwar (Maharashtra), Pachmarhi and Amarkantak (Madhya Pradesh) and Parasnath (Bihar) occupying an area of about 0.3 million ha. Important species occurring in southern hill forests are *Syzygium cumini*, *Ficus* spp. and some species of *Lauraceae*. The northern form generally consists of *Quercus, Castanopsis*, etc.

Sub-tropical Pine Forests

Sub-tropical *chir* pine forests occur in ranges having 1000 Mt to 1850 Mt elevation throughout the central and western Himalayas. *Chir pine* (*Pinus roxburghii*) is a characteristic species in western and central Himalayas, while *Khasi* pine (*Pinus kesiya*) in north-eastern hills in similar elevations. These forests are almost pure throughout their zone of distribution. The understorey is also not pronounced. The total area of these forests is approximately 3.7

million hectare distributed in several Himalayan states.

Sub-tropical Dry Evergreen Forests

These forests occur in low rainfall areas and consist of xerophytic thorny and small leaved evergreen species. Such forests are localised in north-west corner of the country in an area of approximately 0.2 million ha. The typical species are *Olea cuspidata* and *Acacia modesfa* in the top canopy and *Dodonea* shrub in the degraded forests.

Montane Wet Temperate Forests

These forests have characteristic feature of the eastern Himalayas and found between 1800 mt and 3000 mt elevation in high rainfall areas. These forests occupy about 1.6 million ha. Hilltops of some of the southern hills e.g. *Nilgiris* are also occupied by these forests. In northern form of these forests, characteristic species are *Quercus*, *Castanopsis, Machilus*, *Rhododendron*, etc. In southern hills, important species are *Syzygium* and *Ternostroemia* spp. *Rhododendron nilagiricum* is an important component in *Nilgiri* hills. The forests are luxuriant with dense undergrowth.

Himalayan Moist Temperate Forest

These are commercially important forests and found between 1500 mt and 3000 mt elevation in Himalayas, primarily localized in the central and western Himalayas except in area where the mean annual rainfall is below 1000 mm. The total area under these forests is about 2.6 million hectare. These forests are categorized into two forms; the lower form consists of *Quercus leucotrichophora, Q. dilatata, Pinus wallichiana* and *Cedrus deodara*. With increase in altitude, the upper form consisting of *Abies pindrow, Picea smithiana,* and *Quercus semecitrpifolia* become dominant. The east Himalayan hills are occupied by *Q. lineata, Q. lamellosa, Q. Pachyphylla, Tsuga dumosa*, *Picea spinulosa* and *Abies densa*. *Cupressus eorulosa* is a conspicuous species on limestone formation. Alders (*Alnus nepalensis*) and blue pine (*Pinus wallichianu*) colonise the new sites.

Himalayan Dry Temperate Forests

These are open evergreen forests with open scrub undergrowth. These forests occur in the upper ranges of Himalayas, which experience low rainfall. Such forests extends upto an area about 0.2 million hectare. These forests consist of both coniferous and broad-leaved species. In the western Himalayas, the characteristic species are *Pinus gerardiana, Cedrus deodara* and *Queropoda ilex*. At higher elevations, *Juniperus macropoda* forests are also found. In the eastern Himalayas, the common species are from genera *Abies* and *Picea*. In higher hills, *Juniperus wallichiana* is common. Again, in ranges between 2500 to 4000 Mt elevation, some other species e.g. *Larix grsffithiana, Populus euphretica, Salix* spp., *Hippoph* spp., *Myricaria* spp., etc. also occur.

Sub-alpine Forest

These forests occur throughout Himalayas above 3000 mt elevation upto the tree limit in the western Himalayas, *Abies spectalialis* and Betula *utilis,* and in eastern Himalayas *Abies densa* and *Betula* spp. are characteristic species. High level blue pine (*Pinus wallicltiana*) forests occur in exposed sites. *Rhododendron* forms the understorey.

Moist Alpine Scrub

These forests occur above the tree line upto about 5500 m height in Himalayas. The shrubs are rarely more than 1 m in height. Common species are *Rhododendron*, *Juniperous* spp. and *Betula* spp.

Dry Alpine Scrub

These forests consist of open xerophytic dwarf shrub and are common in areas where rainfall is below 400 mm. The common genera present are *Juniperous, Carangana, Eurotia, Salix, Myricaria* etc.

Actual Forest Cover

Though, the recorded forest area in the country is more than 75.0 million ha, but the actual forest cover is much less. The first attempt to determine the actual forest cover in the country by using satellite data was made by the National Remote Sensing Agency (NRSA) in 1984. NRSA interpreted the landsat imagery of 1972-75 and 1980-82 and found that the forest cover has reduced to 46.34 million ha. (NRSA, 1984). However, NRSA estimates suffered from many inadequacies and did not represent the correct picture of forest cover in the country (FSI, 1987). The Forest Survey of India (FSI) made an independent assessment of forest cover by visual interpretation of landsat imagery on 1 : I million scale for the year 1981-83 and removed some of the deficiencies of NRSA estimates. FSI makes the assessment of actual forest cover in the country in every two years by using the landsat imagery. The results of these estimates for different years are given in Tables 1.3, 1.4 and 1.5.

Table 1.3 : Changes of Forest Cover over 10 years.

(Area in Sq. Km.)

Sl. No.	Category	Area as per 1995 assessment	Area as per 1997 assessment	Area as per 1999 assessment
1	Dense forests (crown density 40% and over)	385,037	367,260	377,358
2	Open forests (crown density 10% to less than 40%)	249,309	249,309	255,064
3	Mangove forests	4,533	4,827	4,871
	Total	638,879	633,397	637,293

Source : FSI (1995, 1999).

The assessment made by FSI in 1999 gives the figure of actual forest cover as 63.72 million ha against the previous assessment of 63.88 million ha (1997). Thus, on the basis of these gross figures one can safely conclude that there has been a reduction of 1586 sq. km. of forest cover over a period of 5 years. The annual rate of loss of forest cover thus works out at 31,720 ha. A detailed account of state-wise distribution of recorded forest area and forest cover is presented in Table 1.4.

Table 1.4 : Geographic Area, Recorded Forest Area and Actual Forest Cover of Various States/UT

(sq. km.)

State/UT	Geographic area	Recorded area	Forest per cent	Forest area	Cover per cent
Andhra Pradesh	2,75,068	63,814	23.20	44,229	16.08
Arunachal Pradesh	83,743	51,540	61.54	68.847	82.21
Assam	78,438	30,708	39.15	23,688	30.20
Bihar	1,73,877	29,226	16.81	26,474	15.23
Delhi	1,483	42	2.83	88	5.93
Goa	3,702	1,424	38.46	1,251	33.79
Gujarat	1,96,024	19,393	9.89	12,965	6.61
Haryana	44,212	1,673	3.78	964	2.18

(Contd.)

Himachal Pradesh	55,673	35,407	63.60	13.082	23.50
Jammu & Kashmir	2,22,235	20,182	9.08	20,441	9.20
Karnataka	1,91,791	38,724	20.19	32,467	19.93
Kerala	38,863	11,221	28.87	10,323	26.56
Madhya Pradesh	4,43,446	154,497	34.84	131,830	29.73
Maharashtra	3,07,690	63,842	20.75	46,672	15.17
Manipur	22,327	15,154	67.87	17,384	77.86
Meghalaya	22,429	9,496	42.34	15,633	69.70
Mizoram	21,081	15,935	75.59	18,338	86.99
Nagaland	16,519	8,629	52.04	14,164	85.43
Orissa	1,55,707	57,184	36.73	47,033	30.21
Punjab	50,362	2,901	5.76	1,412	2.80
Rajasthan	3,42,239	31,700	9.26	13,871	4.05
Sikkim	7,096	2,650	37.34	3,118	43.94
Tamil Nadu	1,30,058	22,628	17.40	17,078	13.13
Tripura	10,486	6,292	60.01	5,745	54.79
Uttar Pradesh	2,94,411	51,663	17.54	34,016	11.55
West Bengal	88,752	11,879	13.38	8,362	9.42
A&N Islands	8,249	7,171	86.93	7,606	92.21
Chandigarh	114	31	27.19	7	6.14
Dadra & Nagar Haveli	491	203	41.34	202	41.14
Daman & Diu	112	—	—	3	2.68
Lakshadweep	32	—	—	0	0
Pondicherry	439	—	—	0	0
Total	3,287,263	765,210	23.28	637,293	19.39

Source : FSI (1999)

Table 1.5 : Forest cover as per 1999 Assessment

Class	Area (Sq. Km.)	Percentage of Geographical Area
Dense Forests	377,358	11.48
Open Forests	255,064	7.76
Mangrove	4,871	0.15
Scrub	51,896	1.58
Non-forest	2,598,074	79.03
Total	3,287,263	100

Source : FSI (1999).

The growing stock per ha in the forest of India is about 74.42 m^3, which is far less than the world average of 110 m^3 (FSI, 1995). There is an acute shortage of timber and firewood in the country. The potential productivity of the forests of the country is much higher than the actual production. The sites have larger inherent capabilities and the present production can be increased, if the process of soil degradation and biotic interference are checked completely and technology induction into the defforestation strategey is accelerated. The average annual wood productivity of Indian forest is about 0.7 mt per hactare. This is abysmally lost in comparison with productivity rate prevailing in developed countries (Table 1.6)

Table 1.6 : Average annual production of wood in selected countries

Country	Average annual production in m^3 per ha.
USSR	1.9
USA	3.0
Japan	2.5
UK	3.5
Australia	4.5

Forest Influences

Forests are important for the mankind not because they provide timber and a variety of non-wood products but because they influence various environmental factors. Forests are also popularly known as earth's blanket and world's air conditioner. They are crucial to survival of mankind because they constitute the life-supporting system in the earth.

The following are some of the important protective and ameliorative functions of the forests:

(i) Forests play a significant role in maintaining the CO_2 balance in the atmosphere. Without sufficient forest cover, the CO_2 which is released in the atmosphere is not completely utilized, resulting in higher percent of CO_2 in the atmosphere. This, in turn will result in warming of world temperature, disturbance in the climate, melting of polar ice saps, increasing sea levels, etc. The CO_2 per cent in the atmosphere has already reached 0.042 per cent against the normal point of 0.030 per cent. If this increases continuously, higher temperature will be witnessed in many parts of the globe. Such disturbances on the earth may bring about un-imaginable miseries to the mankind. Forests act as efficient sinks for the carbon dioxide.

(ii) It has been found that forests increase local precipitation by about 5 to 10 per cent due to the orographic and micro-climatic effects resulting in conditions favourable for the condensation of the clouds.

(iii) Forests reduce temperature and increase humidity. Temperature inside the forests is about 3° to 8°C less than the adjoining open areas. Reduced temperature makes the life comfortable. It also reduces evaporation losses. The effect of forests on temperature is not limited to forested areas but the effects are experienced in nearby areas also.

(iv) Forests maintain the productivity of the soil by adding a large quantity of organic matter and re-cycling of nutrients. The dried fallen leaves of trees are used as manure. Supply of firewood from forests releases dung to be used as manure.

(v) Tree crown reduces the erodibility of rain drops and checks splash erosion. Forests increase infiltration rate and water holding capacity of the soil resulting in much lower surface run-off. This helps in checking soil erosion.

(vi) Forests check floods. They intercept 15 to 30 per cent of the total rainfall. They increase infiltration and water holding capacity of the soil. This results in reduced surface run-off checking soil erosion. Mostly, floods are caused due to siltation of river channels, which are formed due to erosion and higher peak discharges caused due to greater surface run-off.

(vii) Forests conserve soil and water both. They prolong the water cycle from its inception to the final disposal into streams and oceans. The longer the water is retained in the land, the greater is the usefulness in nurturing crops and trees and in maintaining regular supply of water in stream throughout the year. Forests increase sub-surface run-off that is much slower than surface run-off and it does not cause erosion.

(viii) Forests and trees reduce wind velocity considerably. Reduction of wind velocity causes considerable reduction of wind erosion, which in turn checks shifting of sand dunes and halts the process of desertification.

(ix) Forests, by reducing erosion check the siltation of irrigation and hydal reservoirs. Rapid siltation of various reservoirs in the country is the result of deforestation in the catchment areas of these reservoirs.

(x) Forests are the storehouse of biological diversity. Several unknown plants may be potential for medicines and food. Deforestation eliminates several species of plants and animals: About 5000 species of plants including 180 species of mammals, many species of birds and several thousand species of insects are already at the verge of extinction due to loss of forests.

(xi) Forests protect from physical, chemical and noise pollution. Dust and other particulate pollutants cause serious problems and forests help in protection from these problems.

(xii) Forests and trees provide shelterbelt and windbreak effect, which is beneficial to agricultural crops, particularly in arid and semi-arid areas. Shelterbelt and windbreak increase agricultural production.

(xiii) Forests provide recreational facilities to the people. A large variety of trees and shrubs, animals and birds attract a large number of people towards them. National parks and sanctuaries rich in flora and fauna are visited by a large number of tourists.

(xiv) Forests provide experimental field and laboratory for college and university students. They provide sites for ecological and environmental studies.

(xv) Forests have a natural healing effect for a number of diseases. As a result a number of Natureapathy Centers are established in well-wooded areas.

Forest Products

Forest products or forest produces may be broadly classified into two main groups:

(i) Wood produce, and (ii) non-wood produce. The wood produce includes timber and firewood and non-wood produce include all produces other than timber and firewood. Non-wood produce are commonly referred to as minor forest produce.

1. Forests are valuable natural resources. The goods obtained from the forests are of immense importance. Timber is a major forest produce and it is used extensively for various purposes. In India, maximum part of timber produced is used for construction of houses,

agricultural implements, bridges, sleepers, etc. In India, about 30 million cubic metres of timber is produced from the forests. Many species e.g. *teak, sal, deodar, sissoo, babul, chir, haldu, axlewood, rose-wood, dipterocarps*, etc. yield valuable timber.

2. Wood is a universal fuel. For thousand of years until the advent of coal, oil, gas, electricity, etc., wood constituted man's chief source of fuel. Even today, more than half of the total world population uses wood for fuel. It remains a major source of domestic fuel in India. Approximately 175 million cubic metres of wood is used as fuel in the country, most of which comes from forests (NCA, 1976).

3. Forest provides raw material to a large number of industries, e.g. paper and pulp, plywood and boards, saw mills, furniture making, and packing cases, match splints and toys, etc.

4. A large number of non-wood products are also obtained from forests. These are commonly called Minor Forest Products (MFP) not because these are of minor significance but since they are harvested in smaller quantities. Some of the important minor forest products are as under:

(i) *Edible products*: Fruits, flowers, seeds, roots, rhizomes, tubers, etc. of several forest species are edible. Fruits and seeds of *Anacardium occidentale, Tamarindus indica, Syzygium cumini, Emblica officinalis, Buchanania lanzan*, etc., flowers of *Madhuca indica*, green pods of *Moringa oleifera*, new shoots of bamboo, etc. are in great demand. There are more than 213 species of trees, 14 species of palms, 128 species of shrubs, 216 species of herbs, 4 species of ferns and 15 species of fungi in our forests which yield edible products.

(ii) *Grasses and grazing*: Forests provide grazing facility to about 30 per cent of the total livestock population of the country. The grasses such as *Cenchrus, Lasiurus, Heteropogon, Bothriochloa, Andropogon, Eragrostis, Iseilema*, etc. are abundantly found in the forest as a ground cover. Some grasses like *Eulaliopsis binata* are also used for rope and paper making.

(iii) *Fodder trees and shrubs*: Forests provide fodder from trees, shrubs and climbers. Leaf fodder of several tree species is almost as nutritious as that of agricultural fodder crops. Good fodder yielding tree species includes: *Ailanthus excelsa, Moringa oleifera, Sesbania* Spp., *Morus alba, Albizia lebbeck, Leucaena leucocephala, Pongamia pinnata, Hardwickia binata,* etc. Several shrubs, herbs and climbers also yield good quality fodder leaves.

(iv) *Bamboo and canes*: India is very rich in bamboo resources. About 120 species of bamboo are found in the country. About 5.0 million tones of bamboo is harvested from country's forests every year. Approximately 10 per cent of the bamboo is used for housing, 25 per cent for rural/agricultural works, 25 per cent for paper-pulp and remaining quantity for packaging and other uses.

(v) *Medicines*: Out of 2000 items of drugs mentioned in different epics, over 800 are of vegetable origin. A large number of these are obtained from plants found in forests. Different parts of various forest plants e.g. roots, shoots, leaves, fruits, seeds, barks, etc. are used for drugs. Some of the drugs obtained from the forests are exported to different parts of the world for their high medicinal value.

(vi) *Oil seeds*: Many tree species produce oil-bearing seeds, which are commercially important. Some of these oils are fit for human consumption. Generally, these seeds are used in soap industry. Tribals use these oils for various purposes. There is a potential of production of about 1 million tones of oil every year from forest tree seeds.

(vii) *Essential oils*: Essential oil is a good source of earning forest revenue. The total yield of essential oils from the forests is about 1500-2000 tones. Many species in the Indian forest yield essential oil, which is used in making perfumes, soaps, cosmetics, etc.

(viii) *Tans and dyes*: A variety of vegetable tanning materials are produced in the forests. Important vegetable tanning materials are the myrobalan nuts and bark of wattles (*Acacia mearnsii, A. decurrens, A. nilotica and Cassia auriculata* etc.). Other tanning materials include leaves of *Emblica officinalis* and *Anogeissus latifolia*, bark of *Cleistanthus collinus,* fruits of *Zizyphus xylocarpa*, bark of *Cassia fistula, Terminalia alata, T. arjuna,* etc.

(ix) *Gums and resins*: Gums and resins are excluded by trees as a result of injury or incision to the bark of wood. Gums are collected from several tree species. Resin is obtained from Pinus spp. Gums and resins find wide-use in general industries and pharmaceutical industries.

(x) *Fibres and flosses*: A wide-range of plants, yielding fibre occur in the Indian forests. Fibres are obtained from tissues of different parts of certain woody plants which are used for making cloth, rope and cordages.

(xi) *Spices and flavourings* : A variety of spices like kalajira (seeds of *Carum carvi*), dalchini (bark of *Cinnamomum zeylanicum*), cardamom (dried capsules of *Ellatteria cardamomum*), tejpat (leaves of *Cinnamomum tamala*), etc. are obtained from forests used locally and throughout the world.

(xii) *Animal products*: Animal products include lac, honey, silk, horns, fur, skins, tusks, etc. Lac is a resinous secretion of the 1ac insects which feed on forest trees, particularly on *Butea monosperma*. Similarly, silk worm is reared on *Terminalia alata* and *Morus alba* plantations. Honey is another product, which is obtained from forests.

(xiii) *Tendu leaves and other leaves*: Tendu leaves are used to prepare bidi and therefore, these are also called bidi leaves. Annual collection of tendu leaves is about 90,000 tonnes in the country. In Madhya Pradesh alone, about 45 per cent of this quantity is produced. Leaves of trees such as *Bauhania* spp., *Butea* spp., etc. are used for making plates, dona, etc.

(xiv) *Miscellaneous products*: Forests yield a variety of other products which have not been included in above groups. Such produce include katha, latex, insecticides, soap nuts, etc.

Forests provide employment to a large number of people. During 1985, forests provided employment to about 10 million people in the primary sector. Much larger employment was generated in secondary and tertiary sectors. Forests and various forest activities help tribals to improve their socio-economic condition by providing gainful employment. Tribals are employed for the collection, processing and marketing of various forest products, which helps them in improving their economic conditions.

Forests provide a good source of revenue to the Government,

which is used for various developmental works. During the year 1990, forests provided revenue worth Rs. 18,500 million.

Forest Productivity

The current productivity of the country's forest is only 0.7 cum per hectare per year. Productivity on the basis of recorded production from forest is only 0.5 or less than that. But if the unrecorded production of firewood which is removed from forests on headloads is also taken into account, then productivity is about 0.7 cum per hectare per year.

Table 1.7 : Actual Productivity in well protected Forests of India

Region	Productivity (cum/ha/yr)
Western Himalayas	2.21
Eastern Himalays	2.03
North-East	1.66
Western Coast and Andaman and Nicobar Islands	3.85
Deccan	1.35
Central India	1.05
Gangetic Plain	0.80
Dry forests of the Indus Plains	0.41

Note: These figures relate to woody biomass above the ground, as against organic productivity, which refers to total plant biomass.

Wood is the most important component of the visible part of the forests. The current average annual production of wood from forests is 30 million cum of which 12 million cum is timber and 18 million cum is firewood. The potential productivity of forests is, however, much more (Table 1.8). The actual productivity assessed by the Survey of India in well-managed and reasonably well-protected forests regions of the country is given in Table 1.7. Since these figures do not represent average productivity but that of forests relatively free from biotic interference, they represent potential productivity of the forests of the country. The potential production from India's forests on the basis of these figures works out at 90.63 m.cum per year. However, on ecological basis, O.N. Kaul worked out the organic productivity of the country's tropical, sub-tropical and temperate forests at 9.0, 8.2, and 6.0 tones per hectare, respectively.

Table 1.8 : Potential Productivity of India's Forests

Region	Forest cover (mha)	Potential production (m cum/yr)
Western Himalayas	5.38	11.89
Eastern Himalayas	6.46	14.85
Northeast	7.33	12.17
Western Coast and Andaman and Nicobar Islands	1.95	7.51
Deccan	15.95	21.46
Central India	17.49	18.36
Gangetic India	4.75	3.80
Dry Forests of the Indus Plains	1.44	0.59
Total	60.70	90.63

Growing stock is the measure of dynamic forest capital. Statewise growing stock of entire country in major forest strata was assessed by the Forest Survey of India (FSI) using available information from (i) Vegetation maps based on 1989-91 satellite data, (ii) Thematic maps, and (iii) Ground inventories of forest carried out by the Forest Survey of India in the past. The rate of annual increment of growing stock of various forest strata was assessed by applying Von Mantel's formula (1 = 2GS/R). Statewise details of Growing Stock, overall volume per hectare and annual increment are given in Table 1.9. The growing stock of the country was estimated to be 4740.858 million cubic metres and the overall volume per hectare to be 74.42 cubic metres. The total annual increment of growing stock has been estimated at 87.622 million cubic metres. These estimates are for forests having more than 10 per cent crown density (FSI, 1995). The data regarding growing stock, volume per hectare and annual increment in forests of India is presented in Table 1.9 and forest stratum wise details of growing stock is given in Table 1.11.

Table 1.9 : Growing Stock, Volume per hectare and Annual Increment in Forests of India

Sl. No.	State/UTs	Growing Stock (million Cum)	Volume/ha (in Cum)	Annual Increment ('000 Cum)
1.	Andhra Pradesh	291.394	61.66	5929
2.	Arunachal Pradesh	775.125	112.8	15243
3.	Assam	304.423	124.2	6061
4.	Bihar	96.72	36.38	1715
5.	Goa, Daman & Diu	12.644	101.16	231
6.	Gujarat	63.921	53.07	1459
7.	Haryana	1.432	27.92	27
8.	Himachal Pradesh	254.41	203.5	1603
9.	Jammu & Kashmir	458.929	224.49	6402
10.	Karnataka	272.411	84.23	5574
11.	Kerala	98.883	95.67	1983
12.	Madhya Pradesh	698.182	51.56	14122
13.	Maharashtra	225.554	51.43	5008
14.	Manipur	95.452	54.1	1889
15.	Meghalaya	104.14	66.0	2150
16.	Mizoram	66.012	35.3	1332
17.	Nagaland	94.887	66.1	1903
18.	Orissa	246.15	52.21	4432
19.	Punjab	0.95	7.07	23
20.	Rajasthan	12.892	9.84	292
21.	Sikim	39.286	125.9	767
22.	Tamil Nadu	69.602	39.31	1394
23.	Tripura	15.452	27.9	316
24.	Uttar Pradesh	338.633	99.71	5818
25.	West Bengal	22.859	37.68	433
26.	Andaman & Nicobar Islands	79.644	119.62	1494
27.	Dadra & Nagar Haveli	0.871	42.2	22
	Total	4740	74.42	87622

Source : FSI, 1995.

The potential production of various minor forest produce other than timber and firewood i.e., NTFP (Non-Timber Forest Produce) has been worked out by Tirath Gupta and Amar Guleria in 1982. It is presented in Table 1.10.

Table 1.10 : Potential Production of Minor Forest Produce

Description	Current production('000t)	Potential production('000t)
Fibre and flosses	6	50
Grasses (other than fodder)	350	525
Bamboo	1932	4309
Canes	14	21
Essential oils	2	3
Non-edible oilseeds	430	6671
Tans and dyes	187	290
Gums and resins	91	175
Lac and tassar silk	22	35
Tendu leaves	210	300
Drugs and spices	3	5

Table 1.11 : Total Estimated Growing Stock and Annual Increment of Forests of India.

(in ,000 Cum)

Sl. No.	Forest Stratum	Total Growing Stock	Annual Increment on Growing Stock
1.	Fir	153033	1722
2.	Spruce	9550	105
3.	Fir-Spruce	31215	347
4.	Blue Pine	81175	1083
5.	Deodar	27473	366
6.	Chir-Pine	111960	1865
7.	Mixed Conifers	383932	3850
8.	Hardwood mixed with conifers	47018	783
9.	Upland Hardwoods	111710	1862
10.	Teak	320546	8013
11.	Sal	515459	8591
12.	Bamboo (tree crop in bamboo)	36371	909
13.	Dipterocarpus	683	1
14.	Khasi Pine	7271	181
15.	Khair	2406	80
16.	Salai	3107	104
17.	Alpine Pasture	619	
18.	Western Ghat Evergreen	47403	790
19.	Western Ghat Semi-evergreen	37560	625
20.	Western Ghat Deciduous	18357	459
21.	Miscellaneous	2794010	55876
	Total	4740858	87622

2

Silvicultural Management of Forest Resources

Introduction

Forest is a dynamic entity. It needs to be managed in such a way that the renewable natural resources would be available to humanity in perpetuity. Therefore, the forests are managed according to sound principle of silviculture. Silviculture is the art and science of cultivating forest crop. A forest crop requires to be carefully tended, harvested and replaced by new crops according to certain silvicutural procedures. When a set of silvicultural procedure is initiated along with a homogenous set of objectives of management then it is called 'silvicutural system'. Thus, silvicultural system has been defined as a method of silvicultural procedure worked out in accordance with accepted sets of silvicultural principles, by which crops constituting forests are tended, harvested and replaced by new crops of distinctive forms. In other words, the silvicultural system is a package of inputs applied to the forest crop throughout its life with an objective of giving the forest a predefined character. The selection of silvicultural system for any type of forests necessitates consideration of an array of factors. These factors influence the choice of silvicultural systems. Some of the important factors are illustrated below:

1. Condition of regeneration and growth in the forests.
2. Nature of terrain and soil.
3. Protection considerations.
4. Nature of forest produces and ecological services required.
5. Economic considerations.
6. Existing rights and privileges.
7. Aesthetic considerations.

Silvicultural Systems in India

Several silvicultural systems practiced in Europe were adopted in Indian forest with little modification. These systems were highly suitable to the temperate forests of Europe. They were not fully compatible with the Indian forest system. However, in order to produce large sized timbers from the forests, certain European systems were followed with some modification. We shall discuss those important systems in relation to their relevance to ecological balance and forest conservation in Indian context.

Clear-Felling System

It is a silvicultural system in which equal or equiproductive areas of mature crop are successfully clear-felled in one operation followed by regeneration, most frequently artificial but sometimes natural also. This system was adopted in India around 1842 for teak forests. Under this system, successive equal areas of forests are clear-felled and regenerated. This suffers from maximum disadvantages from the point of ecological balance of the forests. The complete clearance of forest produces certain adverse condition such as desiccation, and general deterioration of soil fertility, primarily on account of erosion, laterization and destruction of humus. This results into failure of regeneration. The debris left in the forest floor constitutes a fire hazard. Weeds also invade the cleared areas, which generally hampers establishment of main crop. From the point of environment conservation, this system is least beneficial. This system, however, envisages uniform crop of even age gradation, ensuring an assured timber yield.

Selection System

This is a silvicultural system in which felling and regeneration are distributed over the whole of the area and the resultant crop is so uneven-aged that trees of all ages are found mixed together over every part of the area. Such a crop is referred to as 'selection forest' or 'all-aged forest'. This system necessitates removing of mature and over-mature trees singularly over a large area of forests, on a periodic basis, without creating any lasting gap in the forest canopy. This is especially suitable for the temperate and tropical forest areas with hilly terrain. In this system, the removal consists of the following category of trees.

- Dead and dying trees.
- Diseased, and defective trees.
- Trees lacking in vigour and trees of undesirable species intervening with better stems.
- Trees of prescribed exploitable girth.

Removal of aforesaid trees is followed by tending and regeneration operations. This system works well when regeneration in the forests is assured. If the natural regeneration fails, and we keep on removing mature and bad trees then, over the time the forest area becomes open and barren. This is happening in most of our forests today. It is because of the fact that in one hand, majority of our forests lacks natural regeneration due to forest fires and excessive grazing and in the other hand, mature trees are removed. However, there are several advantages in this system. It maintains constant forest cover and so continued soil fertility is assured. There is also a possibility in the system to maintain the original composition of the crop and uneven character of the stand. The system also provides immunity against insect pests to the crop. It helps to restore a suitable balance between pests and predators. Apart from this, the system is most natural one. Operations under it follow the footsteps of nature closely. This system is ideally suited for maintaining the ecological balance of any forest ecosystem. Besides, there is an assured seed source from the older trees, so even if an accidental fire destroys the younger crop or the regeneration in one year, there is a possibility for further recruitment. This is the most flexible system where one can manipulate to obtain a complete and balanced composition of forest trees. The system can also be adapted to grow wood to cater to the local needs. However, marking of selected trees for felling under this selection system requires considerable professional skill and expertise.

Shelter-wood System or Uniform System

This is a system that, aims at concentrated regeneration, in which the canopy is uniformly opened up over the whole area of a compartment to obtain uniform regeneration. Under this system, the mature trees are removed by successive felling. Therefore, Troupe, the famous forester has aptly termed it as a "system of successive regeneration felling". Depending upon the methods of carrying out the regeneration felling and distributing them in time and space, several sub-systems have also been developed. It was widely practiced in European forests. With the incoming of European foresters in India, it was applied to the forests of Utter Pradesh, Himachal Pradesh and Madhya Pradesh. The system has advantages as well as disadvantages. What is significant is the fact that the system, among other things, has the advantages of a selection forests also. However, the system had many disadvantages: Few of such disadvantages are illustrated below:

1. Less concentrated working per unit area compared to clear felling.
2. Younger regeneration takes more time for establishment.
3. It requires tremendous skill in managing successive operations to manipulate overwood and to bring out the younger crop.

Coppice System

In this system, the young crop originates from coppice shoots arising from the adventitious buds on the stumps of the felled trees or by other vegetative means. Here the rotation is usually short. This is ideally suited for those forests, which need to be managed for fuelwood and small timber production. The forest should have the species that has the potentiality to reproduce by means of root suckers and coppice shoots. In the past, many dry deciduous forests in India were managed under this system. The single most demerit in the system lies in the fact that it exhausts the nutrients in the soil. However, over the period, several modifications took place in this system in response to local needs and potentiality of forest. These modifications were incorporated into the main system to form a full-fledged system or sub-system. These systems are known as "coppice-with-standard" and "coppice with reserve". They were widely adopted in dry deciduous forests of Central India. These systems are still in vogue in many parts of India.

Modified System

Most of the silvicultural systems aimed at bringing natural regeneration in the forests. Therefore, all operations carried out under these systems had the focus on natural reproduction. Nevertheless, due to the increasing biotic pressures on our forests, no system ensured adequate natural regeneration. Hence, responding to the need of regeneration in the forest floor, foresters in the past supplemented the system with some special operations. Such operations are called as underplanting, gap planting, enrichment planting, Restoration of Degraded Forests (RDF), etc. Two of such methods or special operations are illustrated below.

Under-planting : This method is usually adopted for shade bearers. Softwood species like *Ailanthus malabaricum, Anthocephalus cadamba, Artocarpus fraxinifolius, Santalam album, Astonia scholaris* are normally raised as *under-planting* species.

Gap planting : In the forests, owing to lack of adequate natural regeneration gaps in the forest floor are formed. In such situations, gaps are planted up with suitable species, generally light demander species. Over the time, this method assumed wide importance and application in many types of forests.

Choice of Silvicultural System

Every forest requires a silvicultural system to rejuvenate itself. The system helps the forests to maintain its diversity, composition, stocking and productivity in perpetuity. If a forest is to continue, then it needs to regenerate, otherwise, it will vanish. Therefore, while formulating management plan, it is the duty of the forest planners to propose appropriate silvicultural system to regenerate forests on a sustainable basis. In fact, there are many silvicultural systems, which are in practice in different parts of our country. Therefore, it is imperative to choose the suitable silvicultural systems for regenerating the forest crop. While selecting the suitable silvicultural systems for the forests a wide range of factors are considered. Few of such factors are illustrated below.

State of conditions required for regeneration : While deciding about the system, it is necessary to assess the state of conditions that are fundamental to rejuvenation of forest crop. The points generally considered are light requirement of the crop and availability of seedlings in the forest floor and ease of regeneration. Over the years, the ground conditions required for natural regeneration in the forest has deteriorated. Frequent fire incidences and high grazing pressures have alarmingly marred the prospect of natural regeneration. In such situations, there is one option available to forest planners. That is the effort to make the forest floor conditions compatible for natural regeneration.

Topography and soil conditions : The state of regeneration in the forests growing on rocky and precipitous slopes is never uniform. Regeneration in such forests is found mostly in some pockets of the hills, cliff, or valleys. Generally, such rocky and mountainous areas constitute the catchment of river basins. In such areas selection system is generally applied. At present, this system is being adopted widely in most of our forest areas. Therefore, topography also influences the choice of silvicultural system for a forest.

Compatibility with the external factors of locality : While selecting the silvicultural system, it is imperative to examine the compatibility of the system with external factors of locality like frost, wind, snow, draught and proneness for pest and insect damage. If it is found after evaluation that the silvicultural system shall not brook the onslaught of draught, frost, etc., then the silvicultural system does not stand for consideration. For example, if a forest is found prone to frost or draught, then it is appropriate and judicious to adopt selection or shelterwood silvicultural system. It is because of the fact that these systems provide protection to the seedling crop from adverse effects of locality factors.

Objects of management : Objects of management determine the selection of silvicultural system. If the objective of forest management is to produce large sized timbers, then the high forest system will be the choice. However, when the object is to produce small timber, pole and fuel for the local communities, then the coppice system is adopted. In all such situations, other factors are also considered while selecting the system. Where the primary aim of forest management is to conserve the bio-diversity and to produce timber as a subsidiary output, then the selection system with focus on conservation of bio-diversity is adopted. In a hilly and mountainous forest area, the goal of management is to protect the watershed. Therefore, in such condition, maintenance of existing forest cover remains the sole aim of forest management. Considered in this backdrop, it is the objective of management which primarily influences the selection of silvicultural system.

Conditions of Applicability of Silvicultural Systems : Advantages and Disadvantages

Forests are managed under different sivicultural systems. However, before deciding about the system, it is worthwhile to consider the conditions of applicability and merits and demerits of the said system threadbare. Unless this exercise is undertaken honestly during the planning process, then the objectives of management are bound to fail. Therefore, while considering different management options for a forest, it is indispensable to examine the prospect of a silvicultural system in the context of predecided goal. Following are some of the factors of consideration with respect to different silvicultural systems.

1. *Clear-felling System*

— Favourable factors of locality especially terrain and other ground a condition.
— Susceptibility of the site for soil erosion and loss of biological diversity.
— Socio-political compatibility towards clear-felling and subsequent plantation of a single species.
— Site potentialities for producing large sized timber.
— Species suitability for the site.
— Prospect and constraints of artificial regeneration.
— Operational feasibility.

2. *Uniform System or Conversion to Uniform System*

— Possibility of obtaining regeneration at will during the regeneration period irrespective of the fact that the area constitutes hills or plains.
— Prospect of protection for the regeneration crop in the initial stages of development against adverse climatic factors.
— Prospect of soil erosion and consequent land deterioration in the event of having the area clear-felled.
— Possibility of invasion of weeds and grass.

Generally, this system has the potentiality of producing large sized timbers if the site is a productive one. However, it is not suitable in a number of situations. Few of such situations are as follows :

— When the slopes of the forest areas are very steep and rugged in nature.
— When the pressure of cattle grazing in forests is so high that it is not possible to close the area for a long period.

— Where regeneration cannot be ensured at will during a reasonable period of time.

3. *Selection System*

Selection system has more merits than demerits in Indian context. Few of advantages are as follows:

— The system helps the forests to maintain a continuous canopy at the top and some layer of cover immediate below the canopy. This not only helps to conserve soil and moisture but also uses the site optimally without degrading the environment.
— The crop has an in-built defensive capacity to resist injuries by insects, pests and adverse climatic factors. It prevents invasion of grass and weeds.
— The forest regenerates itself continuously without any limit if biotic factors are kept within the threshold limit.
— It results in producing more growing stock and large sized trees per unit area than the uniform system. This is the best silvicultural system to produce large sized timbers.
— It provides insurance against loss of bio-diversity.
— This system provides a trade-off among various forest products and services.

However, this system has some demerits. Few of such disadvantages are illustrated below:

— The system suffers from some operational disadvantages. Marking and felling in annual coupes requires considerable technical skills and professional expertise.
— Cost of timber harvesting is high on account of the fact that production operation is carried out over a vast area with low intensity of yield.
— Protection of the forest area after main felling is a difficult proposition. It is because in the present socio-economic condition it is nearly impossible to protect the forest for the regeneration process to set in. Apart from this, it is also difficult to protect the forests from fire and grazing.
— Sometimes, while executing the system, it becomes difficult

to maintain the required mix of valuable and invaluable species. Beside this, the concept of valuable and invaluable species and their mix keeps on changing. In the absence of supportive research data, it becomes intractable to decide about the right mix of species composition, density and the growing stock.

However, the merits of selection system outplay the demerits. The system is widely applied in our forest management practices. In this context, few of important applicabilities of the system are enumerated below:

1. The system is preferred for those areas where protection against soil erosion, landslides and moisture loss is the paramount objective. Therefore, this system is adopted for those forest which are situated in catchment areas.

2. This system is chosen when other systems do not hold promises because of changing environmental and socio-political conditions.

3. The system is adopted when conservation of stratified forest ecosystem is a necessity from the point of productivity of the forest, in terms of wood as well as non-wood forest products. This system has also the potentiality to respond to the needs of the local people.

4. This system also ensures best possible conservation of biological diversity of the forests.

Accessory Silvicultural System

Two-Storied High Forest System : Application of this system leads to formation of two-storyed forests, *i.e.* crop of trees in which the canopy can be differentiated into two strata. The crop in each story is even aged. This system has never been subjected to practice seriously.

High Forest with Reserve System : This system is also almost similar with the pervious one. It was scarcely put to practice.

Improvement felling : It is not a silvicultural system in pure sense. It is because of the fact that it does not aim at regenerating the crop or producing a crop of distinct characteristics. Trevor (1923) has suggested that in view of the fact that large areas of degraded and poor forests in India are likely to remain under improvement felling to prepare them for eventual management under a more intensive system, and that as the treatment, which aims at bringing these forests to the normal state, implies and entails encouragement of regeneration, it should be treated as an accessary system. (Ram Prakash and L.S. Khanna (1979).

Improvement felling is a set of operations aimed at improving forest crop. Unlike that of other silvicultural systems, it does not aim at initiating a regeneration process. In other words, it is a method of treatment, which involves essentially the removal of inferior growing stock in the interest of better growth of the more valuable species and individuals. It is usually applied to mixed uneven-aged forests. The treatment comprises of the following operations:

1. Felling of dead, dying and diseased trees.
3. Felling of unsound mature trees if they are silviculturally available.
4. Thinning of congested group of poles.
5. Cutting back of badly-shaped saplings and advance growth expected to give better coppice shoots.
6. Removal of undesirable undergrowth or trees of inferior species which are likely to prevent regeneration of valuable species.

Selection-Cum-Improvement felling system

In good productive sites, selection felling is carried out with pronounced treatment of improvement felling. In this hybrid silvicultural system, the operation comprises two distinct components. They are as follows:

— Selection felling operation.
— Improvement felling operation.

In selection felling, an exploitable diameter or girth is determined with respect to different species or group of species. Trees above the specified diameter or girth limit, if found silviculturally available, are felled. The basic aim is to encourage regeneration. Crop improvement is subsidiary to it.

In improvement felling, all the aforesaid operations are aimed at improving the growing stock by removing the inferior part.

Coppice with Standard

This system was also widely applied to many forest areas to grow small timber for the local people. Advantages and disadvantages of the system are enumerated below:

Advantages

1. The overwood of standard retained under the said system protects the soil and conserves the moisture.
2. The standard serves the purpose of seed bearers and provides seedling regeneration. It helps in switching to other high forest system in future.
3. The system has the potentiality to produce large sized timber.

Disadvantages

1. It is difficult to put the system into practice. It requires great skill in maintaining correct balance between standard and coppice. Above all, it is difficult to effect proper distribution of standards in the whole of the forests. It has been adopted in many forests. However, the problem lies in the fact that the standards left under the system are not uniformly distributed.
2. The system has some intrinsic silvicultural incongruities. The standards left in the forest during felling operation develop tendency to become branchy because of profuse lateral light they receive during few years after main felling.

Coppice with Reserve

This system is also widely applied to our forests. In the past, this was adopted in many parts of our miscellaneous forests. This system has its own advantages and disadvantages:

Advantages:

1. It helps to protect the forest ecosystem.
2. It helps to maintain crop mixture.
3. It helps to improve the condition and composition of the crop.
4. It fulfils the needs of the local people.
5. Reserves induce seeding and so it helps in regeneration process in the forest.

Disadvantages:

1. Its practice requires high degree of skill.
2. It does not produce large sized timber.

This system has a wide range of applicability. It can be applied into the forest, which shows the following conditions :

1. When the crop varies greatly in density, composition, quality, growing stock and proportion of the valuable species.
2. When most of the species are good coppicers and the coppice power of the species is vigorous.
3. When it is possible and socially feasible to protect forest areas against uncontrolled fire and grazing at least for 5 to 10 years after main felling.

Practice of Silvicultural Systems in different Forests of India: A Retrospect

As discussed earlier, Silvicultural systems are adopted on the basis of the following criteria:

— Objectives of management.
— Requirement of the site and compatibility of forests to the system.
— Silvicultural requirement of the species constituting the forest composition.
— The intensity of demand for the forest produce like timber, poles, fuels, etc. locally as well as nationally.
— The degree of development of the transportation systems and other infrastructure in the forest areas.

Over the years, the criteria for selecting silvicultural systems have changed. Presently, the objectives of management critically influence the selection of silvicultural system for a forest. Forest policy directions and guidelines generally decide objectives of management. Now, all parts of Indian forests are accessible and they have been brought under ambit of the management plan or working plan.

Different forests of India were worked under different silvicultural systems. A synopsis of silvicultural system adopted with other management attributes in different forests across the country is presented below to provide a perspective with regard to management of Indian forests in the post-Independence era.

1. Deodar Forests

Jammu and Kashmir

1. Objective of management:
   To meet the local and commercial wood demands.
2. Silvicultural system:
   — Shelterwood (Compartment) System for accessible forests.
   — Indian Selection System for forests situated in steep and difficult terrain.
3. Exploitable diameter: 75cm.
4. Rotation: 150 years.
5. Yield Regulation: By volume and calculated by various standard methods.

Uttar Pradesh :

1. Silvicultural system:
   — Group selection system.
   — Indian irregular shelterwood with floating periodic blocks.
   — Uniform system.
2. Yield Regulation:
   — The yield was variously fixed in different working circles by volume units or by number of selection trees.

2. Spruce and Silver Fir Forests

Jammu & Kashmir

1. Silvicultural System: Indian selection system.
2. Exploitable diameter: 75 cm.
3. Rotation: 50 to 200 years.
4. Felling cycle: 30 to 40 years.
5. Yield Regulation: By number of trees in accordance with Brandish method.

Himachal Pradesh

1. Silvicultural System : Indian irregular shelterwood system (Punjab shelterwood system).
2. Exploitable diameter : 70 cm.
3. Rotation : 180 years with 6 periodic blocks.

Uttar Pradesh (Chakratta Division)

1. Selection System: Selection System.
2. Exploitable diameter : 60 cm.
3. Rotation : 160 to 180 years.

Himachal Pradesh

1. Silvicultural System: Punjab Shelterwood System (For Deodar and Kail forests).
2. Rotation: 150 years with 5 periodic blocks.
3. Exploitable Diameter: 60 cm.

3. Pine Forests

Assam (Forests of Khasi and Jaintia Hills)

1. Silvicultural System: Clear-felling with or without retention of the standards.
2. Rotation: 120 years.
3. Yield: Controlled by area.
4. Regeneration: Artificial methods mainly by 'Taungya method'.

Himachal Pradesh

1. Silvicultural System: Punjab Shelterwood System.
2. Rotation: 120 years with 4 periodic blocks.
3. Exploitable Diameter: 60 cm.

Uttar Pradesh

1. Silvicultural system: Uniform system with floating periodic block.
2. Rotation: 120 years.

4. Evergreen Forests

Assam (Mesua—Dipterocarp formation — semi-evergreen forests)

1. Silvicultural system : I. Indian irregular shelterwood with floating periodic block. (Upper Assam—Mesua—Dipterocarpus Forests)
   II. Conversion to Uniform system. (Assam valley—semi-evergreen formation).
2. Yield regulation: By volume with area check and girth limits for exploitation.

Kerala

1. Silvicultural system: Selection system.
2. Exploitable girth: 210 cm.

Tamilnadu/Karnataka

1. Silvicultural system: Selection system.
2. Rotation: 150 years.
3. Yield Regulation: By area.
4. Felling cycle: 30 years.

Andaman

A. Non-tidal forests of the Andamans :

1. Silvicultural system : Conversion to uniform with concentrated natural regeneration.
2. Rotation: 150 years (hardwood).

B. Mangrove forests of Andamans:

1. Silvicultural system: Clear-felling in strips (with three periodic blocks).
2. Rotation: 99 years with a regeneration period of 33 years.

5. Moist Deciduous Forests

Sal Forests

A. Assam:

1. Silvicultural system: Uniform system. Alternate strip system, clear-felling and selection-cum-improvement.
2. Rotation: 100 to 150 years.
3. Yield regulation: By volume in Uniform system and by number of trees above certain maximum exploitable girth limit with an area check in Selection-cum-Improvement (SCI) System.
4. Regeneration: Natural regeneration or Assisted natural regeneration in case of uniform system. Artificial regeneration in case of clear-felling system.

B. Bihar :

Sal Forests :

1. Silvicultural system : Conversion to uniform (for the site ranging from quality I to III).

2. Conversion period: 120 years (with 6 periodic blocks of 20 years each).

3. Yield Regulation : By volume and area.

Sal forests in the upper slopes of the hills:

1. Silviculture system : Selection system.
2. Felling cycle : 20 years.
3. Exploitable girth : 90 cm.
4. Yield Regulation : By area and by the percentage of selection trees to be removed.

C. Madhya Pradesh :

Bastar and Bilaspur : Sal forests

1. Silvicultural system : Selection-cum-Improvement Felling systems.
2. Exploitable girth : 150 to 180 cm.
3. Yield regulation : By number of trees and by Sagreiya's modification of Brandish method.

South Raipur, Mandla and Bilaspur : Sal forests

1. Silvicultural system: Conversion to uniform .
2. Conversion period: 120 years with four period blocks of 30 years each.
3. Yield regulation : By area.

Raigarh and Sarguja : Sal forest

1. Silvicultural system : Coppice with Reserve system.
2. Felling cycle : 20 to 30 years.

6. Moist Mixed Forests with Teak

Forests of Raipur, Rajnandgaon, Balaghat, Seoni, Chindwara, Betul and Hosangabad.

Site quality = I areas

1. Silvicultural system : Improvement and slection-cum-improvement felling system.
2. Exploitable diameter : 60 cm to 120 cm.
3. Felling cycle : 20 to 40 years.
4. Yield Regulation: 50 % of selection trees are removed.

Site quality = II (Moist teak forests)

1. Silvicultural system: Conversion to Uniform
2. Conversion period : 80 to 120 years, with 4 to 6 periodic blocks of 20 to 30 years each.
3. Yield Regulation : By area in all divisions except Raipur division where a volume check was applied.

7. Moist teak forests of Maharashtra

1. Silvicultural system : I. Conversion to uniform (in Chanda division).
   II. Selection-cum-improvement felling (in Thane division).
2. Regeneration: Natural regeneration: in Chanda forests.

8. Moist mixed forests without teak

1. Silvicultural system : I. Selection system (Forest of Gujarat, Kerala and Tamil Nadu).
   II. Improvement Felling (M.P. forests).
   III. Selection-cum-Improvement (Maharastra Forest).

Felling cycle : 40 to 50 years.

9. Dry deciduous Forests

A. Dry Mixed Forests with teak or sal:

Forests of Andhra Pradesh, Bihar, Gujarat and Andhra Pradesh

1.Silvicultural system: I. Andhra Pradesh. Selection-cum-Improvement.

II. Bihar. Coppices with standard system (About 80% of forests of Bihar state were worked with coppice rotation varying from 20 to 40 years felling cycle).

III. Gujarat. (a) Simple coppice with artificial regeneration in patches,
(b) Coppice with reserve,
(c) Selection coppice,
(d) Selection-cum-Improvement.

2. Yield Regulation : By area method.
3. Felling Cycle: 30 to 35 years.

B. Dry sal and teak Forests of Madhya Pradesh (Bastar, Raipur, Rajnandgaon, Mandla, Balaghat and Hoshangabad)

1. Silvicultural system: Coppice with reserves in Dry sal forests.
2. Felling Cycle: 20 to 40 years.
3. Yield Regulation: By area method.

C. Dry teak forests of Maharashtra, Karnataka, Punjab and U.P.

1. Silvicultural system: I. Coppice with Reserve system with enrichment planting (Maharastra),
II. Coppice system followed by artificial regeneration of valuable species (Karnataka), and
III. Coppice with standard (with rotation of coppice 30 years and of standard 90 years) (Punjab and U.P.).
2. Yield Regulation : By area method.

D. Dry teak forests of Orissa (Kalahandi, Bolangir, Rayaganda and Jaypore forest division)

1. Silvicultural system : Selection system and coppice system.
2. Yield Regulation : By an area method.

E. Mixed Forests without teak or sal (Dry deciduous)

1. Silvicultural system : I. Selection system (A.P. and Orissa).
II. Coppice with reserve (M.P.).
III. Coppice with standard (Rajasthan).
IV. Coppice system (Orissa).

2. Felling Cycle : 40 years.

9. Thorn forests

1. Silvicultural system : I. Simple coppice system (Andhra Pradesh).
II. Improvement Felling (Gujarat).
III. Clear-felling with plantation of economically important species (Tamil Nadu).

10. Bamboo forests

Different bamboo species are found in different forests of India. A list of bamboo species is provided in Table 2.2. Cutting cycle of bamboo practiced in different parts of India is presented in Table 2.1.

Table 2.1 : Bamboo Forests

1. Dendrocalamus Strictus

S.No	State	Cutting Cycle	Other Remarks
1.	Andhra Pradesh	3 years	—
2.	Assam	4 years	—
3.	Bihar	3 to 4 years	A minimum number of mature culms are cut and all the immature culms are retained.
4.	Gujarat	4 years	—
5.	Jammu & Kashmir	4 years	—
6.	Tamil Nadu	3 years	Removing only mature culms from each clump.
7.	Madhya Pradesh	4 years	—
8.	Maharashtra and Karnataka	3 to 4 years	For dendrocalamus — 4 years. For Bambusa aurindinacea—3 years.
9.	Punjab	2 to 4 years	Leaving three to six old healthy straight culms in each clump. Current years' culms are not cut.
10.	Uttar Pradesh	4 years	
11.	West Bengal	4 years	—
12.	Tripura	3 years	—
13.	Orissa	4 years	—

Table 2.2 : Distribution of different species of Bamboo in different Forests of India

S. No.	Name of the States	Name of bamboo species	Other Remarks
1.	Andhra Pradesh	Dendrocalamus strictus	Pure formation mixed with trees.
2.	Assam	Dendrocalamus strictus	
3.	Bihar	Dendrocalamus strictus	
4.	Gujarat	Dendrocalamus strictus	
5.	J & K	Dendrocalamus strictus	
6.	Kerala	Bembusa aurundinacea	Pure formation occurs in limited extent. Occurs along with tree associates of ever-green and semi-evergreen forests.
		Dendrocalamus strictus	Limited extent.
		Ochlandra travancorica	Occurs along with evergreen and semi-ever-green forests.
		Ochlandra rheedii	Occurs along banks of rivers and streams.
7.	Maharashtra and Karnataka	Dendrocalamus strictus	Dry deciduous zone.
		Bambusa aurundinacea	Dry deciduous forests along with water courses.
8.	Uttar Pradesh	Dendrocalamus Strictus	-do-
9.	Punjab	Dendrocalamus Strictus	-do-
10.	West Bengal	Dendrocalamus hamiltonii	Along the gorges of Tista and its tributaries and outer foothills.
11.	Orissa	Dendrocalamus strictus and Bambusa aurundinacea	—

Yield Regulation and its Principles

Yield regulation is an important aspect of forest management. It has influenced the management practices of our forests in the pre as well as post-Independence period. However, over the years, the stocking in our forest depleted and extent of dense forests declined. The productivity of our forest is on a decline. Considered in this backdrop, the conservation of existing dense forests and restoration of degraded forests assumes stupendous significance in the practice of forest management. The primacy of timber over the management practices has come to an end. The aspects of conservation of forest ecosystem have taken precedence over everything in forest practices. This has resulted into a situation in forest management where the aspects of yield regulation lost its importance and relevance. However, in areas of dense forests, it is still necessary to develop suitable methods of yield regulation while evolving the strategy for sustainable forest management.

The silvicultural systems like clear-felling system, shelterwood system or uniform system are not generally practiced in Indian forests at present. Instead, the selection and coppice systems are in vogue now-a-days. The obvious reason of change over of silvicultural system is attributed to fast depletion of our forest resource base. Now with a new awakening, the forestry management has widely adopted the selection system with minor modification to suit the silvicultural requirements of our forests. Nevertheless, we shall discuss in this chapter the yield regulation in selection forest on a border framework.

The methods of yield regulation that generally applied to a selection forest is based on volume and/or increment of growing stock (G.S.). This is also called the formula method. This method perfectly applies to normal forest where there is balanced distribution of age gradation and also, a normal increment. The two aspects of balanced distribution of age gradation and increment are complementary to each other. Such forests are rarely found in India. Though, attainment of normal forests remained the objective of management practices in good old days, but it was never attained in its true sense. The concept of normal forests remained a dream unfulfilled in the Indian history of forest management. However, in such selection forests which exhibits varying degree of near normalcy in stem distribution, this formula method can be applied with caution. In spite of all flaws, this method is widely used to determine the trend of timber productivity and to regulate wood production. There are two such popular methods. They are illustrated below:

Von Mantle's Formula

This method is widely used in calculating the yield of timber in such forests, which are worked under selection system. In an un-evenaged or selection forests, we generally find representation of stems almost of all age classes. In such forest, the formula indicates the relationship among growing stock, increment and rotation to estimate the following parameters.

1. Growing stock of the forests.
2. Rotation.

In a selection forest it is always difficult to measure rotation. However, on an average, the rotation of dominant trees is taken as the rotation of the crop. This may be taken as 100 years or near to it, basing on the composition of the crop. The formula aims at estimating the increment of the forests. The formula goes in this way:

I = 2GS / r
I = Increment (in the shape of final yield)
GS = Growing stock
r = Rotation

This Von-Mantle's Formula is a popular method of measuring the increment in selection forests. This is owing to the fact that the method is simple to use and it applies to Indian forest conditions. In the past also this method was widely adopted for regulating the timber yield. Now-a-days, while evolving forest management practices, inventory of tree resources is generally prepared to estimate the growing stock of the forests. Therefore, with a growing stock calculated on the basis of sound principles of random sampling, it becomes very easy to calculate the increment of the forests. However, lack of proper distribution of young, middle and old-aged crop in the forests leads to unrealistic forecast of yield. Therefore, considered in this perspective, it is best suited to even-aged forests than to normal forests. However, in the absence of any other reliable and easy method, this is generally used for yield regulation in the selection forests along with some supplementary check method.

Brandis Method

There are many methods to calculate the yield of selection forests based on the volume. Brandis method is one of those methods which was developed in the middle of nineteenth century. It was in vogue for a long period. It was applied to Indian selection forests in those times when there was no practice of inventorying the tree resources of the forests. The Brandis method was based on some data generated from strip enumeration and stump analysis in teak forests. The method was contingency method to exploit the merchantable trees in selection forests. The yield was calculated based on average number of trees reaching exploitable size annually plus a fraction of surplus number of trees in the girth class above the exploitable class or size. Again, there was a flexibility of removing the surplus trees in the first felling cycle or in the future felling cycle. In other words, with this method, the system of reservation of trees in the selection girth classes started with an aim to sustain a future yield. However, with an introduction of a system of resource survey and of inventorying of trees, the Brandis method lost its relevance and applicability.

Smithie's Safeguarding Formula

This system of yield calculation based on volume was developed by Smithie in the thirties. This method was evolved to cater to the sal forests of U.P. which were worked under selection system. Application of other methods of yield regulation led to removal of selection trees as surplus stock. This gave rise to a situation where the yield of the forests kept on decreasing over successive felling cycles. Against this backdrop, Smithie considered it necessary to evolve a system to calculate the yield based on the principle of sustained yield in selection forests. This is how this method of safeguarding formula was evolved. At present, the method is still widely used for yield calculation in selection forests all over India.

This method is based on the following premises:

1. It is necessary to maintain sustained yield, mostly from "Selection class of trees" at every felling cycle. This sustained yield was termed as "selection yield".
2. The sustained yield of selection trees is based on the trees having girth above exploitable limit.
3. No method of yield calculation can safeguard the selection yield fully.
4. Area yield with a check on the selection trees can be applied to all types of forests to safeguard the selection yield to the extent possible.

This method is applied to such selection forests where the following conditions are met:

1. The exploitable or selection trees are fairly distributed over every unit area of the working circle.
2. The exploitable trees are regularly being recruited in every unit area of the working circle within the time frame of the predecided felling cycle.
3. The middle-aged and younger diameter or girth classes are

well represented and the present rate of recruitment into the two highest diameter/girth classes (I and II) will not decrease for quite a few decades.

Methodology of yield regulation under Smithie's Safeguarding Method

The methodology of yield calculation in this method is as follows:

1. First decide the felling cycle and exploitable diameter or girth of economically important trees. Exploitable girth for teak is decided on the basis of stump analysis or on the secondary data with regard to growth. Exploitable girth for other species are generally decided on the growth parameters recorded in preservation plots or on the basis of secondary data regarding growth statistics.
2. Fix the diameter or girth classes. These classes are indicated with symbols like I, II, III classes of trees. Class I symbolizes trees above exploitable diameter and others stand for trees below it until the youngest.
3. Enumerate the number of trees available in various diameter or girth classes. More importantly trees, which stand for classes I, II and III are required to be enumerated for calculation of yield. However, in order to assess the representation of middle and younger age classes in the crop, it is also necessary to enumerate the trees falling in all classes. This also helps in ensuring sustained yield in subsequent felling cycles. The youngest diameter class usually below 10 cm diameter is not enumerated. In Madhya Pradesh, there is a practice of enumerating all trees above 20 cm girth.
4. Find out the "x" number of trees per hectare. This 'x' indicates the number of existing class II trees, which pass into class I during the felling cycle. This "x" number is determined from the formula given below:

$x = f / t$ (II - z % II),

$x =$ Number of trees selected for felling.

$f =$ felling cycle in years.

$t =$ Time taken by a class II trees to pass into class I stage.

$z =$ Percentage of class II trees that do not pass into class I stage in 't' years on account of mortality or removal in thinning, etc.

3
Forest Management Plan : A Historical Perspective

Introduction

In the Pre-British period, Indian subcontinent had plenty of forests. The population at that period was also low. Forest resources were only used locally by the rural folk. There was only a semblance of regulation in the administration of forest. With the advent of British Raj in India, forest resources received great attention. With the increase in communication facilities in most parts of India, exploitation of forest started. Such exploitation of forest was very irregular and unsystematic. British Government wanted to convert the forests of India into a sustained source of timber to European markets. Towards this end, the British Government embarked on programme to organize and exploit the forests on some scientific principles. Management of Indian forests on the scientific principles started with the introduction of a management plan. The management plan for a forest area was called as 'working plan'. Working plans are the written documents enunciating the management strategies for a particular forest area.

The history of management of forests began in the year of 1837. Gradually, the practice of managing forest resources by some sort of plans and estimates evolved. Finally, the system of managing forest by the method of working plan emerged. Therefore, the saga of forest management in India is a long one. A better understanding of the nuances of the evolution story is possible if we study the different phases separately. Hence, we divide the historical account of forest management into the following phases:

1. Early Period: Before 1884
2. Post Early Period: 1884 to 1914
3. War Period: 1914 to 1939
4. Pre-Independence Period: 1939 to 1947

5. Post-Independence (Early Period): 1947 to 1961
6. Post-Independence (Late Period): 1961 to till today.

The fact that planning is the *sine qua non* of forest management was realized long ago. Planning functions began with activities like survey, settlement, reservation, demarcation, area consolidation, mapping etc. With the passage of time, it was further realized that in such preliminary activities it was not possible to take appropriate strategic decision with regard to forest management. Hence, other planning activities like stock mapping, regeneration study and demand assessment were added to those earlier exercises. Because of these studies, it was felt necessary not only to control felling of timber trees in the forests but also to make efforts for promoting regeneration and improving the crop. This laid the foundation for necessity of a concept like 'forest management plan' or 'working plan'. As the term implies, it is a plan for working a particular forest over a certain period, usually not less than 10 and not more than 20 years, under the pre-determined objects of management. (Anon, 1961).

Early Period : Before 1884

Forest management in real terms began with the concept of forecasting annual timber yield on the basis of some form of estimate of the crop and the rate of growth of trees. The man who initiated this exercise was Mr. Munro, the then Superintendent of Forests in Travancore in 1837. He estimated that about 1 lakh teak trees would be required to be felled in the forest of Travancore. His way of computation of number of trees required to be felled was primarily based on "personal observation and experiences of nearly twenty years". Therefore, this concept of estimating the yield on the basis of growth of trees and stand structure and evolving felling pattern based on that study heralded a beginning in the field of forest management. Therefore, Stebbing was perhaps right in assuming that "the credit of having been the first to introduce a simple form of forest working plan in India must be ascribed to Mr. Munro". (Anon, 1961).

Mr. Munro's effort in estimating yield and regularizing felling, was in fact a precursor to the evolution of working plan in its present form. Afterwards the German trained forester Dr. Brandis made a true beginning on his appointment as Superintendent of Forest in Pegu (Burma) in 1856. It was Brandish who propounded for the first time the fundamental principle that in any forest to be worked as many first class trees (trees over a prescribed diameter) could and should be felled in one year as would be replaced during the year by the growing stock of the second class trees. In order to apply his method, he estimated the growing stock in tree forests with detailed size class wise distribution. He assessed the growing stock by a method of "Linear Valuation Survey". He determined the growth of teak trees by the method of ring counting. Finally, Brandis prepared an estimate of the yield of Pegu forest and presented it in his first Working Plan. The plan was prepared for a brief period of 5 years. This method of yield calculation developed by Dr. Brandis was latter known as "the Indian Method" of yield regulation. Thereafter, it received international acceptance and was incorporated into the syllabus of forest management.

With the appointment of Brandis as the first Inspector General of India, a concerted effort was made to prepare working plan in India. Brandis visited many forests and took the necessary initiative to prepare simple working plan for selected forest areas. In the year 1864 Brandis, Steward and Wood prepared a "Valuation Survey Report on Bushahr forest in Punjab. Mr. Webster conducted similar surveys in the hills of Kumaon and Garwal and prepared a plan on the basis of the survey findings. In the year 1864, Dr. Brandis visited the sal forests of Central Provinces and undertook a valuation survey in those forests. This valuation survey method consisted of counting trees in a given small portion of the area and estimating the aggregate number of trees over the whole area by simple process of multiplication. This valuation method was used to be conducted without any stock map. An account of initial working plans is presented in Table 3.1.

In addition to Dr. Brandis, two other trained foresters Mr. Schlich and Ribbentrop continued the practice of formulating working plans for different areas to a great extent. However, the forest area for which the working plans were prepared constituted a very small part.

Table 3.1 : An account of initial working plans

S. No.	Name of the Forests	Name of the forester	Year	Nature of working plan exercise
1.	Singhbhum forests (Bihar)	Mr. Schlich	1873	Yield Estimation
2.	Palamau Forests (Bihar)	Mr. Johnston	1874	Valuation Survey
3.	Kanasar, Konain, Deoban and Ekra Block of Chakrata Division (U.P.)	Dr. Brandis	1875	Preliminary Report

Till the year 1884-85, only 282 square kilometre of forest in India was covered under different management plans or working plans. This tardy progress was primarily attributed to initial heavy work associated in forest settlement, survey, demarcation, fire protection and most importantly, to paucity of trained staff.

In the year 1882, decentralization of forest administration took place. Until this, the affairs of the forest departments were controlled by the Central Government. Inspector General of Forest exercised control over the conservators of forests in the provinces. Therefore, with the decentralization, the revenues from the forest went to local governments. This created a situation where revenue raising irrespective of its impact on forests became the guiding principle for forest management. "As a result, even where working plans existed, the prescriptions thereof, could be easily deviated from, and even dodged by the local forest officers, under the direct influence of the Local Government", (Anon, 1961). With the beginning of this phase of decentralization, the importance and role of working plan in the forest management got diluted gradually. More focus was given on the felling of trees and propping up revenue. Where the working plan did not exist the need for their preparation was conveniently lost sight of (Anon, 1961).

However, this phase of undoing the initiative for preparing working plan continued for a short spell only. In the year 1984, Mr. Schlich as the Inspector General of Forests, took an epoch making step in reversing the retrogressive trend in the field of forest management. He centralized not only the process of formulating working plans but also the management of forests as per the prescriptions of the working plans. For this, he after sustained efforts obtained Government of India's sanction to his scheme of centralization of forest planning, management and control. He opened a working plan branch to accelerate the process of making working plans for different areas. The sagacious and visionary act of Mr. Schlich was a landmark in the history of forest management in India. Mr. Ribbentrop, in his memorial report of 1900, referred to the above step, saying: "This was one of the epoch-making events in our forest history."

Post early Period : 1884 to 1914

Centralization of working plan process for different forests and the management of these forests as per the plan prescriptions heralded a new beginning in the history of working plan preparation in India. This intensified the drive to undertake activities, which facilitated the preparation of working plan in many states.

Table 3.2 : List of early working plans and the officer who prepared them

S.No.	States	Forest area	Year	Forest officer who prepared
1.	Central Province (Presently M.P.)	Hoshangabad (Bori forests)	1894	Caccia
2.	Punjab	Kalesar	1892	D'Arcy
3.	Himachal Pradesh	Kulu	1894	Fisher
4.	Bihar	Singhbhum	1890	Haines
5.	West Bengal	Darjeeling	1892	Mason
6.	West Bengal	Sunderbans	1893	Heinig
7.	Kerala	Nilambur	1895	Cushington
8.	Assam	Goalpara	1993	Perrce

Few of such pioneer working plans are indicated in Table 3.2. During this period from 1884 to 1914, the exercises of working plan preparation continued through the agency of the local forest officers. But, what is noteworthy is the fact that during this period there was a forest policy in force, which guided the formation of working plan prescriptions. Besides, there was an official practice to prepare working plans under the strict technical advice of the Inspector-General of Forests who had the necessary authority to scrutinize the draft prescriptions and to ask the local government to amend it accordingly, if found not compatible with the principles of forestry. In such situation, where the local government disagreed to the recommendation of Inspector-General of Forests, such working plan prescriptions were required to be referred to Government of India for a final decision. Such mechanism of control and supervision, which aimed at improving the quality and the spread of working plan formulation process, paid rich dividends. As a result, forest areas under working plan increased in many states. In Utter Pradesh, by the end of the year 1899, over 75 per cent of the forests were brought under the cover of working plans. In Maharashtra and Gujarat, by the end of year 1897-98, about 6434 square kilometre of forest area was brought under the purview of the working plan management system.

Apart from the increase in extent of forest area, the quality of the plan also improved a lot. Unlike that of the initial years when plans were very sketchy and simple, during this phase working plans began to be written systematically. After 1890, the practice of writing working plans in a systematic format began. The forester D'Arcy laid the foundation for a standardized format for working plan document in his monumental treatise on Forest Working Plans in 1891.

In spite of all advancements in the field of working plan, the phase was marked with many disadvantages. They are as follows:

1. Working plans were brief and simple.
2. The forest areas were not divided into compartments.
3. Stock mapping was not done usually, because of lack of topographical sheet on suitable scale.
4. Yields were regulated by the area method. In stray cases, volume control method was put into practice.

With the beginning of twentieth century, few of plans that were prepared in earlier years came up for revision. The process began in the existing set-up. Nevertheless, from 1906 onwards, the task of scrutinizing the plan was assigned to the Superintendent of Working Plans stationed at the Forest Research Institute, Dehradun. This practice facilitated the process of incorporating advancements in the field of forest research into the making of working plans. Thereby, it began a new era in the process of working plan formulation in India.

War Period: 1914 to 1939

This is a brief phase, which marks a difference from the previous one. The momentum gained in the previous phase died out, primarily on account of out-break of First World War. Apart from this, other factors also played an important role in dampening of the enthusiasm that characterized the first decade of this century. Those factors were inaccessibility of forest areas, poor communications, lack local demand for forest produce and diversion of trained forest staff to such activities like logging in accessible areas to meet timber demand for the war. After cessation of war, however, the situation improved and preparation of working plans again gained momentum in the country. The phase between the end of first world war and the beginning of second world war had a lot of distinctive characteristics.

1. The plans were comprehensive in nature.
2. Detailed stock maps were prepared on suitable topographical

sheets on appropriate scale. Such topographical sheets were made available by Survey of India.

3. Working plan areas were divided into blocks and compartments.
4. Research findings made by Forest Research Institute, Dehradun began to be incorporated into the final preparation or revision of working plans. Forest Research Institute, by this time identified the different forest types of India and provided a framework for classification of forests. This step helped in evolving proper management regime for different forests. Apart from this, the fact that forest research created a data base enabled the working plan officer to gather definite information about not only growth and silivcultural characteristics of most of the timber tree species, but also the technique of obtaining regeneration and their probable response to different set of silvicultural treatments. During this period, another dimension was also added to the working plan formulation exercise. This was the aspect of financial calculation of yield on account of working in the forests. Mr. Bourne did an exemplary work by adding to the existing process of working plan formulation.

This period marked the application of intensive method of silviculture. In many working plans, there was a prescription of conversion of existing irregular forests into regular crop. This method worked successfully in teak and sal area where regeneration was not a problem. This system also worked with some good result in Himalayan forests, which responded favourably to the introduction of successive regeneration felling. However, the introduction of intensive methods of management met with failures in the deciduous sal forests as well as moist and dry teak forests. Such forests were beset with problems in obtaining sufficient natural regeneration within a given period. Adoption of silvicultural system like conversion to uniform system played havoc with such forests.

Recruitment of officers to the Indian Forest Service started after the cessation of First World War. This facilitated the process of working plan preparation to a great extent in this phase. However, this recruitment only lasted up to the year 1939. This event again slowed the pace of efforts to bring fresh forest areas under the fold of working plan. Added to it, another important factor virtually gave a set-back to the progress in this direction. That is decentralization of control of Government of India on working plan preparation. The Inspector General of Forest lost hold over the drive for bringing forests area under working plan.

Pre-Independence : Period 1939 to 1945

The world event of Second World War gave serious body blow to the working plan process in the country. The working plans, which expired, could not be revised. In many forests areas, felling were carried out in total disregard to the working plan prescriptions. Irregular and destructive working known as 'advance working' was carried in many parts of the country. The main objective of forest management pursued by the local government was to exploit the forests to cater to the war needs of British Empire. Exploitation of more accessible forest areas began on a larger scale not known in the past. The surge in exploitative activity in the forests primarily aimed at supplying wood to the ship building sector of Britain. This resulted into depletion of timber trees of higher diameter sizes in accessible forests.

However, with the end of Second World War, the work on preparing working plan for left out areas and revising the lapsed working plan was again restored. More serious attempt was made to survey and assess the damage on account of 'advance working' in the accessible forest and to suggest ways and means to improve the growing stock. The Central Provinces created a separate circle for working plans in the year 1945. Revision of many working plans was taken up on a war-footing. In the year 1945-46, a special conference was held in Dehradun to standardize the technique of partial enumeration, required for yield estimate. Yield Table and Volume Tables were further improved to suit the management requirement of different forests.

Post-Independence Period (Early 1947 to 1961)

Independence heralded a new beginning in the history of forest management in India. The importance of working plan was restored in the post-independence era. During this phase more detailed information on the growth and yield were collected to constitute the base for working plan preparation. Apart from this, the practice of preparing detailed stock maps began.

Success of silvicultural system in many deciduous forests largely depended on natural regeneration. But, such forests failed to have natural regeneration within a given period. Given this, the strategy of retention of advance growth as a part of the future crop emerged as an answer to the problem of regeneration. This ethos gradually crept into the forest management practices in many parts of India.

In many states like Madhya Pradesh, Karnataka, Jammu and Kashmir, forest areas which were under the princely states were added to the forest departments in the period from 1948 to 1950. In such forests, there was virtually no forest-working plan in force. Likewise, there was no organization to carry out any systematic management in such princely states. After the merger of such forest area into the state forests, the concerned forest departments undertook the task of preparing working plans for such areas afresh.

Abolition of proprietary rights over forests in the post-Independence period increased forests areas for drawing up forest working plan. Such forests which were burdened with proprietary rights were never surveyed and demarcated properly. Apart from this, no preliminary planning work had ever been undertaken with respect to forest management in such areas. Therefore, all necessary spade works like survey, demarcation, stock mapping, notifying started after their merger with the forests of the states.

The work on bringing most of the fresh forest area under the fold of working plan received a fillip in the Post-Independence phase. When the country got freedom, about 37 per cent of the forest area were under the operation of working plan. But, in the year 1957-58, the forest area under the scientific management increased to 45 per cent and an additional 11 per cent of the forest area was taken up for preparing working plan. In this phase, there was a distinct focus on scientific management of forest.

The increasingly large areas taken for the preparation of working plans year after year are indicative of the realization on all sides of the vast headway that has to be made in the scientific management of our forests, (Annon, 1961). The global position of forests with respect to the practice of working under a management plan or working plan is presented in Table 3.3. It is clear from this Table that in the year 1958, the percentage of forest area in most of the developed countries, which worked under a system of management plan, varied from 22 to 50 per cent. During this period, India also had brought about 40 per cent of forest areas under the fold of scientific management.

In the sixties and seventies of this century, the country made rapid strides in the field of bringing more forest area under the fold of working plan. Every state developed its own organization to take up the work of formulating working plan. Utter Pradesh had been the pioneer state in India in the field of establishing suitable organization for preparing and revising working plans. The state also evolved a working plan code to systematize the planning process. Gradually, other states also followed the path and built up suitable planning organization. In the state of Madhya Prasesh, then was an elaborate and efficient organization that undertook preparation and revision of working plan. Here, there is a tradition of assigning the job of preparing or revising the working plan to a senior forest officer and printing the plan in his personal name. This has created a condition, which encouraged and facilitated the forest officer to devote full attention to formulate the plan prescription compatible with the requirement of the crop and the people living near the forests. In states like Maharashtra, there is a practice of permanent working plan division attached to territorial circles. Such permanent divisions kept on making or revising working plans for the forest under the circle.

Table 3.3 : Management of forests in selected areas

S. No.	Name of the areas	Total closed forest area (000 ha)	Area under management plan (000 ha.) 1963	1983	Percentage of coverage under working plan
1.	Europe	145486	93010	83484	57 %
2.	Soviet Union	791600	299965	791600	100 %
3.	United States	195256	84378	102362	52 %
4.	Tropical Asia	305510	37370	39790	13 %
5.	Tropical Africa	216634	10610	1735	1 %
6.	Tropical America	678855	2444	522	—

Source : Based on data from WRI (1986).

Post-Independence Period (late) 1961 to till today

In this phase, almost all forest areas leaving aside some major pockets in North-eastern part of the country, were brought under the operation of working plan prescriptions. Revision of working plans also gained momentum. In this period, two exercises were added to the working plan formulation process. They are forest tree inventory or resource survey and the socio-economic survey. The tree inventory exercise provides the detailed information about stem distribution in the forests and the condition of the crop. This helped in making correct estimate of timber yield as well as regeneration status in floor of the forests. The inventory also provides a benchmark in the field of forest resources. The socio-economic survey helped to assess the quantum of dependence of people on the forests and its resultant impact on the resources. This also helped in understanding management and utilization pattern of available forest resources.

With the Wildlife Ordnance 1972, more focus was laid on wildlife conservation in India. In sanctuaries and national parks, wildlife management plans were prepared. After putting forestry in the concurrent list, Forest Conservation Act was promulgated in the year 1980. This heralded the beginning of legislative control of Government of India over the forestry matters. Otherwise, forest conservation was absolutely a state affair before this epoch making constitutional amendment in India. Again, the enunciation of New Forest Policy in 1988 proved another milestone in the field of forest management in India. The policy necessitates formulation of management plan for every forest area in our country. This was again emboldened by the directive issued by the honorable Supreme Court in the year 1996 through a Public Interest Litigation (PIL) case, which made it mandatory that no forest in India will be exploited or worked without any working plan approved by Central Government. This again gave further impetus to the process of working forest with a definite management plan.

Forest Management — A Global Perspective

In general, state ownership dominates the world forest resources. However, a vast stretch of forest is still under private ownership. In the United States, the federal government owns 34 per cent of the total forest area (Mather, 1990). In Canada, the federal forest accounts for 27 per cent of the wood volume (Dunster, 1988). In European community countries, the average holding size for private, commercial and state forests are respectively 4.5, 199 and 910 hectares (Mather, 1990).

State ownership or communal or private ownership does not ensure implementation of any management or working plan. Ownership has no relation with management control, Mather (1990) analyzing the relationship of ownership with management, said that, management of forests over the years is declining. "A number of firm conclusions can be reached about the nature of management control. First, it is clear that state ownership does not necessarily correlate with state management, nor indeed with any form of management. Second, it is apparent that much of the world forest area remains unmanaged in any formal sense. Third, there is disquieting evidence that the extent of forest management has actually decreased in much of the world forest area in recent decades." (Mather, 1990)

A small proportion of the world forest area is under formal management plan. Data collected by the FAO for the year 1963, World Forest Inventory indicated that 975 million hectares out of a total forest area of 4,229 million hectares were under management plans. Therefore, globally, around 23 per cent of the forest area is subjected to working plans. World Resource Institute has opined that less than 25 per cent of the world's forests were actively managed in the early 1980 (WRI, 1986). But, the data with regard to coverage of forest area under working plan in different continents or countries varies contrastingly and does not constitute any pattern. In Europe and North America, the proportions of the forest area under management plans were respectively 55 and 67 per cent, but in Latin America and Africa, on the other hand, the proportions were under 2 per cent (FAO, 1966). However, the World Resource Institute data with regard to areas under management plan in different continents is slightly different. In old Soviet Union countries, almost all forest areas are under working plan (Table 3.3).

If we analyze the extent of forests under management plan in different continents, the picture is more disheartening in most of the area. In Europe, most of the publicly owned forest is covered either by plans or by controls. In the EEC countries, just over 75 per cent of the publicly owned area is covered by management plans and the reminder is subject to controls (UNFCE/FAO 1985). Of the privately owned forests in the same region, however, only 6 per cent are covered by management plans. (Mather, 1990)

In Africa, the area subject to management plans has fallen from over ten million to under two million hectares, or less than 1 per cent of the closed forest area. Furthermore, much of the managed area is concentrated in a small number of countries, (Mather, 1990). In the early 1980, two-thirds of the managed areas in Africa was in Ghana (Lanley, 1982) and three quarters in Ghana and Uganda (WRI,

1986). Political turmoil in the latter has severely reduced the extent and effectiveness of management and illegal agricultural encroachment and increase in unlicensed exploitation of forest produce has followed (Hamilton, 1984). In former Belgian and British colonies, working plans, often complemented by silvicultural treatments for forest regeneration and enrichment, were introduced early, and intensive forest management in these areas amounted to 4 to 4.5 million hectares of closed forests by 1960. Thereafter, management plans were abandoned in many countries, as a result of population pressure on forests and staff shortages in forest services (Mather 1990). In Latin American countries, there is a marked decline with respect to coverage of forests under working plans. However, the case with Asian countries is different. The pattern of coverage of management plan somewhat tallies with that of African situation. World Resource Institute in the year 1986, has reported that India accounted for 80 per cent of the forests under working plan (WRI, 1986). However, as things stand today, almost all forests in India, excepting few pockets in North-eastern region, come under working plans.

In African countries, management plans have been abandoned or have not been followed up in many areas, on account of population pressure on the forests and also inadequate funding and staffing of forest services. In other words, non-compliance of working plans or deviation of working plan prescriptions takes place primarily not only because of adverse socio-economic conditions but also due to lack of necessary political will. In a number of countries such as Nigeria, Zaire and Tanzania, the extent of management has contracted considerably in recent decades and intensively managed forests no longer existed in these countries (Mather, 1990 and Lanly, 1982).

Therefore, the state and pattern of forest management differs from state to state. The ownership pattern, and the public opinion and the socio-economic conditions are primarily influencing the emergence of forest management plans. The countries ridden with mounting poverty, unemployment and weak political structure are still lagging behind in the front on scientific management of forests. However, in the global front, old Soviet Union countries and India stand distinctly in having achieved the task of bringing about 100 per cent of the forests under management plan. But, what is disquieting is the fact that large areas of state-owned forest in the tropics except India, are now still in different stages of exploitation and destruction by not only the logging companies but also government. If this pace of forest destruction continues unabatedly, then it will spell doom to the tropical forest resources.

PART - II

FOREST PLANNING : PRINCIPLES, RATIONALE AND PRACTICES

4

Elements of Planning

Introduction

Planning is 'number one task' of forest management. It symbolizes the determination of course of action to achieve a desired result. It aims at setting objectives of an organization. It is like deciding in advance what to do, how to do, when to do and who is to do. It is also an institutional process; it is characterized as the process of thinking before acting and to act only on the basis of facts and realities. Planning is the precursor of all other managerial functions. In fact, planning is the foundation of forest management.

Planning is a way of organizational life. Every organization in today's world plans its activity. Every plan is unique to the planner as well as to the situation. Planning begins with a goal or objective or targetted outcome that an organization wishes to achieve. A plan is a means to achieve the said goal. From the point of development plans are of two types. The first one is informal type and the second one is formal type. Informal planning is a product of intuitive and instinctive reaction to a situation. Formal planning is a structured, systematic and scientific process of investigation, assessment and action.

Definition of Planning

Planning is deciding in advance what is to be done. It involves the selection of objectives, policies, procedures and programmes from among the available alternatives. A plan is a pre-determined course of action to achieve a specified goal. In short, it is a blue print for action. Different authors, however, have defined planning differently. Few of such definitions are as follows:

1. *David Cleland and William King:* "Planning is the process of thinking through and making explicit the strategy, action and relationship necessary to accomplish an overall objective

or purpose."

2. *George R. Terry:* "Planning is the selecting and relating of facts and the making and using of assumptions regarding the future in the visualization and formation of proposed activities believed necessary to achieve desired results."
3. *Theo Haimann:* "Planning is deciding in advance what is to be done. When a manager plans, he projects a course of action for the future, attempting to achieve a consistent, coordinated structure of operations aimed at the desired results."
4. *Koontz O'Donnell:* "Planning is an intellectual process, the conscious determination of courses of action, the basing of decisions on purpose, acts and considered estimates."
5. *George B. Galloway:* "Planning is the opposite of improvising. In simple words, it is an organized foresight plus corrective hindsight... Conceived as a process, planning embraces a series of steps: (i) the determination of objectives to be sought, (ii) research to understand the problem, (iii) the discovery of alternative solutions, (iv) policy making and chosing between alternatives, including the frequent choice of doing nothing, (v) the detailed execution of the chosen alternative known in physical planning as layout or design."
6. *Alfrod and Beatty* : "Planning is the thinking process, the organised foresight, the vision based on facts and experience that is required for intelligent action.

Characteristics of Planning

Planning has the following characteristics:

1. Planning is closely associated with the goals or objectives of the organization.
2. Planning is mainly concerned with looking ahead in the future.
3. Planning involves the selection of best alternative.
4. Planning is required at all levels of management.
5. Planning is a continuous and unending process.

Advantages of Planning

There are many advantages of planning. Few of such advantages are summarized below:

1. It enables a purposeful and orderly set of activities instead of random action.
2. It facilitates the process of decision-making in the organization.
3. It helps in identifying specific objectives and strategies, which guides future activities.
4. It provides opportunities to understand and determine resource status, constraints, and other problems.
5. It provides opportunities to weigh alternative courses of action and to evolve the best possible option.
6. It helps in continuing existing successful activities and management practices.
7. It provides opportunities to adapt to unpredictable changes in the environment.
8. It helps in cementing and synergizing the organization towards achieving the common goal. Plans are linkages that connect the work activities of individual and groups at different organization levels to a common purpose. Planning is the activity that clarifies the higher level objectives as the end towards which lower level objectives act as means.
9. It helps in improving the performance at each level of the organization.

In nutshell, planning has been found to help performance in five ways: (1) planning focuses management attention on objectives, that can generate results; (2) planning helps management establish priorities; (3) planning assists management to emphasize organizational strengths; (4) planning helps management to cope with ever-changing external environment; and (5) planning facilitates control.

Features of a Good Plan

According to L.F. Urwick, a good plan is that which (i) is based on a clearly defined objective, (ii) is simple, (iii) provides for a proper analysis and classification of action, i.e., which establishes standards, (iv) is flexible, (v) is balanced, and (vi) uses available resources to the utmost before creating new authorities and new resources. The best test of effective planning is its realistic and viable nature. A good plan must be pragmatic and must lead the organization forward on the path of progress and prosperity. A good plan opens up new avenues, and ways of doing things and reveals specific opportunities previously unknown to the planner.

Objectives of Planning

The objectives of forest management planning are as follows:

— The basic objective of forest planning is sustainable development of forest resources.

— It aims at optimum utilization of forest resources and other natural endowments. In other words planning aims at sustainable and best use of resources.

— It aims at serving the village community and the society at large by extending the tangible and intangible benefits. It aims at using the resources in such a way that it helps in maximizing the benefits to the society on a sustainable basis. Benefits encompass wood, non-wood benefits and life supporting elements.

Planning Approach

Planning, by simple definition, is to 'devise detailed methods for doing, arranging, and making something'. For doing different things different approach should be adopted.

Some useful approaches employed in forest management planning are explained below :

Bottom up Approach

This approach necessitates planning from the bottom or base level. The problems at the base level should be studied first and data at the bottom level should be collected. On the basis of the problems at the bottom level the plan should be evolved to address it. Therefore, the approach which looks at the bottom level for identification of problem and for generation of supporting data is called bottom up approach. Many management plans failed because such an approach was not adopted in the past. Generally, local communities are not adequately involved in forest management planning process. As a result, such plan miserably fails. Therefore, this approach has immense relevance to the field of forest management planning. There are several ways that can be applied to involve people in the planning process. The present practice of joint forest management is one of them. Involving villagers and communities for planning development of forests, pasture and protected areas constitute bottom up approach.

Iterative Approach

Planning basically is an iterative approach. Before a final plan is prepared, many surveys, studies, assessments, alternative considerations and revisions will have to be made. Generally, in the field of forest management planning one or two preliminary reports are prepared on the basis of management need. Such reports are prepared by gathering, analyzing and interpreting data from the field. After a thorough discussion on such reports in the review committee, generally consisting of senior officers from different branches of forestry, an ultimate strategy for sustainable development of forests is evolved. Therefore, in the process of finalization of management plan many steps are repeated to hammer out the best appropriate strategy. Such steps may consist of different field study, analysis and presentation of reports.

Flexible Approach

An approach is called flexible when it leaves enough room for future adjustments, modification and revision. A management plan is the starting point of forest management. The plan is drawn for a period of 10 to 15 years. On account of long span of the plan, it is necessary to adopt a flexible approach in formulating any forest management plan. This approach will help to cope with new problems and policies and to induct new technique in the management. Forest management is a complex task dealing with social, economical, cultural, legal, institutional, physical and ecological problems. Difficulties arise during implementation. It has been found that in order to fine tune the plan, it becomes necessary to revise and recast original strategy and objective. Therefore, it is imperative to make the plan flexible; so that any additional experience or knowledge gained in the process of executing the plan can effectively be incorporated in the management plan. In order to achieve this, the component of monitoring and evaluation process, in addition to other elements, should be built into the plan.

Planning Levels

Planning process begins with identification and determination of goal and objectives. Such an exercise depicts the purposes, premises, values and direction of planning. Once the goals and objectives are decided, the next stage of planning follows. Planning in an organization requires well-defined and delineated frameworks. Different organizations have different framework for planning. Generally, the framework for planning in a forest department has three levels of management hierarchy. The planning functions of these three levels are as follows:

1. Top level management (State level). It determines strategic goal and formulates strategic plan for the forestry sector at the state level.
2. Middle level management (Territorial, divisional and circle levels). It determines tactical goal and formulates tactical plan.
3. Lower level management (Range level). It determines operational goal and formulates operational plan.

The national and state forest policy determines the organizational goal and it sets stage for drawing up a strategic plan for the state. Now, this plan becomes the inputs for the establishment of tactical goals. Based on the tactical goals, the tactical plan is formulated. Similarly, tactical goal determines operational goals. Based on these gaols, the operational plan is drawn.

Types of Plans

Organization dealing with management defines goals at three different levels. Therefore, in order to achieve those avowed goals at different levels the organization requires evolving plans at those three levels. Such plans are called as strategic, tactical and operational plans.

Strategic plan aims at achieving strategic goals. It illustrates the ways in which the organization will utilize its resources and conduct its activities in order to achieve the strategic goals. The top functionaries of the organization formulate strategic goals. Basically it covers the long-term perspective planning.

Tactical plans aims at achieving tactical goals. It encompasses specific segments and aspects of the strategic plan. Suppose, the strategic plan comprises a component of rehabilitation of degraded forests. Then the tactical plan identifies it as one of the objectives of the plan. It aims at evolving suitable strategy to address it. Therefore, tactical plan defines major problems already identified in the strategic plan and draws treatment to improve on the situation.

Presently, in the forest departmental organization forest management plan (working plan) constitutes the tactical plan. Middle management functionaries prepare the tactical or working plan in the forest department. The middle level manager plays the role of tactical planner who is called as working plan officer in the department.

Operational plan aims at achieving operational goals. It supports the tactical plan. It comprises actions that are necessary for the achievement of operational goals. Lower management functionaries like forest range officer prepares such plans. Operational plan covers a short time frame and it serves as a guide for day-to-day operations. The harvest plan, plantation project, fire management plan, thinning plan etc. are some of the examples of operational plan.

As already explained, any major programme is a plan. Apart from this, other courses of action are also called as plans. Therefore, any course of action can be called as plans. Effective planning requires proper understanding of existing different courses of action. Practically, plans are classified as: (1) purposes or missions, (2) objectives, (3) strategies, (4) policies, (5) procedures, (6) rules, (7) programmes, (8) budgets. To a great extent, they all constitute a distinct hierarchy.

Mission

Mission identifies the basic function or task of an enterprise. Every kind of organized operation necessitates a mission. Every organization or enterprise has a basic function or task, which is assigned to them by the society or government. Sustainable management of forest resources is the mission before the forest departments.

Objectives

Objectives or goals are the ends towards which activity is aimed at. In other words, they are the results to be achieved. They not only represent the end of planning, but also symbolize the purposes towards which other management functions are aimed at.

Strategies

Strategies have been defined variously. However, the most comprehensive definitions are presented below:

— General programmes of action and deployment of resources to attain comprehensive objectives.

— The determination of basic long-term objectives of an enterprise and adoption of courses and allocation of resources necessary to achieve their goals.

Generally, strategies provide a framework for guiding, thinking and action. Strategies are put into action by means of plans. Such plans are called as strategic plans. Tactics are the action plans through which strategies are executed. Therefore, strategic plans are always supported by tactical plans.

Policies

Policies are general statements, which influences thinking. They define an area with which a decision is to be made. They also ensure that decisions will be consistent with the avowed objectives. They are guides to decision making and so, allow certain amount of discretion.

Procedures

Procedures are also a kind of plan that establishes a required method of understanding future activities. They are guides to actions and practices rather than thinking. They generally, constitute the manner and pattern in which certain activities should be accomplished to achieve an objective. In other words, procedures are nothing but chronological sequences of required actions.

Rules

Rules are the simplest types of plan. They indicate a specified action in a required manner. No discretion is allowed in such rules. Therefore, rules guide action and activities.

Programmes

Programmes are complex packages of goals, policies, procedure, rules, task assignments, steps to be taken, resources to be employed and other basic elements necessary to carry out a given course of action. Programmes are supported by budget.

Budgets

A budget is a statement of expected results expressed in numerical terms. Generally, a budget is expressed in financial terms or numerically measurable term. It is planning as well as a control instrument. In an organization, programme budgets are prepared. Such budgets identify goals and develop detailed programmes to meet the goals. Besides, they estimate the cost of each programme. In some organization, zero based budget programmes are prepared. Here, the goals and objectives are determined first and programmes required to achieve them are formulated. In other words, the approach is to evolve budget programme from scratch or 'base zero'.

Planning Process

Planning is the beginning of management process. Management as a whole is defined as the process of planning, organizing, leading and controlling the resources of an organization in the effective and efficient pursuit of specified organizational goals (Fig. 4.1 and 4.2). This definition of management amply illustrates the critical relationship between planning process and other management process. (Fig. 4.3). Therefore, management of any resources begins with planning. Planning helps in identifying what must be done and how it must be done. In other words, planning sets the stage and becomes central to other key management practices of the organization.

Planning process comprises seven closely related steps or activities (Fig 4.4). These steps are as follows:

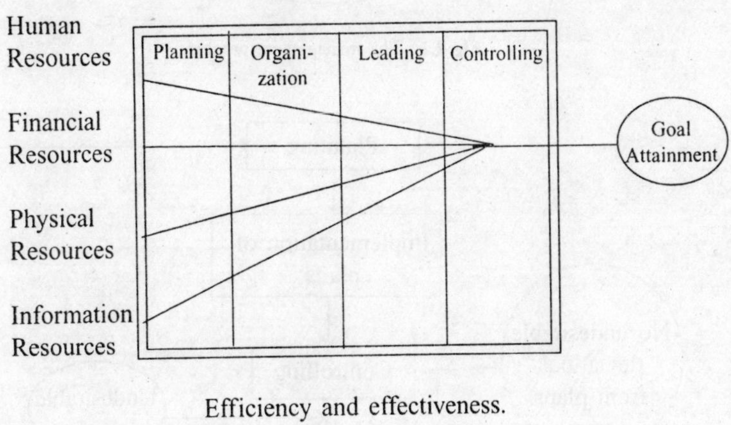

Fig 4.1 : Management in organization.

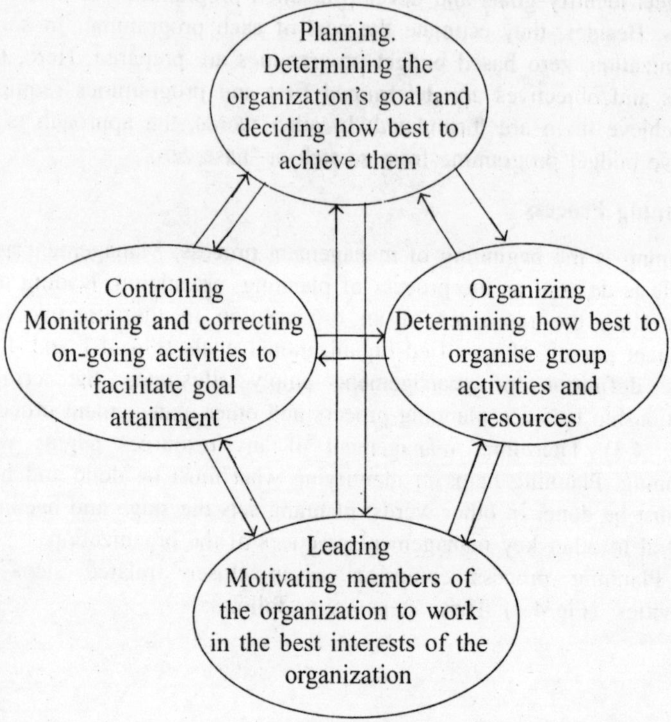

Fig 4.2 : The management process.

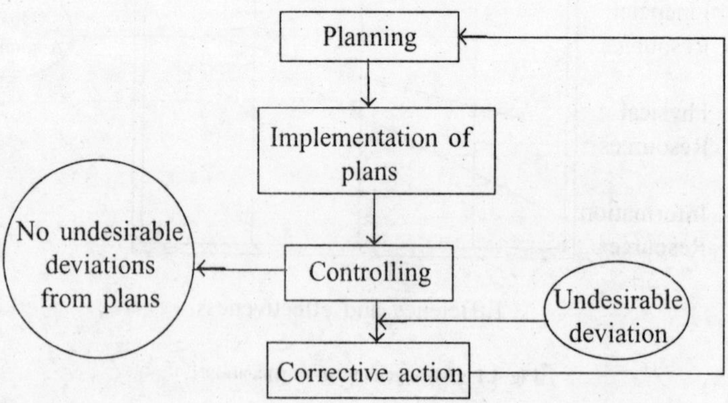

Fig 4.3 : Close relationship of Planning, Implementing and Controlling.

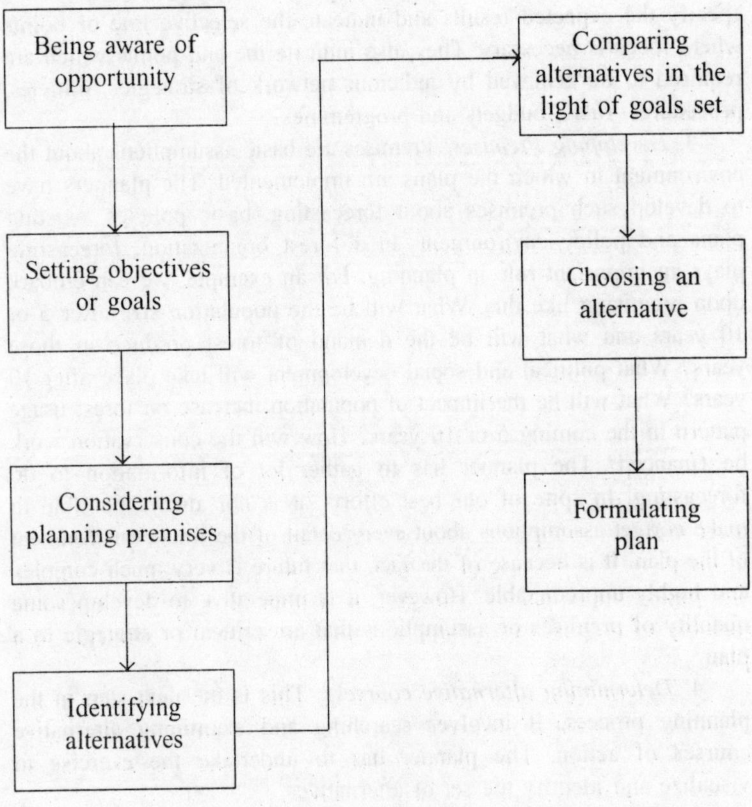

Fig. 4.4 : Planning Process.

1. *Being aware of opportunities:* This exercise of making oneself aware of opportunities in the internal as well as external environment is the precursor of the real planning process. A planner must know where his organization stands and what are its strengths and weaknesses. He should develop awareness within himself about the opportunities and constraints, problems inherent in the environment. Determining pragmatic objectives depends on such awareness. Therefore, effective planning requires realistic diagnosis of the opportunity and problem situation.

2. *Establishing objectives:* Establishing objectives in an organization is the second step in the planning process. Objectives are established on long term as well as short term basis. Objectives

specify the expected results and indicate the selective line or points where focus is necessary. They also indicate the end points which are required to be achieved by judicious network of strategies, policies, procedures, rules, budgets and programmes.

3. *Developing Premises:* Premises are basic assumptions about the environment in which the plans are implemented. The planners have to develop such premises about forecasting, basic policies, existing plans and policy environment. In a forest organization, forecasting plays an important role in planning. For an example, we can embark upon premising like this. What will be the population size after 5 or 10 years and what will be the demand of forest produce in those years? What political and social development will take place after 10 years? What will be the impact of population increase on forest usage pattern in the coming 5 or 10 years? How will the conservation work be financed? The planner has to gather lot of information to do forecasting. In spite of our best efforts, it is not always possible to make correct assumptions about every detail of the future environment of the plan. It is because of the fact, that future is very much complex and highly unpredictable. However, it is imperative to develop some quantity of premises or assumptions that are critical or strategic to a plan.

4. *Determining alternative courses:* This is the next step in the planning process. It involves searching and examining alternative courses of action. The planner has to undertake the exercise to visualize and identify the set of alternatives.

5. *Evaluating alternative courses:* This is the next step in the planning process. After identifying an array of alternatives, it is necessary to evaluate each alternative by weighing them in the context of premises and goals.

6. *Selecting a course*: This is the next step in the planning process. This is the stage where major decision taking process takes place. Analysis and evaluation of alternative courses expose the merits and demerits of each course. The planner has to select a single course or a group of compatible courses in the context of avowed goals and objectives.

7. *Formulating plan:* This is the last step in the planning process. After taking proper and rational decision about the set of alternatives, the plan is formulated.

Strategic Planning

Strategic planning is the process by which the strategy of an organization is analyzed and decided on. In general sense, it is usually perceived as a mediating force between the organization and its environment. Strategic planning is sometime referred to as 'framework planning'. It means, examining problems, opportunities, constraints, resources and abilities and taking decisions about general priorities and courses of action (Dale, 1998).

Today, every organization does strategic planning. The degree of strategic planning may differ. In the field of natural resource management, strategic planning assumes stupendous relevance and significance. Conceptually, strategic planning is simple. It is analyzing the current and expected future situation, determining the direction of actions for the organization and developing means of achieving the goal or mission. But, in practice, it is an extremely complex process which necessitates scientific and systematic approach to identify and analyze factors external to the organization and meticulous matching them with the capabilities of the enterprise. It encompasses wide areas. It goes beyond the allocation of resources to achieve organizational objectives. It inculcates and promotes strategic thinking. It comprises an appropriate organizational structure, an effective management information system and a budgeting system.

Strategic planning symbolizes interaction among three main elements. These elements are:

- Organization and resources.
- Opportunities, problems and constraints.
- Objectives and outputs.

The organization-specified objectives identify the types and determine the magnitudes for output. Such output leads to fulfilment of objectives. The organization mobilizes resources like manpower, information, finance and physical assets to convert them into inputs for the activities that the organization intends to carry out. As already said that while formulating the strategy, there is an imperative need for analyzing opportunity and problems/constraints.

Tasks in Strategic Planning

The main tasks in strategic planning are as follows:

1. *Problem analysis:* It deals with identification of problems, their causes and effects. It explores the extent to which it is feasible to resolve and decide about the prioritization of problems.

2. *Stake-holder analysis:* It deals with an assessment of needs of all people who have some interest or connection with the project, plan or programes. During the course of such analysis, all aspects, which may impinge on the interests of the people or stake-holders are taken upon. Interests of these people who are likely to be affected are analyzed. Likewise, interest of the people who are likely to promote the programmes will also be analyzed.

3. *Analysis of objectives:* This exercise encompasses discussion, analysis and determination of all goals and objectives.

4. *Analysis of inputs:* This exercise helps in determining the need and feasibility of arranging resources in one hand and decides about the quantity of resources required to use.

5. *Analysis of external influencing factors:* This task pertains to identification of factors that are likely to promote the opportunities and enhances constraints separately. Apart from this, it also helps to develop future course of action to deal with such factors in future.

6. *Analysis of organization:* This task involves assessment of present organizational capabilities. Besides, it prescribes necessary changes in the organizational structure and process to improve the organizational strength.

Goal Setting Process in Forest Management Planning

Goal setting is the beginning of the planning process. There are three basic management related issues in the goal setting process. These issues are critical to successful delineation of goal and objectives. The issues are illustrated below:

1. Dealing with multiple goal and objectives.
2. Barriers to goal and objective setting process.
3. Improving the goal setting process.

Forest organization has many goals and objectives to achieve. Most of the time these goals and objectives are not complimentary to each other. Quite often these are in conflict to each other. For example, biodiversity conservation and timber production are two conflicting goals. Activities that aim at conserving bio-diversity and wildlife will be at the cost of timber production. Such conflict and contradiction appear on the way of determining goals and objectives.

However, such problems are resolved if the planner addresses himself to assessing the intensity and implication of such conflict and contradiction. After understanding the conflicts and constraints in its perspective, the next task is goal optimizing or balancing or reconciling of possible conflicts among different goals and objectives. The planner has to decide whether it is appropriate to undertake the fullest attainment of one goal or objective or a group of goals and objectives to the exclusion of others. In a forest organization which is fully controlled by the government, the best choice is to pursue a course of action designed to attain a modest mix of all the goals and objectives.

The goal setting process in the forest organization suffers from two important infirmities. They are: (i) selection of unattainable goals and objectives, and (ii) overemphasis on either quantitative or qualitative goals or objectives. However, the goal setting process can be improved by adopting the following guidelines (Griffin 1990):

- Planner should understand the purpose of goal.
- Goal should be properly stated:

  1. They should be specific.
  2. They should be concise.
  3. They should be time tested.

- Goals should be horizontally and vertically consistent.
- Planner must accept and be committed to the goals.

5

Sustainable Forest Management: Principles and Policy Environment

Introduction

Forests are the planet's largest and most important terrestrial ecosystem. They have profound influence on the structure and functions of human habitat locally and globally. They constitute the largest reservoir of plants and animals on the land. They sustain much of world's diversity of life embodying stupendous genetic wealth developed over evolutionary time. They provide innumerable goods and services to human society. They provide the means for their livelihoods as well as succor to millions of people who live close to forests. Forests affect the quality of atmosphere and temperature regime, which in turn regulate global and local climate. They provide spiritual, aesthetic and symbolic value for millions of people.

All these benefits from forests can continue to flow to society, present and future, on a sustained basis. But this is possible if the forest resources are managed on the basis of sustainability principles. Managing forests sustainably necessitates an elaborate management system. Which aims at maximizing the achievement of aforesaid objectives for the benefits of society at large.

— Defining Forest Management
— Changing Perspectives
— Multiple Management of Forest
— Sustained Yield

Defining Forest Management

Forest management has been defined differently by many authors embodying basic ingredients in different proportion and verying emphasis. Some of these different definitions are presented below:

1. Forest Management is defined as the practical application of

the scientific, technical and economic principles of forests. (Anon, 1953)

2. Forest management is the application of business methods and technical forestry principle to the operation of a forest property.
3. Forest management is that branch of forestry whose function is the organization of a forest property for management and maintenance, by ordering in time and place the various operations necessary for the conservation, protection and improvement of the forest on the one hand and the controlled harvesting of the forest on the other.
4. Forest management, by implication, is not a basic subject in itself; it is the practical application of science, technology and economics to a forest estate for the achievement of certain objectives—mainly production of wood timber and industrial raw material, and other forest products such as resin, gum, tan, bark, etc. It is based on the knowledge of a number of basic subjects' science, such as silviculture, ecology, geology, pedology, botany, mensuration, pathology, economics, finance, etc. (Ram Prakash *et al.* 1979).
5. Forest management is defined as dealing with "... the overall administrative, economic, legal, social, technical and scientific aspects involved with the handling of conservation and use of forests. It implies various degrees of deliberate human interventions, ranging from action aimed at safeguarding and maintaining the forest ecosystem and its functions, to favouring given socially or economically valuable species or groups of species for the improved production of goods and environmental services (FAO. 1991).

Changing Perspectives

In the past the country had good dense forest. The population growth was not alarming. The demand for the forest produces remained within the limit to which the existing forest could sustainably produce. Deforestation was never perceived as a problem. Therefore, management of forest in the pre-Independence period broadly focused on three main tasks. They were:

1. Control of composition and structure of the growing stock.
2. Harvesting and marketing of forest produce.

3. Administration of forest property and personal.

The forest management practices prevailing in the pre-Independence period continued till eighties. Apart from this, a sea change has taken place in the socio-economic environment pertaining to forest of the country. In one hand, the anthropogenic pressure on the existing forests increased manifold and community control over forest use system weakened. In the other, more unsustainable harvest and use of forest resources increased by leaps and bounds. Both these factors combined to form a situation where the growing stock of our forest started declining alarmingly. As of today, the country has 76 million hectare of forest land. Out of it, we have only 39 million hectare of dense forest, which satisfies the parameter of a forest ecosystem. The rest of forest land is degraded and barren. Therefore, considering the forest scenario of the country in the aforesaid backdrop, the focus of forest management at present time lays on the following issues:

1. Restoration of degraded forest.
2. Development of medium and dense forests.
3. Conservation of existing dense forests and its resources.

Restoration of degraded forests requires protection and afforestation. Again development and conservation of existing good forests necessitates evolving mechanism not only for controlling composition and structure of the forest crop but also for sustainable harvesting and use of forest produce.

Multiple Management of Forests

Forests are managed for a multiplicity of purposes. In a welfare society like ours, the role of forest in serving people assumes utmost significance. The multiple purposes of forest management are generally protective, climatic, productive, scientific, recreational. But while managing a unit area of forest, all such purposes cannot be equally synchronized. One purpose has to take precedence over other. However, in an array of objectives one would be principal and others become subsidiary to the principal one. Apart from this, the subsidiary multiple objectives have to be prioritized and then they should be integrated and synchronized with the principal one. Till today, in case of the management of dense forests, the timber production has received the utmost primacy. However, now the time has come for the forest management to address itself to the task of prioritization of objectives basing on a synchronized assessment of community and national needs.

Likewise, the degraded forests are required to be tackled for rehabilitation work on a war footing; so that we shall start arresting the process of depletion of soil resources in one hand and sustainably producing benefits from the forest areas for the local community in the other.

Sustained Yield

Sustained yield has been an age-old principle of forest management world over. It pertains to production of timber and other measurable forest produce on a sustainable basis. This principle has been accepted as the basic for evolving timber management practices in the working plan in the past. However, sustained yield has been defined and explained variously by different agencies or authors. Few of such definitions/explanations are as follows:

1. Sustained yield: (a) "The material that a forest can yield annually or periodically in perpetuity. (b) As applied to policy, method or plan of management, it implies continuous production with the aim of achieving at the earliest practical time at the highest practical level an approximate balance between net worth and harvest by annual or somewhat longer periods" (Anon, 1953).
2. "The regular, continuous supply of the desired produce to the full capacity of the forest".
3. "The yield of timber or other forest produce from a forest which is managed in such a way as to permit the removal of approximately equal volume or quantity of timber or other forest produce annually or periodically in perpetuity".

The concept of sustained yield has been evolved from the basic consideration that the latter generations may derive from the forest at least as much of the benefits as the present generation. The principle of sustained yield envisages that a forest should be so exploited that the annual or periodic felling do not exceed the annual or periodic growth, as the case may be. (Ram Prakash *et al*. 1979).

Sustainable Forest Management

Forest provides social, economic and environmental benefits. Therefore, evaluation of forests from a narrow angle does not portray the totality of forestry. In order to explore forestry on a holistic basis, it is necessary to comprehend the concept and principles involved in "sustainable development".

Sustainable development is defined as "... the management and conservation of natural resource base and the orientation of technological and institutional change in such manner as to ensure the attainment and continued satisfaction of human needs for the present and future generations. Such sustainable development (in the agriculture, forestry and fisheries sectors) genetic resources, is environmentally non-degrading, technologically appropriate, economically viable and socially acceptable" (FAO, 1991).

The World Commission on Environment and Development in 1997 produced one of the most widely accepted definitions of sustainable development. This definition of sustainable development is presented below :

> "Development that meets the needs of the present without comprising the ability of future generations to meet their own needs."

The concept of sustainable development recognizes that utilization of resources will change natural ecosystems, but that conservation is also important. It also recognizes that utilization of forests is important for achieving social goals, such as poverty alleviation. But such uses must take care of the fact that the ability of the future to meet their own needs is maintained. In other words, the paradigm of development should be based on a trade-off between the needs of present and future. However, over the years, many governments have started adhering to the concept of sustainable development and integrating it into their policies.

Sustainable forest management is forestry component of sustainable development. This is development which is economically viable, environmentally benign and socially beneficial and which balances present and future needs. More precisely speaking, sustainable forest management is the process of managing forests to achieve one or more clearly specified objectives of management with regard to the production of a continuous flow of desired forest products and services, with out undue reduction of its inherent values and future productivity and without undue undersirable effects on the physical and social environment (ITTO, 1998)

Sustainable forest management aims at ensuring that the values derived from the forest meet present-day needs while at the same time ensuring their continued availability and contribution to long-term development needs. The values can be forest products like wood, fruits, etc. and services like soil and water conservation and life supporting services. Forest management involves planning the production of wood for commercial purposes as well as meeting local needs for fuelwood, poles, food, fodder and other purposes. It includes the protection or setting aside of areas to be managed as plant or wildlife reserves, or for recreational or environmental purposes. (FAO, 1993). "Sustainable management is considered under the headings of wood conservation, wild-life management, the conservation of biological diversity, forest protection and climate change; policy, institutional and legal aspects, socio-economic aspects and research needs." (FAO,1994)

In sustainable forest management, timber production should not be the sole aim and that the sustained provision of a wide range of environmental, social and economic needs must be included. In the past system of working, the gap between sustained yield for wood and multipurpose sustainability was glaring. The science and practice of forestry is largely simultaneous conservation and use of forests for the welfare and development of human societies. The difficulties arise mainly from attempts to combine conservation with potentially destructive uses. Using forest for wood production only is potentially one of the most destructive uses of forests.

However, wood is the most important product of the forests. The sustainable management of forests for wood production is based on the simple principle that harvesting the wood should be done at an average annual rate that is no greater than at which the forest in question can grow it and without harming the forest's capacity to supply other goods and services. Elucidating this principle is easy, but the problem lies in its application. The increment being physically inseparable from the trees, with which it is produced, must be harvested by felling a part of the forest equal in amount. Here lies the complexity of sustainable management.

The first two elements of determining the annual increment and identifying mature trees having volume equal to increment have been taken care of in the past system of yield management. The practice continues today also. The third element is concerned with regeneration or silvicultural part. This has gone beyond the control of the present system of management. The fourth is the harvesting or using the forest. In practice, it is not easy to devise, measure, apply or monitor any one of these, and the difficulties are compounded because all parts have to work together and simultaneously. The difficulties increase with the ecological and structural complexity of the forests to be managed and their economic and social environment (Janzen,1975).

Sustained Timber Yield to Sustainable Forest Management

The transition

The transition from sustained yield management for wood to successful sustainable forest management in a broader sense is now the main challenge. It is, as some says, an impossibility (Jacob, 1988). There is no gainsaying of the fact that sustained yield management for wood has been an uphill task. In this backdrop, the task of combining and synchronizing not only sustained yield of wood with the unimpaired production and protection of the many other non-wood products, but also services and values of forests on a holistic basis is the most daunting and challenging job today. Therefore, it is high time to address to this task.

To sum up, sustainable forest management encompasses an array of activities or objectives whose significance changes with the change of socio-economic and political environment. These activities are as follows:

- The production of wood on a sustained basis for commercial purposes.
- Conservation and harvesting of Non-Wood Forest Products (NWFP) and other genetic resources on sustainable basis.
- Meeting local needs for fuel wood, poles, food, fodder and other purposes.
- The protection or setting aside of areas to be managed as plant or wildlife reserves (PAs) or for recreational or environmental purposes.
- Regeneration of wastelands and degraded forests.

In order to make forest management sustainable, it is also indispensable that the aforesaid objectives or activities must maintain or improve the forests, ecological integrity and contribute to people's well being both now and in future. Ecosystem integrity has been defined as the ability to support and maintain a balance, integrated, adaptive biological community having a species composition, diversity and functional organization, comparable to that of natural habitat in the region. Likewise, well-being encompasses the economic, social and cultural aspects of people lives as influenced by forest management. Therefore, apart from carrying on activities or determining objectives for forest management, it is necessary to ensure that two conditions are satisfied. These conditions are: (1) ecosystem integrity is maintained now and in the future, (2) well-being of people is maintained or enhanced now and in future.

Sustainable Forest Management : The Policy Environment

The country had a great tradition of environmental conservation which taught us to respect nature and to take cognizance of the fact that all forms of life—human, animal and plant—are closely interlocked and that disturbance in any one gives rise to an imbalance in others. Today also, our Constitution lays down provision for environment conservation. The Part IV of the Constitution encompasses the directive principles of state policy which lays down the duties for the state to protect and improve the environment and to safeguard the forests and wildlife of the country. Apart from this, the Constitution, under article 51(g) states that "it shall be the duty of every citizen of India, 'to protect and improve the natural environment including forests, lake, rivers and wildlife and to have compassion for living creatures'."

Objectives of forest management are normally determined on the basis of constitutional provision, various legislations, current forest policy (1988) and national conservation strategy and development 1992. Besides, Stockholm Conference (1974) on environment and Rio Summit (1991) on environment and development, other international meetings, conventions, and workshops held in the last decade influenced forest management practices immensely. Such policy statements released after prolonged deliberation at the national levels, have profoundly influenced the course of forest management in the country.

The country has a great tradition of managing forest resources on the basis of avowed forest policy. The first National Forest Policy was passed in the year 1894. Thereafter, in the post-Independence era, the second forest policy was announced in 1952. Now, the third Forest Policy, i.e. New Forest Policy was passed in 1988. The New Forest Policy heralds a new beginning in many areas of forest management.

New Forest Policy, 1988

The New Forest Policy was promulgated on 7th December, 1988. In its preamble, it has been emphasized that forests in the country have suffered serious depletion. The reasons cited in the document are relentless pressures arising from ever inadequacy of protection measures; diversion of forest lands to non-forest uses and tendency to look upon forests as revenue earning resource. Therefore, there arose a need to review the situation and to evolve a new strategy of forest conservation. Forest conservation includes preservation, maintenance, sustainable utilization, restoration and enhancement of natural environment. The new strategy has come in the form of National Forest Policy, 1988.

The basic objectives that govern the National Forest Policy are as follows:

1. Maintenance of environmental stability through preservation, where necessary restoration of the ecological balances that has been adversely dısturbed by serious depletion of the forests of the country.
2. Conserving the natural heritage of the country by preserving the remaining natural forests with the vast variety of flora and fauna, which represent the remarkable biological diversity and genetic resources of the country.

— Checking soil erosion and denudation in the catchment areas of rivers, lakes, reservoirs in the interest of soil and water conservation, for mitigating floods, droughts and for the retardation of siltation of reservoirs.

— Checking the extension of sand dunes in the desert areas of Rajasthan and along the costal tracts.

— Increasing substantially the forest, tree cover in the country through massive afforestation and social forestry programmes.

— Meeting the requirements of fuelwood, fodder, minor forest produce and small timber of the rural and tribal populations.

— Increasing the productivity of forests to meet essential national needs.

— Encouraging efficient utilization of forest produce and maximizing substitution of wood.

— Creasing a massive people's management with the involvement of women, for achieving these objectives and to minimize pressures on existing forests.

The principal aim of forest policy is to ensure environmental stability and maintenance of ecological balance including atmospheric equilibrium, which are vital for sustenance of all life forms, human, animal and plant. The derivation of direct economic benefit must be subordinated to this principal aim.

The strategy advocated to achieve the objective of the New Forest Policy has been elucidated as follows:

- Increase in forest cover.

  The national goal is to have one-third of the land area under forest or tree cover. In hills and mountainous region, two-third of the area should be brought under forest as tree cover.
- Massive afforestation and tree planting programme. A massive time bound programme of afforestation and tree planting with particular emphasis on fuel wood and fodder development on all denuded and degraded forest areas should be the national imperatives.
- Scientific management of state forests. All forests should be worked with an approved management plan. Scientific and technical inputs should be applied to enhance the productivity of the existing forests.
- Linking rights and concessions with the carrying capacity of forests.

The rights and concessions, including grazing always remain related to the carrying capacity of the forests. The capacity itself should be optimized by increased investment, silvicultural research and development of the area. The holders of customary rights and concessions in forest areas should be motivated to identify themselves with the protection and development of a forest from which they derive benefits.

- Discouraging diversion of forest lands. Diversion of forest lands for non-forest purpose should be discouraged and if at all it is necessary, then the purpose of non-forest use should be subject to most careful examination.

- Promoting wildlife conservation.
- Forest management should take special care of the needs of wildlife conservation, and forest management plan should include prescriptions for this purpose.
- To associate tribal people closely in the protection, regeneration and development of forests as well as to provide gainful employments to people living in and around forests.
- To discourage shifting cultivation.
- Efforts should be made to contain the menace of shifting cultivation.
- To prevent damages to forests from encroachments, fires and grazing.
- Encroachment on forest land has been on the increase. This trend has to be arrested. The incidence of fire and grazing has to be minimized.
- To promote forest extension, education research, data base management and inflow of finances into the forest sector.
- Forestry can be promoted with the improvement in its supporting environment. So, the fields like forestry extension, research, education, data base management should be addressed forthwith. Apart from this, the objectives of this policy cannot be achieved without the investment of financial and other resources on a sustainable basis.

National Conservation Strategy and Policy Statement on Environment and Development, 1992

Apart from the New Forest Policy (1988), the National Conservation Strategy and Policy Statement on Environment and Development (1992) also influences the forest management regime. This latest policy statement embodies the core issues of forest policy, among other aspects pertaining to all sectors having relevance to environment.

The survival and well-being of a nation depend on sustainable development. It is a process of social and economic betterment that satisfies the needs and values of all interest groups without foreclosing future options. To this end, we must ensure that the demand on the environment, from which we derive our sustenance, doesn't exceed its carrying capacity for the present, as well as future generations (NCSPSED, 1992).

Over the years, there has been progressive pressure on the

environment and the natural resources, the alarming consequences of which are becoming evident in increasing propositions. These consequences detract from the gains of development and worsen the standard of living of the poor who are directly dependent on natural resources. It is in this backdrop, the country has heralded a new era in enunciating a new strategy and policy on conservation and sustainable development.

In June 1992, the Ministry of Environment and Forest released a policy statement outlining India's National Conservation Statement on Environment and Development. "Sustainable development" was the key phrase in the agenda for action. The basic aim of the agenda has "to ensure sustainable and equitable use of resources for meeting the basic needs of the present and future generation without causing damage to the environment". Fig. 5.1 depicts the different features of sustainable management of natural resources.

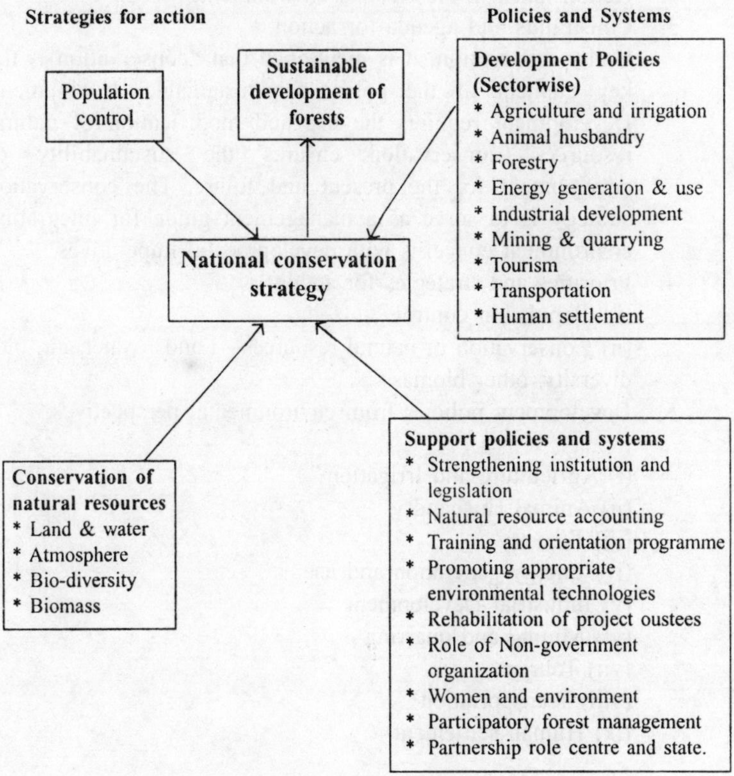

Fig. 5.1 : Sustainable Management of Natural Resources.

Government of India has already adopted legislative and administrative action in several key areas. Environment is a state subject and forest is in concurrent list. The National Conservation Strategy and the Policy Statement on Environment and Development is in response to the need for laying down the guidelines that will help to weave environmental consideration into the fabric of our national life and of our development process. It is an expression of our commitment for reorienting policies and action in unison with the environmental perspective (NCSPSED, 1992). Therefore, in order to understand the contours of policy statement pertaining to sustainable development, it is necessary to discuss it threadbare.

The National Conservation Strategy and the Policy Statement on Environment and Development have the following sections.

1. Environmental problems: Nature and dimension.
2. Action taken in the field of environment.
3. Constraints and agenda for action.

   Under this section, it is mentioned that "conservation is the key element of the policy for sustainable development. Development requires the use and modification of natural resources, conservation ensures the sustainability of development for the present and future. The conservation strategy is to serve as a management guide for integrating environment concerns with developmental imperatives.
4. Priorities and strategies for action.

   (i) Population control.

   (ii) Conservation of natural resources—Lands, water, air, bio-diversity, other biomass.
5. Development policies from environmental perspectives.

   (i) Agriculture and Irrigation
   (ii) Animal Husbandry
   (iii) Forestry
   (iv) Energy generation and use
   (v) Industrial Development
   (vi) Mining and quarring
   (vii) Tourism
   (viii) Transportation
   (ix) Human settlement

6. International cooperation
7. Supply Policies and Systems
8. Conclusion.

From the policy document, we shall discuss about three relevant aspects related to forest resource management. They are presented below:

1. Agenda for action
2. Conservation of Natural resources like land and water, bio-diversity, and biomass.
3. Development policies for forestry.

The Agenda for Action

The agenda for action calls for an integrated action on the part of the society, government institutions and individuals. These are as follows:

1. To ensure sustainable and equitable use of resources for meeting the basic needs of the present and future generations without causing damage to the environment;
2. To prevent and control future deterioration in land, water and air which constitute our life-supporting systems;
3. To take steps for restoration of ecologically degraded area;
4. To prevent further damage to and consume natural heritage;
5. To ensure that the environment and productivity of coastal areas and marine ecosystems are protected;
6. To conserve and restore the biological diversity, gene pool and other resources through environmentally sustainable development and management of ecosystems, with special emphasis on our montane, marine and coastal, desert, wetland, riverine and land ecosystems; and
7. To protect the scenic landscapes, areas of geomorphological significance, unique and representative biomass and ecosystems and wildlife habitats, heritage sites/structures and areas of cultural heritage importance.

Conservation of Natural Resources

(1) *Land and Water*

Sustainable use of land and water requires the following steps:

- Classification, zoning and apportionment of land for uses such as agriculture, forestry, grass land, green areas, industrial activities, human settlements, etc.
- Enactment of laws for appropriate land uses to protect the soil from erosion, pollution and degradation.
- Measures to ensure equitable access to and responsibility for sustainable use of land and water resources.
- Micro land planning to develop appropriate methodology and implementation of action plan by involving the people at the village land in social forestry programmes, landuse planning, afforestation etc.
- Country-wide campaign to minimize soil erosion and runoff by carrying out extensive works like contour trenching, contour bunding, terracing, construction of small storages, catchment treatment and protection of the vegetational cover in the catchment and watersheds.
- Restoration and reclamation of degraded areas.
- Measures for preventing wind erosion by undertaking special programmes of conservation and afforestation in desert areas.
- Development of suitable agro-silvicultural techniques with special emphasis on hilly areas and in arid and semi-arid zones.
- Conservation of wetlands for ensuring sustainable ecological and economic benefits.

(2) *Bio-diversity*

India's biological diversity is very rich but unfortunately its wealth is being eroded due to various reasons. This diversity needs to be preserved. Action for conservation must be directed to:

- Intensification of surveys and inventorization of biological resources in different parts of the country including the island ecosystems. The survey should include information on distribution pattern of particular species, population, communities and the status of ethnobiologically important groups;
- Conservation of bio-diversity through a network of protected areas including Biosphere Reserves, Marine Reserves, National Parks, Sanctuaries, Gene conservation centres, Wetlands, Coral reefs and such other natural habitats of bio-

diversity. This should include taxonomic and ecological studies on the flora and fauna with adequate emphasis placed on the lower vertebrate, invertebrate and micro flora which are important in contributing to the healthy maintenance of ecosystems;

- Full and correct rehabilitation of rural poor/tribals displaced due to creation of national parks/biosphere reserves/tiger reserves;
- Protection and sustainable use of plants and animals genetic resources through appropriate laws and practices;
- Protection of domesticated species/varieties of plants and animals in order to conserve indigenous genetic diversity;
- Maintenance of corridors between national parks, sanctuaries, forests and other protected areas;
- Emulating and support for protecting traditional skills and knowledge for conservation;
- Development of methodology to multiply, breed and conserve the threatened and endangered species through modern techniques of tissue culture and biotechnology;
- Discouragement of monoculture and plantation of dominating and exotic species, in areas unsuited for them and without sufficient experimentation; and
- Restriction on introduction of exotic species of animals without adequate investigations.

(3) *Biomass*

For the vast majority of our people, the foremost need is for fuelwood, timber, fodder, fibre, etc. The issue of sustainable resource utilization, therefore, has to be specially addressed first from the point of view of the biomass requirements of the rural people. Action must be directed to:

- Devising ways and means by which local people can conserve and use those resources of the common lands and degraded forests, so that they have a stake in the continuing productivity of the resources;
- Encouraging private individuals and institutions to regenerate and develop their waste-lands;
- Raising fuelwood species and provision of alternatives to reduce dependence on fuelwood;

- Taking measures to increase the production of fodder and grasses to bridge the wide gap between supply and demand;
- Raising of bamboo and species providing small timber for local house construction and agricultural implements;
- Increasing biomass to meet essential requirement of biomass based industry;
- Promoting direct relationship between forest based industry and farmers to raise needed raw materials, provided this does not result in diversion of prime agricultural lands and displacement of small and marginal farmers;
- Extensive research and development in forestry for better regeneration and improved productivity;
- Development of technologies for enhancing the productivity and efficiency of use of all biomass resources;
- Institutional and technological systems to enable rural artisans to sustain biomass based crafts; and
- Curtailment of the supply of subsidized biomass resources to industrial consumers.

Development Policies for Forestry

Implementation of the policy and realization of objectives of conservation and sustainable development will require integration and internalization of environmental considerations in the policies and programmes of development in various sectors.

Curtailment of consumerism and shift towards use of environment friendly products and processes, and low waste generating technologies through conscious efforts and appropriate economic policies including pricing of natural resources as well as fiscal incentives and disincentives will be guiding factors for ensuring conservation and sustainable development.

For environmental conservation and sustainable development, the steps, which need to be taken with respect to forestry, are outlined below:

Concerted efforts should be made for raising the forest cover and for conservation of existing forests which constitute an essential life support system and an important source of food, fibre, fodder, fuel and medicines etc. For attaining the goal of having at least one-third of our land area under forest cover, intensified measures on a mission mode are required to be taken along with commensurate mobilization of resources for this purpose. As outlined in the National Forest Policies (1988), the action points should include the following:

- Maintenance of environmental stability through preservation and, where necessary, restoration of ecological balance that has been adversely disturbed by serious depletion of forests of the country;
- Conserving the natural heritage of the country by preserving the remaining natural forests with the vast majority of flora and fauna, which represent the biological diversity and genetic resources of the country;
- Increasing substantially the forest, tree cover in the country through massive afforestation and social forestry programmes, especially on all denuded, degraded and unproductive lands involving the local people in this endeavour by giving them tangible economic motives and employment opportunities;
- Meeting the rights and concessions for requirements of fuel wood, fodder, minor forest produce and small timber of the rural and tribal population with due cognizance of the carrying capacity of forests;
- Increasing the productivity of forests to meet the essential national needs;
- Encouraging efficient utilization of forest produce;
- Restriction on diversion of forest lands for non-forest use and compensatory afforestation in case where diversion is unavoidable;
- Afforestation on common lands by the local communities through usufruct sharing schemes;
- Motivation of farmers, land owners to resort to tree farming in similar manner of crop based farming;
- Substitution of wood by other materials, alternative sources of energy and fuel efficient stores;
- Permission to forest based enterprises after a thorough scrutiny regarding the availability of raw materials;
- Supply of forest produce to the industrial consumers only at its true market value and not at concessional prices;
- Involvement of local people and dedicated grassroots non-governmental organization in the afforestation programme and for protection of existing forests;
- Creation of land banks for compensatory afforestation;
- Promotion of staffs feeding and rotational grazing;
- Restoration and protection of grazing lands;
- Involvement of local people in the policy planning on pasture lands and stalk feeding to avoid fodder scarcity; and
- Incentive for growing fodder crops and establishment of fodder banks.

6
Forest Management Plan: An Introduction

Introduction

Forests in India are owned and managed by government. Therefore, forest ownership is public in nature. Public ownership of forests enjoins upon government a responsibility to manage forests in such a way that it maximizes generation of public goods. Over the years, government has managed forest keeping the interests of the public at large. More specifically, the forests in India are managed for production as well as protective purpose. In the past, the production role took precedence over protection. However, at present, the role reversed. Forests are now managed with a primary objective to protect the entire ecosystem where man exists.

Besides, with the introduction of New Forest Policy in the year 1988 a new era dawned in the forestry sector. Forest management focused on conservation of forest resources with the help of community participation. The place of production forestry and revenue yield has been relegated to a position of low priority. With New Forest Policy in force the instrumentality of management has changed. There is a shift from unilateral management to participatory form of management. Such changes in the paradigm of forest management necessitate proper determination of goals and objectives of management for a forest division. Forest management plan or forest working plan is an instrument by which the goals and objectives are delineated on the basis of policy environment and resource status.

Forest Management Plan or Working Plan: Definition

Management of any resource requires a management plan. Forest, being an important resource since time immemorial necessitated a management plan for generation of sustained timber yield. This management plan in forest parlance is called "Working Plan". A management plan is largely about resolving conflicts, choosing goals, objectives and making decisions. If conflicts cannot be resolved, objectives are chosen wrongly or out of context and if decisions are wrongly taken, the plan will be a failure (Sawarkar, V.B., 1995). Working Plan had been defined variously by different authors in different contexts and at different point of time. This is the reason why there is wide variation in their content and tone. However few of such definitions are given below:

1. Working Plan is a written scheme of management aiming at continuity of policy, controlling the treatment of a forest. (Anon, 1983).
2. Working Plan is the simplest possible statement of what is known about the working plan areas; its configuration, soil climate, vegetation, its possibilities; what has been done in the past, what should be done in future, how it should be done and what records should be kept.
3. A Working Plan sets fourth the purpose with which a forest should be managed so as to best meet the interests, and therefore, the wishes of the owner; and indicates the means by which this purpose may be accomplished. In other words, it is a forest regulation prescribing the application of certain cultural rules and the execution of certain works, in order to produce a given result (D'Arcy, W.E., 1910).

Functions of Working Plan

Working Plan performs variegated functions. Few of such functions as enumerated by the Forestry Commission on Working Plan are:

— To ensure that National Forest Policy is made effective at all levels of forest management.

— To ensure the orderly and economic development of the forest within framework of existing policy and to prevent un-co-ordinated adhoc management decisions being taken without understanding their wider implication.

— To ensure that the aims of management are clearly defined and that the methods of achieving them are realistic and are the results of the careful study of all relevant factors.

— To ensure that all supervisory staffs concerned with the forest are aware of its intended future development and thereby achieve continuity of development when there are changes in staff.

— To bring together in a readily accessible form such information as necessary for the efficient management of the forests.

Besides, there are other important functions of working plan. Such functions are:

— To coordinate all the planning activities for the development and growth of the forests.
— To assess and compute status and potentialities of a forest.
— To provide a convenient repository for the data required in the day-to-day management of forests.
— To examine the constraints imposed upon the forests, the various possible courses of action and to work out which particular combination of possibilities most successfully achieves the management objectives.
— To convert objectives of management into achievement by means of annual plan of operations and allocation of budget, which relies on the needs, identified during the planning process.
— To provide a mechanism by which the higher management may exercise a fair degree of control over the type of operation undertaken.
— To provide an instrument to effectively monitor and evaluate the forestry operation.
— To induct modern technology and tools of analysis, monitoring and evaluation into the operation system.
— To understand the effects and impact of past management on the forest resources and the people living around forests.

Relevance of a Written Plan

Planning for forest management is a process of observation, appraisal and analysis with respect to resources, organizational strength, problems, constraints and compatible courses of actions. This process is learnt much better if the results of thinking and reasoning are written down in a simple but organized way. Apart from this, the fact that sustainable development of forestry, which needs a long term planning, calls for a systematic, logical, scientific and analytical plan. Such plans are executed for a longer period of 10 to 20 years. Therefore, considered in the backdrop of those two factors, it becomes imperative to have an elaborate written plan in the field of forestry.

Changing Role of Working Plan

Concept of managing forests changes with the passage of time. Since the inception of forest management the primary objective of drawing a management plan was to maximize timber production from the forests. Therefore working plan aimed at evolving a felling programme so that the entire accessible part of the forest would be taken up for harvesting wood in a definite period. In other words, the main focus was to hammer out a harvesting plan. However, over the years, in order to obtain sustained yield of timber on a long-term basis, regeneration strategy was amalgamated into the working plan. This led to the emergence of working plan of sort. During this period the methodology of computing the timber projection was basically very archaic and was based on thumb rules and cursory observation. Employing this methodology the working plan used to draw timber yield projection. This practice continued till the beginning of twentieth century. Afterwards, the practice of partial enumeration of trees and other improved methods to calculate timber yield began in the forest management. The exercise of computing timber yield projection on the basis of partial enumeration symbolized an improvement in technique and methodology. This method of estimation of timber continued till 1960, when the methodology of preparing detailed inventory of tree resources came into practice. Presently, this practice of conducting tree resource survey during management planning process is in vogue in all over the country.

Over the years, the role of forest changed radically. Timber producing function gradually relegated to a place of low priority. Societal and protective objectives of forest management took recedence over other production and revenue earning objectives. Societal values of forests gradually changed with the passage of time. Considered in this backdrop, the necessity for adjusting plans to meet changing circumstances has assumed immense relevance and significance today than in the past. But what is disconcerting is the fact that today the situational framework where forest management works is more conditioned by a host of intractable problems and constraints. Therefore, there may be little room for manoeuvre and consequently little choice as to the future course of action.

Objectives of Management

Forest management at divisional level is the sum total of three distinct and separate activities. These activities are:

1. Translation of forest policy into plan,
2. Execution and implementation of plans, and
3. Exercising administrative control over the plans.

Objectives occupy an important place in forest management process. They determine the shape of future events. Lack of objectives or failure to keep them clearly in the mind makes the task of management more difficult. By establishing goals and objectives the planner provides the spotlight over the courses along which management practices and activities are to be organized. In fact, an objective has an inherent power to stimulate action. Objectives help us to secure a clear vision and unity of purpose.

Management is the means by which objectives are achieved. If an objective is not reached, then there is either:

— Poor or wrong management inputs, or
— Poor or wrong choice of objectives which is not reachable given the potential resources and the existing ground realities (Sawarkar, V.B., 1995).

Determination of objectives and their optimization provides a framework for evolving strategy for sustainable forest management. While determining the objectives of management for a forest division, emphasis should be given to clarity and distinctiveness. Besides, it should also be ensured that the objectives are in consonance with the established forest and environmental policy and socio-economic tradition of the locality. Deciding of objectives provides a basic framework for evolving comprehensive strategy for sustainable forest management. While determining the objectives of working plan, it is done at two stages. One is at the stage of beginning and second is at the stage of forming working circles. The objectives so determined in the outset are called general objectives of management. The objectives set in the second stage are called special or specific objectives of management. General or creed-type objectives symbolize social viewpoint of forest management. They have less significance, so far operational part of management is concerned. Special or specific objectives are the real objectives that concern the management very much. But specific or special objectives of management are always subordinate to the general objective of management of a working plan (Table 6.1). The general objectives of forest management are enumerated below:

1. Maintenance of environmental stability.
2. Preserving the natural forests along with biological diversity and genetic resources.
3. Checking soil erosion and conserving moisture.
4. Increasing forest area by afforestation.
5. Meeting the requirements of fuelwood, fodder, minor forest produce and small timber of rural and tribal population.
6. Increasing the productivity of forests to meet essential national needs.
7. Encouraging efficient utilization of forests produce and maximizing substitution of wood.
8. Managing forests with people's participation, for achieving those aforesaid objectives and to minimize pressure on existing forests.

Table 6.1 : Index for special objectives of management.

S. No.	Description of working circle areas	Few special objectives of management	Working circle
1.	Badly eroded areas; steep slopes	1. Soil Conservation	Protection Working Circle.
2.	Areas with good site quality and dense forests	1. Sustained timber management. 2. Normalcy in age gradation.	Selection-cum-improvement working circle.
3.	Areas with rich habitat for wildlife.	1. Conservation of wildlife and their habitat	Wildlife conservation working circle.
4.	Areas replete with rich bio-diversity	1. Bio-diversity conservation	Bio-diversity conservation working circle.
5.	Areas with degraded forest	1. Rehabilitation of degraded forest.	Rehabilitation working circle.

Forestry is a long-term enterprise. It necessitates long term planning. The instrument that epitomizes long term planning is called Working Plan. Basically, this is a management plan evolved for the sustainable management of forest resources of a division. For sake of analysis and conceptualization, sustainable forest management connotes different things for different forests. The examples of degraded forest and well-stocked forest are taken for illustration.

1. Sustainable management of degraded forests encompasses the following activities:

- — Soil and water conservation.
- — Protection and conservation of all available lifeforms including the vegetational growth.
- — Development of all physical resources.
- — Restricted and minimal uses of resources till the point of optimal recovery.
- — Developing supporting socio-economic and political envoironment necessary for promotion of sustainable resource management.

2. Similarly sustainable management of well-stocked forest encompasses the following activities:

- — Silvicultural management of forest resources either for sustainable timber yield or Non-Wood Forest Products or botn with optimal mix on the basis of site potentialities and community requirement.
- — Growth and improvement of growing stock.
- — Promotion of natural regeneration.
- — Controlled and equitable use of resources according to the carrying capacity of forests.
- — Soil and water conservation, where necessary.
- — Developing supporting socio-economic and political environment necessary for promotion of sustainable resource management.

Planning Process

Formulation of management plan involves a systematic planning process, which culminates into a document. This document is called as 'Forest Management Plan' or 'Forest Working Plan'. The major steps involved in planning process are enumerated as below:

1. Analysis of external environment and determination of resource status of the forest.
2. Evaluation of results of past management on the forests.
3. Identification of problems and constraints in the last plan and their assessment.
4. Determination of objectives.
5. Forecasting broad planning premises.
6. Developing appropriate strategy.
7. Formulation of management plan or working plan.

Analysis of external environment and determination of forest resource status

Analyzing the external environment followed with determination of the forest resource status of the division is the beginning of the planning process. Before embarking on the actual planning, the organization should carefully analyses the external environment prevailing in the district and the state. The term external environment connotes a wide spectrum of social, economic, cultural and political fields. Generally socio-economic and political conditions prevailing in the district should be included in it i.e., social and economic status of different classes of the society, social aspirations, forest produce market conditions, intensity of anthropogenic pressures on the forests, dependence of village communities on forests with relation to different forest produce. Although, these conditions or factors are uncontrollable and unpredictable, yet the forest organization has to adapt its plans, strategies and programmes to changing trends in the external environment.

Objective analysis and estimation of forest resource status of the division is an important precursor to actual planning. It is technically known as 'Resource Audit'. Plan and strategies are to be adjusted according to prevailing forest and other resources. Therefore, such an analysis to determine the resources will give an exact idea about the strength and weakness of the forests. Therefore, resource audit is the real starting point of the planning process.

The resource status of the division is determined by conducting many field studies and surveys like inventorying, stock mapping, soil survey bio-diversity study etc. The qualitative status of the resource can also be adjudged by field observations. The data collected from such methods are processed, analyzed and interpreted to provide knowledge about the resource status of the forest division.

Evaluation of results of past management

Evaluation of results of past management is the second step in the revisionary process of working plan. Every forest division in India is managed under some management plan or working plan. Every plan has a definite time horizon. Before the plan period is completed revision of the plan is taken up. The process of revision necessitates evaluation of the results of the last plan on the forests. As a convention, evaluation is always done with respect to working circles. Separate measurable and easily identifiable indicators are selected for each circles and a comparison is made between two similar indicators: one at the beginning of the plan period and another in the end of the plan period. Generally the indicators taken are easily quantifiable. Such indicators can be regeneration status, growing stock, tree density, stem distribution, crop condition, density wise areas, etc.

Identification of problems and constraints and their assessment

Identification of problems and constraints is the next stage in the reversionary process of working plan. Special tasks are taken up to identify main problems and constraints. Plan should be centred on solving or alleviating those identified problems and constraints. However this exercise should be undertaken in the overall context of potentiality of the forest resources.

Major forest problems varies from once place to another. Generally, the problems are interrelated and cannot easily be separated. However, for the sake of presentation and understanding, we may group some of the common problems in the following way:

— *Technical Problems:* Lack of regeneration in the floor of the forests, lack of balanced stand structure, dominance of young crop, soil erosion, etc.

— *Socioeconomic Problems:* Shortage of fuel-wood, illicit felling, poaching, grazing beyond the carrying capacity, encroachment on forest land, shifting cultivation, increase in anthropogenic pressure, privileges and rights of people over forests, lack of adequate employment opportunity, etc.

— *Institutional Problem:* Lack of institutional capability, erosion of forest department's role, non-functioning of village level committees, etc.

— *Management Problems:* Tardy implementation of prescribed treatment, lack of closed supervision of work, lack of sufficient financial support to past prescription, lack of trained staff, lack of coordination, etc.

Planning aims at examining available resources and their constraints. In a forest setting, the forest resources, the manpower, the finance and the science and information technology constitute the resources. In the course of planning process, one comes through an array of constraints facing the forest management plan. They are as under:

— Poor policy guidelines.
— Insufficient manpower, especially at professional level.
— Lack of fund to support different strategies.
— Lack of data and research for continuous development.
— Lack of priority for technical forest management in comparison to social security schemes or projects.
— More emphasis on utilization of resources and less emphasis on conservation and development.
— Other sociol-economic, institutional and managerial and constraints.

After identifying the problems and constraints, it is time to make an assessment. Assessment is an appraisal exercise. It is done from the point of social, economical, and financial feasibility and ecological sustainability. The purpose of assessing is to evolve appropriate strategies to overcome them. However, assessments of all aspects are equally important from the point of drawing a pragmatic management plan.

Forest management plan relates to multiple use of forest resources. Use of one kind of resources may affect others. Increasing benefits of one type may impair or damage others. For instance, timber management may lead to underproduction of non-timber forest products and degradation of bio-diversity. For effective assessment, not only the production function, demand and price structure of each product need to be understood, but also the relationship and conflicts of resource use should be fully comprehended. Taking into consideration one aspect and forgetting others, or counting only benefits and neglecting its negative impacts can mislead the planner in determining objectives and strategies.

Execution of plan generates a host of effects, which are not accounted for in the analysis of impacts. These are commonly referred to as 'indirect', 'non-market spill over' or 'externalities'. Hence, externalities should be included in the analysis and assessment because they affect the viability of the plan. In many cases externalities are difficult to identify and quantify. Many of them cannot be meaningfully valued. However, an attempt should be made to describe them at least in qualitative terms, if quantification and valuation are not possible. An important point to remember is that when a positive externality has been identified, it is also necessary to search carefully for any corresponding negative externality.

Determination of objectives

The next step in planning is determination of objectives to be achieved by the organization. Objectives are determined at two stages. One at the general plan level and second at the working circle level. The objective at the general level is called 'objectives of management' or 'general objective of management' and objectives at the working circle level are called special or specific objective of management. Objective indicates what basically is to be done, where the primary emphasis is to be placed, what is to be accomplished by the rules, procedures, programmes and strategies. The objectives are usually set against possible resources, institutional capabilities and local needs.

Forecasting of broad planning premises

In preparing plans for the future, the management has to make some general prediction about the future shape of things. This is done through forecasting. Forecasting is a systematic attempt to probe into the future by inference from known facts relating to the resource environment. Forecasting exercise is necessary for evolving variable planning premises or assumptions. Planning premises are actually assumption and prediction about the future. They constitute the basic framework for planning. The process of making premises consists of two stages. One is developing the basis through systematic investigation and the second is through visualization of future.

Developing strategies

Developing strategy is an important step in the planning process. This task is accomplished in different stages. These stages are as follows:

— Determining alternative course of action.

— Evaluating alternative courses.

— Selecting the best course.

It is a common knowledge that an action can be performed in several ways; but there is a particular way, which is the most suitable for the organization as well as for that point of time. The planner should try to find out these alternatives and examine them carefully in the light of planning premises and assumption. Having sought out alternative courses of action and examined their strong and weak points, the next step is to evolve them by weighing the various factors in the light of premises and objectives and goals. The next step—selecting the course of action—is the point at which the plan is adopted. It is the real point of decision-making. After evaluating the various alternatives, the best and the appropriate alternative is selected. This process is called 'Strategy Development'. In forest resource management, it is generally found that a combination of more than one alternative is the best course. In such situation the planner should select several appropriate alternatives and combine them in an integrated fashion.

Formulation of management plan

After the best programme is decided upon, the next task is to work out details of management plan and formulate the component parts with proper priority and sequence. Now detailed prescriptions are formulated for different working circles and aspect pertaining to miscellaneous regulation for forest practices.

Preparing the final management plan or working plan is the marks the end of plan process. All the attributes of the external environment and resource status, problems, constraints, management options are analyzed, discussed and presented in an organized fashion in the final working plan report. The descriptive and prescriptive parts of the report should be arranged in a logical sequence. This would facilitate smooth implementation. Priority areas should be spelt out according to their ground situation, management readiness and feasibility. A clear determination of priorities and sequence of work at the planning stage would benefit not only future implementation but also budget allocation. An orderly allocation of funds for different thrust areas is possible only with good and realistic planning.

Components of Working Plan

A working plan has to describe the situation, present the argument for and against different courses of action and prescribe for the future. Therefore, traditionally the plan document is organized into different parts accommodating separate aspects.

The first part comprises a number of chapters, each of dealing with a particular subject. This part is brief but deals in total relevance to the subjects. In other words, the chapters in this part should display a great deal of error-free data collection and analysis. The second part of the plan document is the place where various possibilities are discussed and important prescriptions are recorded with its accompanying analysis and justification. It is, therefore, the heart of the plan. The third part encompasses the entire appendix comprising data and maps prepared during the course of working plan process. Apart from these traditional parts, other reports like forest resource survey (Inventory), forest soil resource inventory, joint forest management evaluation report and case studies on specific practices may be appended with as the additional part of the report.

Stages in Working Plan Formulation

The final Working Plan document is the product of meticulous planning process. In the process of formation of plan, there are many important and distinct stages, which assumes significance (Fig. 6.2). Such stages are elucidated below:

- Basic information Report stage: This is the initial stage of the working plan process. In this stage basic secondary information regarding forest division are collected.
- Preliminary working plan report stage: This is the second stage of working plan process. This stage again divided into two sub-stages. They are :
    - First Preliminary Report stage
    - Second Preliminary Report stage.

Appointment of Working Plan Officer marks the beginning of this stage. Generally, this stage spans about 3 to 6 months. The initial phase of this stage culminates in formulation of First Preliminary (working plan) report. The report is then presented to a review committee constituting senior officers of the state forest department and a representative officer of Ministry of Environment and Forests, Government of India.

A conclusive discussion and decision on the issues raised in the First Preliminary Report marks the beginning of the next phase of this stage. This phase is characterized with extensive data collection, their analysis and interpretation. Apart from this, based on interpretations, the future strategy for sustainable forest management is drawn in the form of concrete proposals. Therefore, the Second Preliminary Report encompasses analysis of social-political environment, resources status, problems/constraints and future strategy for forest management. Formulation of the said report and its presentation to the review committee culminates this phase. This phase spans about 12 to 18 months. The important tasks involved in working plan preparation process are presented in Fig. 6.1.

— The Final Working Plan Report Stage: As the name indicates, this is the final stage of the working plan process. A conclusive discussion on the issues raised and strategy proposed in the Second Preliminary report marks the beginning of this final stage of planing process. Generally, this stage spans about 6 to 8 months. Submission of final Working Plan Report marks the end of planning process.

Organization of Working Plan Areas

Forests are territorially classified into different categories for the purpose of description, administration, management and record. However, these sub-divisions are made primarily for the following purposes:

— For organizing forest operations.
— For controlling yield.
— For defining territorial responsibility.
— For providing, a permanent basis for locating forests and maintaining records.

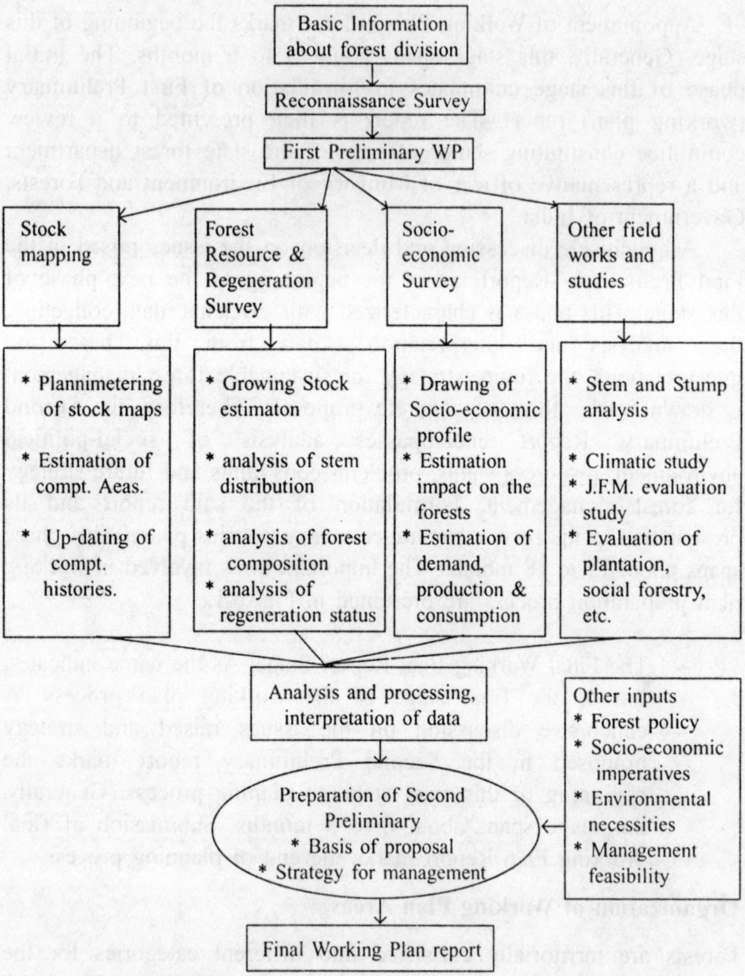

Fig. 6.1 : Tasks in Working Plan Preparation Process.

The former two types are known as management and the latter two types are known as administrative classification. Both these categories, the management and administrative classification of forests assumes significance in the context of working plan (Table 6.2.). However, the management types of classification assume more importance in the course of working plan formulation process. Therefore, sub-divisions of forest from the management point of view are elucidated below:

1. Working circle or Management circle.
2. Felling or Treatment Series.
3. Compartment.
4. Periodic block.
5. Coupes.

Table 6.2 : Sub-division of Forests

Administrative	Management	
	Sub-division	Function
* Working plan areas * Range * Circle * Beat * Compartment	Working circles * Felling series * Treatment Series * Coupes	Organization of work * Silviculture * Yield regulation

Working Circles

The forests of a division constitute a unit. The area of the unit is normally large. The unit contains heterogeneity in terms of site condition and crop composition. Therefore, in order to improve the growing stock, different silvicultural treatments are required for different parts of the plan area. As a result different sets of prescription or working rules are required to be drawn up for different parts of the forests. It is because of this reason that the forest is divided into different categories of areas on the basis of their required management practices. Such categories of forest areas are called working circles.

A working circle is an area of similar or more or less uniform crop, which may form the whole, or part of a working plan area. In other words, it is one of the sub-divisions of the working plan area. Generally, a reasonably similar crops are put together to form a working circle where a common treatment can be applied. In the text-books of forestry, working circle has been defined as, 'a forest area (forming the whole or part of the working plan area) organized with a particular object and under one silvicultural system and one set of working plan prescriptions. In certain circumstances the working circle may overlap.' As for example, the bamboo forest generally constitutes an overlapping bamboo working circle. However, over the years, with the decline in forest crop, the concepts of working circle formation has changed. Now, generally, working circle constitutes forest areas which generally requires a common treatment. For example, all forest areas that need rehabilitation with varying intensities are grouped in one or two rehabilitation circles. Likewise all forest areas, which need artificial regeneration, are grouped in one working circle popularly called as plantation working circle.

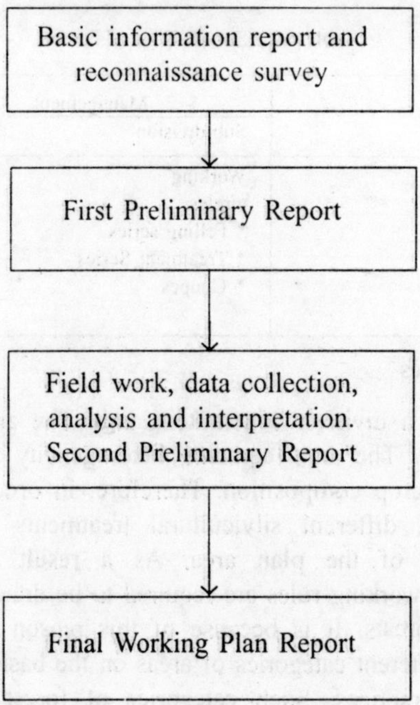

Fig. 6.2 : Stages of working plan formulation process.

Felling Series

Concentrated felling in one forest area in a year shall create labour, transportation, social, marketing, staff, and other operational problems. On the contrary distributed felling emits continuous benefits in terms of employment, to rural folk and precludes prohibition of grazing in the forest area on a continuous time period. Therefore, it is desirable to distribute the felling all over the working circle areas. Such a scheme goes in consonance with social, political, economic and operational requirement. In order to achieve this objective the working circle areas are divided into different felling series areas.

Felling Series (FS) is a forest area forming the whole or part of a working circle and delimited so as: (1) to distribute felling and regeneration to suit local condition, and (2) to maintain or create a normal distribution of age classes. The yield is calculated separately for each felling series, which should have an independent representation of age classes (Anon, 1953). Each felling series is deemed as a self-contained unit of management with a separate calculation of yield and a separate series of silvicultural operations. Division of working circle areas into such felling series facilitates effective control and distribution of work into different ranges.

When the forest area does not have enough growing stock for felling, then in such cases the working circle area is divided into different treatment series. Treatment series are group of compartments, which requires analogous treatment in terms of rehabilitation and regeneration. Such treatment series generally provides a framework for systematic and planned treatment of degraded forest areas.

Compartment

Compartment is a territorial unit of a forest permanently defined for purposes of administration, description and record. It is also the unit of management area. Generally, a compartment is given one set of treatment. This is why commonly the compartment is kept in one working circle. However, in special situation the compartment is divided into sub-compartments. Sub-compartment is a sub-division of a compartment, generally of a temporary nature, differentiated for special description and treatment (Anon, 1966).

Coupe

Coupe is the unit area in a felling or treatment series taken up for felling or treatment every year. A group of compartments or a compartment or a part of compartment may form a coupe. While determining area of coup, it is necessary to decide the quantum of work the treatment requires and its operational feasibility. In a working circle the number of annual coupes would be equal to number of felling or treatment series.

Periodic Block

In the past uniform silvicultural system was in vogue in many parts of the country. Under this system, there was a practice of dividing the forest area into periodic blocks, which symbolized different age classes. Periodic block has been defined as the part or parts of forest set aside to be regenerated, or otherwise treated, during a specified period. The regeneration block is called "floating" or "single" when it is the only periodic block allocated at each working plan revision. When all periodic blocks are allocated and retain their territorial identity at working plan revision, they are termed "fixed" or "permanent" (Anon, 1966).

Notwithstanding the territorial units in preparing a management plan, there are many important management terms, which are extensively used to define management practices. Few of such terms are illustrated below:

1. Felling Cycle

In a selection forest the usual practice is to divide the area of the working circle into a number of coupes. Such coupes are worked at an interval of definite period of years known as "Felling Cycle". Felling cycle has been defined as the time that elapses between successive main feelings on the same area (Anon, 1966). It may vary from 10 to 30 years or more. Its determination depends on the silvicultural needs of the crop as well as the objectives of the management. If the objective is to produce timber of large size, then a longer felling cycle is opted. A smaller felling cycle produces small timber.

2. Exploitable Diameter or Girth

In a selection forest a minimum girth or diameter of the tree is considered suitable for exploitation. This exercise aims at optimizing the yield of timber crop. In timber management practice the exploitable diameter or girth symbolizes rotation of maximum volume yield.

Formation of Working Circles

The strategy for forest resource management begins with the formation of working circles and miscellaneous regulation. Therefore, formation of working circle is an important step in the planning process. By tradition, Indian forests are managed by a system of working circles. Formation of working circles involves the many steps (Fig. 6.3). These steps are as follows:

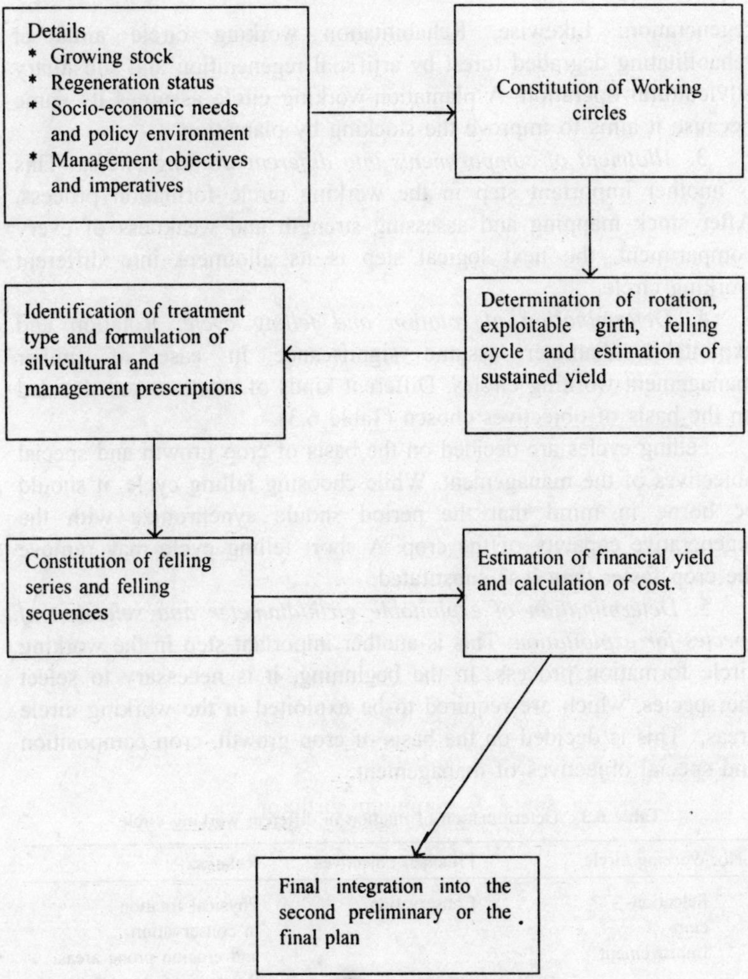

Fig. 6.3 : Process of Working Circle Formation.

1. *Determination of working circles*: Working circles are determined on the basis of resource status, ground requirement and objectives decided.

2. *Naming of working circles*: Working circles are generally named on the basis of regeneration method adopted. However, over the years, nomenclature process has adopted a functional approach. Both the practices are still in vogue. Selection working circle aims at regenerating the crop on the basis of selection system of

regeneration. Likewise, Rehabilitation working circle aims at rehabilitating degraded forest by artificial regeneration and subsidiary silvicultural operation. A plantation-working circle assumes its name because it aims to improve the stocking by plantation.

3. *Allotment of compartments into different working circles*: This is another important step in the working circle formation process. After stock mapping and assessing strength and weakness of every compartment, the next logical step is its allotment into different working circle.

4. *Determination of rotation and felling cycle*: Rotation and exploitable diameter assume significance in case of timber management working circles. Different kinds of rotations are decided on the basis of objectives chosen (Table 6.3).

Felling cycles are decided on the basis of crop growth and special objectives of the management. While choosing felling cycle, it should be borne in mind that the period should synchronize with the regenerative capacity of the crop. A short felling cycle may remove the crop faster than it is substituted.

5. *Determination of exploitable girth/diameter and selection of species for exploitation*: This is another important step in the working circle formation process. In the beginning, it is necessary to select the species, which are required to be exploited in the working circle areas. This is decided on the basis of crop growth, crop composition and special objectives of management.

Table 6.3 : Determination of rotation in different working circle

S.No.	Working circle	Principal objectives	Rotation
1	Selection-cum-Improvement	Conservation	Physical rotation in conservation and erosion prone areas.
		Timber production	Rotation of maximum volume production.
2	Coppice with Reserve (CWR)	Conservation	Physical rotation in conservation and erosion prone areas.
		Timber production	Rotation of maximum volume production.
3.	Rehabilitation of degraded forests	Conservation	Physical rotation.
4.	Biodiversity Working Circle	Conservation	Physical rotation.

6. *Estimation of annual increment or yield*: This is the next step in the working circle formation process. The probable sustained yield with respect to each working circle or felling series is calculated on the basis of yield estimation formulas.

7. *Identification of treatment type and silvicultural prescription*: As we know, forest management pursues multiple objectives. A single unit of forest can have more than one objectives. Each compartment in the forest contains different opportunities and problems. Therefore, each such unit, be it a compartment or coupe, is divided into different treatment areas. Such treatment areas are geographically demarcated and prescription for each such treatment type is formulated. Different treatment types used in management practices are presented in Table 6.4.

Table 6.4 : Treatment types in coupe or compartment

Treatment type	Description	Objectives
Type-A	* Erosion areas * River banks * Sloppy areas * Conservation areas	Protection
Type-B	Dense areas	Production/ bio-diversity conservation
Type-C	Congested pole crop areas	Improvement felling and thinning
Type-D	Degraded areas	Rehabilitation

8. *Constitution of felling series and felling sequence*: After having allotted compartments into different working circles, the next step is to divide the working circle areas into felling series. Felling series are formed on the basis of (1) operational feasibility for carrying out silvicultural works and (2) yield prospects. Once the felling series areas are delineated, it is time to decide on felling sequence. Determination of felling sequence depends upon crop condition, yield potentiality and time when the coupe was harvested in the last plan. After this process, all the annual coupes are again plannimetered and management maps are prepared.

7

Wildlife Management Plan : Principles and Planning Process

Introduction

Conservation of natural resources necessitates sustainable management of such resources and the ecosystem, which constitute the resource base. Apart from this, it is imperative to control adverse anthropogenic pressures and activity that affect the ecosystem. Therefore, sustainable management of natural resources can be accomplished within a pragmatic framework of planning which purports to draw programme for their sustainable development. Planning is the beginning of sustainable management regime. It is also the prerequisite for good management.

Forestry in India has an old tradition of initiating management through planning. This is how "the working plans or management plans" are in vogue in the forest departments since a long time. The practice of preparing and implementing such management plan generally coveres all aspect of the forests. However, over the time, the practice of preparing separate management plan in the field of wildlife conservation has gained curency. A wildlife management plan is a product of planning process that sets out the values and objectives of management for a Protected Area (PA). The plan provides ways and means to achieve the avowed objectives with in a fixed time frame.

History of Wildlife Management Plan

The tradition of managing forest with a working plan or management plan dates back to about 150 years. The need for systematically prepared management plans that provided for management and development of protected areas in accordance with an appropriate range of conservation objectives is a clause of the Bali Action Plan of the World Park Congress held in 1982. The development of guidelines and a manual for the preparation of wildlife management plan forms a major recommendation of the National Wildlife Action Plan 1983. A management plan institutionalises the management approach, thus ensuring realisation of national policies and objectives and obviating whims and fancies of individuals. Therefore, management plans ensure continuity of policy and objectives in the field of forest resource and wildlife management.

A Wildlife Management Plan (WMP) is analogous to a forest management or working plan. The first working plan for India was prepared in 1837 by Munro for the forests of Travancore. Nevertheless, the wildlife management plans are of recent origin. An attempt to devise a code for preparing wildlife management plan was made by Project Tiger Cell in 1972 in the shape of guidelines. Generally, wildlife management planning did not have an independent discipline in the forest organisation. The onus of formulating a wildlife plan rests with the manager of the Protected Area (PA). However, many states, having a distinct focus on wildlife conservation, have started deploying separate officer for wildlife management planning in P.A.

At present, there are three kinds of management plans that exists in the forest department:

1. An exclusive Wildlife Management Plan for a National Park.
2. A Wildlife Management Plan concurrently with a forest-working plan, both relating to the same area.
3. A forest working plan for a wildlife sanctuary that incorporates wildlife and forest management objectives.

In the formulation of the said plans, with respect to the first two categories of plan, there is no problem in formulating objectives, premises and strategies. However, with respect to third category of plan, there is a predominant consideration of timber management with only theoretical recognition to wildlife management principles. The plan is supposed to aim at securing conservation values as its primary functions.

Functions of Wildlife Management Plan

Makinnon *et al.,* (1986) elucidated the role of management plans in the field of wildlife. "It is now accepted as a basic principle of protected area management, that every protected area should have a management plan. The management plans guides and controls the management of protected area resources, the uses of the area and development of facilities needed to support that management and use. It facilitates all development activities and all management actions to be implemented in an area."

Wildlife management plans have definite functions. They are as follows:

- It reveals the resource status.
- It reveals the trends of resource use.
- It lays bare the objectives of resource management.
- It expresses the problems and constraints.
- It provides an array of management options.
- It helps in evolving the optimal strategy.
- It helps in monitoring the resource use and accomplishment of objectives.

Goals, Objectives and Strategies

Goals and objectives are central to the management plan. They constitute the framework for determining what action to take when they will be taken and the budget and personnel required to implement it. A management plan is a valuable tool for assessing the resource strength identifying management needs, setting objectives and priorities and organising resources to achieve the goal. The management plan helps the manager to allocate and make best use of resources available at hand. The plan can also be used to assess the requirement of financial resources. A management plan can also serve as a communication tool to gain understanding and support from both the public and relevant government agencies. Such two-way communication helps in gaining the co-operation of local communities and the political goodwill. The planning process helps the management personal to comprehend the full spectrum of management needs. This ultimately provides a rational base for administration and implementation.

Objectives or goals are the ends towards which activities are aimed. In other words, they are the results to be achieved. They represent not only the end of planning but towards which all management functions are aimed at. Strategies are general programmes of action and deployment of resources used to attain comprehensive objectives. Management is the means by which the objectives are achieved. If any of the avowed objectives are not redeemed, then there is a need to go for a review and evaluation. In such situation, there is either

— Poor or wrong management inputs, or
— Poor or wrong choice of objective, which is not achievable given the resource status and the existing ground realities.

Characteristics of Objective

Determination of objective constitutes the most important facet in a planning process. Before embarking on the task, it is imperative to understand the characteristics of objectives. Objectives have the following characteristics:

- The objectives should be clearly defined. They must be reduced into black and white. A clearly defined objective provides the clear direction for managerial efforts.
- The objectives must be realistic i.e., they must be within the reach of the organisation.
- The objective has to be written in the realm of possible achievement through strategies.
- Objective must be measurable.
- Objectives must have social sanction and acceptability.

Conservation necessitates long-term planning perspective. That means, the plan document should deal with two or three time scale or temporal entities. The short-term life of a plan, usually ten years, draws detailed scale of implementation. The medium term scale, covering two or three planning periods, requires different planning inputs. However, in the context of long term perspective planning, the conservation goals and objectives are identified and determined and broad strategies are evolved to reach those objectives at a very long time scale.

Sustainable forest management is the general goal or mission of the forest department. Mission is the central guiding concept describing the fundamental reason for the existence of the organisation. Again, such mission can be broken down into objectives. The following are examples of some of the objectives in wildlife management:

- Rehabilitating habitat for conservation of tiger.
- Establishment of viable population of guar, covering the specified habitat.
- Creation of in-situ gene conservation area.

Objectives are achieved by implementing strategies. Generally, many strategies are drawn to achieve the objectives. It is always imperative to evolve strategy compatible to ecological, social and political niceties of the said environment.

Management is an art of trade off. Forest management throws a host of conflicts of objectives as well as interests. Therefore, management aims at resolving such conflicts and settling controversies. Generally, in order to achieve this objective, it is necessary to develop alternate resources and sustainable land practices for people depending on forests. Therefore, a management plan should not be written in isolation of other land use, conservation and development policies.

National policies always set broad conservation goals. Legislation provides the legal basis for resolving conflicts that emerge during the course of management. Therefore, state policy and legislation sets guidelines for evolving management practices and resolving conflicts. Such guidelines cover issues remained untouched in the central policy. State policy also takes cognisance of the resources of the state and proposes major strategy. The state policy deals with area like tourism, conservation, education, crop compensation, resources sharing, social security measures in protected areas etc.

Goal Setting Process

The process of setting goals and objectives opens up an arena where two or more objectives are juxtaposed in conflict with each other. For example, the objective of permitting people to collect fuelwood and fodder in riparian zone clashes with the objective of maintaining and restoring the micro-locality in a dry deciduous forest ecosystem. If one objective is selected, then the other one becomes incompatible. Such situations are galore in forest and wildlife management practices. One has to deal with such situations in the course of determining objectives. However, this problem appears intractable, but applying the technique of ranking and evaluation can solve it to great extent.

First, systems of priorities amongst the value on which objectives are based need to be established. Depending on their nature, the value and system falls into different categories. Now, such values are ranked in order of importance. Scaling of values depends on perception of society, and the importance provided by policies. Once, the objectives are ranked in order of priority, then the objectives with lower priority will be allowed to the extent where it does not circumvent the higher priority objectives. Non-compatible goals and objectives can be separated or accommodated, if necessary, by specially or temporally or by both means. Temporal methods are separation methods based on time.

The Need to Involve People

Management decisions are bound to affect different groups of people in different ways depending on what their interests are. If the management of natural resources is to succeed, it must be built on development process, a mechanism to enlist support from the local communities.

Participation of people means involving people in natural resource management and understanding their problem and perceptions and developing strategies that aim at reducing the stress on people and pressure on forests. With reference to management, it is said that good wildlife management is synonymous with effective management of people's pressure on natural resources. If the pressures on forest resources are not reduced to the ecologically sustainable level and to the satisfaction of the social, economic and political aspirations of people, there would be no habitat left to manage. Legal measures and rigid protection are important but they generally lead to social confrontations. Therefore, there is a great need to tread cautiously while resolving conflicts. The art of management is to enhance management capability through people friendly, pressure dissipating strategies, which combine public education with actions that bring about mutually satisfying results.

Planning Horizon

Generally, a minimum period of year and half is assigned for completion of a plan. The first six months is devoted to the preparation of a preliminary plan. The planner should collect the available information, identifies the data gap, and spells out the modalities to remedy it. The Management Planning Officer (MPO) should prepare a summarised part one and should set his objectives and draw strategies

Committees appointed by government to revise the preliminary plan document. The shape, in which the preliminary wildlife management plan is accepted, will form the base for further investigation and preparation of the final plan document. The said committee again revises the final document. Then, the document will be placed before the competent authority for approval.

A management plan is largely about choosing the appropriate goals, objectives and resolving conflicts by making apt decisions to achieve the objectives. The wildlife planning officer faces a challenging situation where the conflicts between the wildlife conservation objectives and forest-based dependency in a forest setting are increasing alarmingly. Compromise solutions must be based on mutually acceptable trade off. However, to evolve a trade off is not an easy task. Conflict resolution always requires creating of alternative resources, skills, vocations and evolving sustainable land use practices for people dependent on forests. The general tendency of people towards wildlife is not encouraging in the wake of increasing incidences of man-animal conflicts and general apathy towards conservation movement.

Hierarchies in Wildlife Management Planning

A wildlife management plan is always formulated in consonance with other land use management policies. However, the planner has to evolve definite management prescriptions for a protected area. While assessing different management options, it may be possible that many decisions concerning plan, falls within the purview of state or national policies. Policy environment for centre as well as state is always complimentary. National policies set broad conservation goals and shows relationship between different users. State level policy sets guidelines pertaining to protected areas and general wildlife conservation in the state. Generally, such policy deals with items like social conflicts in the field of wildlife conservation, tourism, education, crop compensation etc. State policy deals with the specific areas of wildlife conservation in detail and integrate it with the general development policies and programmes.

While evolving a management plan for protected areas, the planner takes a cue from the national as well state policies and develops an array of plans like degraded habitat rehabilitation plan, a buffer zone development plan, a human resource development plan, a watershed development plan etc. In other words, a wildlife management plan is an amalgamation of many plans prepared in consonance with national as well as state policies/action plans.

Components of Wildlife Management Plan

A wildlife management plan comprises three distinct components (Sawarkar, 1995). These components are:

1. *A Description:*
- Of area and surrounding.
- Of the physical resources.
- Of biological resources.
- The artefacts.
- The uses of resources.
- The problems.
- Backward and forward linkages.

2. *An Evaluation:*
- Of area and resources.
- Significant values of area and resources.
- Role of Protected Areas.
- Objectives of Protected Areas.
- Problems and constraints of Protected Areas.
- Backward and forward linkages.

3. *Strategies:*
(Pertaining to Work Programme, Budget and Financial Schedules)
- Staffing pattern and training.
- Development inputs and research.
- Maintenance inputs.
- Utilisation input.
- Work schedules.
- Plan of Operation.
- Monitoring and evaluation.

Organisation of the Wildlife Management Plan Document

A management plan comprises different chapters dealing with different subjects. These chapters are organised in three parts. The first part deals with description and evaluation of resources. The second part deals with the proposed management practices for the resources, and the third part consists of maps and appendices. The organisation of parts, chapters and subjects is illustrated below:

PART I
Description and Evaluation

The Preamble
The Protected Areas: The Existing situation

Chapter 1. *Introduction to the area*
Name, location, constitution and extent.
Approach and access.
The statement of significance.
Chapter 2. *Background information and attributes*
Boundaries, geology, rock, soil, terrain, climate, water resources, range of wildlife, status, distribution and habitat.
Chapter 3. *History of management and present practice*
General/Timber operations including bamboo and firewood harvest.
Silvicultural systems and tending operations, bamboo working.
Collection of Non-Wood Forest Products (NWFP).
Lease/Other programmes and activities; Forest protection/Domestic livestock grassing; Tourism/Research, monitoring and training; Wildlife conservation strategies and their valuation; Administrative set-up/ Communication.
Summary of threats to wildlife.
Chapter 4: *The Protected Area and the interface land use situation*
The existing situation in the zone of influence.
The development programmes and conservation issues.

PART II
The Proposed Management

Preamble
Chapter 5. *Plan objectives and problems*
Objectives of management
Problems and constraints in achieving objectives.
Chapter 6. *The strategies*
The principles.
Boundaries, zonations, zone plans, theme plans.
Chapter 7: *Tourism, Interpretation and Conservation education*

General, Objectives, Problems. The strategies; Identification of zone; Infrastructure development; Regulation; Monitoring and Evaluation.

Chapter 8. *Eco-development*

Objectives; Specific issues; Broad strategies; Village level site-specific strategies.

Monitoring and evaluation.

Chapter 9. *Research, Monitoring and Training*

Research and monitoring; Training including on the job training; Formal training courses; establishing a learning centre.

Chapter 10. *Organisation and Administration*

Structure and responsibilities; Staff amenities.

Chapter 11. *The budget*

The plan budget.

Chapter 12. *The schedule of operations and miscellaneous regulations*

The schedules; Record of deviation and implementation; The record of employment potentials; Control forms; Maintenance of compartment histories; A product field guide for plan implementation; Others like maps, annexes, glossary, list of abbreviation used , etc.

PART III

Appendices and Maps

8

Working Plan Officer: Job Description and Responsibilities

Introduction

Forest mangement plan or forest working plan, as it is popularly known, is prepared by a senior officer of the forest department. The plan lasts for a definite period say 10 to 15 years. Before the said plan expires, the working plan is revised. Here also, the job of revising working plan is assigned to a senior officer. The officer who undertakes the preparation of revision of working plan is called as working plan officer. The working plan officer is generally an officer of the forest department who has put ten to fifteen years in the higher echelon of the forest service, especially in the forest administration and management. It is a general practice that a senior deputy conservator of forests is posted as the working plan officer. The basic aim of assigning the job of drawing a working plan to a senior officer is to ensure that the technical and management expertise gained by the senior officer over the period of his incumbency in the forest department, is applied towards formulating a pragmatic working plan based on the principles of sustainable forest management.

Job Description

Preparing a plan is a challenging job. In management parlance, a job is defined as 'a collection or aggregation of tasks, duties and responsiblities, which as a whole is regarded as a regular assignment to an individual employee'. Therefore, formulating a management plan involves accomplishment of numerous tasks, duties and responsibilities. As we know that formulation of working plan is basically a planning job. The job has the following management functions:

1. Planning.
2. Execution.
3. Control.
4. Monitoring and evaluation.

The job description of a working plan officer contains two different types of work. One is field work and another is office work. The field work encompapasses all activities, which are carried out in the field to generate necessary data. Inventory, stock mapping, compartment history writing, bio-diversity study are some of the few important field works. Office works encompasses planning of field works, map preparation, compilation, analysis and interpretation of data generated during the course of field work, report writing, monitoring and evaluation of all office and field works.

Management of the whole work of preparing plan consists mainly of identification of required type of field works, determination of their methodologies, prepartion of time-table and schedules for different office and field works, report writing, monitoring and evaluation of field and office work. This is done in three distinct phases. First one is planning of all works required for the process of preparation. Second one is execution, which consists of implementation of all the plans. The third one is the function of control that takes care of monitoring and evaluation of all the works accomplished. Therefore, the functions of planning, execution and control play an important role in management of working plan process. The working plan officer at the field as well as at the office level performs all these, functions.

Few important aspects of the job of the working plan officer are explained below :

1. Tasks for the working plan officer.
2. Duties and responsiblities.
3. Monitoring and evaluation of progress of working plan exercise.

Tasks for the Working Plan Officer

The working plan officer has an array of complex tasks to accomiplish (Fig. 8.1). These tasks are as follows :

- Collection of basic Information Report from the concerned territorial division.
- Reconnaissance survey of the forest area for which the management plan is prepared.
- Preparation of First Preliminary (Working Plan) Report.
- Conducting field works like stock mapping, inventory, socio-economic survey, stump and stem analysis, bio-diversity

studies, evaluation of Joint Forest Management (JFM) and plantation works, other studies and generation of data.

- Analysis, processing and interpretation of data.
- Preparation of Second Preliminary (Working Plan) Report along with preparation of maps.
- Preparation of Final (Working Plan) Report and Maps.

The aforesaid tasks have been dealt separately in various relevant chapters in the book. Therefore, no further discussion about the tasks is taken up here.

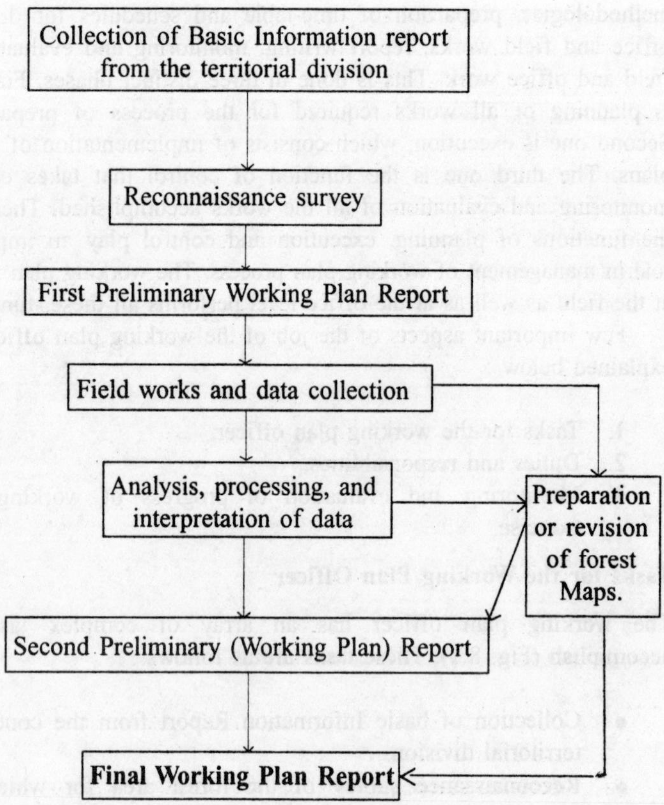

Fig. 8.1: Tasks for the working plan officer.

Duties and Responsibilities of Working Plan Officer

The following are some of the important duties and responsibilities of the working plan officer.

1. *Checking the forest area*

After having been posted as working plan officer for a forest division, the foremost duty of the said officer is to check the forest area proposed to be included under the plan. This will be done by means of perusal of form No. 1 of area register and old stock maps in the scale of 1:15,000 (4" = 1 mile) blockwise. A list of such blocks is generally appended to the old working plans. The territorial division maintains the area register. Any change in area of the blocks is always shown distinvctively in the area register. Hence, while examining the area register, the working plan officer should ensure that all area changes are shown in the area register with details like notification, etc. It is also required that map sheet areas will be added and all block areas as shown in area register will be vertically added. After doing that, it is necessary to ensure that two sets of area aggregate must tally. While carrying out this exercise, it is also essential to take into consideration the diversion of forest areas under Forest Conservation Act of 1980.

2. *Compartment numbering*

Because of the process of continuous reorganization of Reserved Forest and Protectd Forest blocks and divisional boundaries during the course of administration, there is always a need to renumber the compartments of the entire division. While allotting a number, it should be ensured that every compartment gets a separate number. It also happens in certain situation that there is a repetition of compartment numbers in different two divisions situated in one district. In such situation, it is necessary that there should be no overlapping of compartment numbers in one revenue district. Therefore, apart from other reason, it is also necessary to renumber the compartments of the division.

Compartments are generally serially numbered with Arabic numerals as 166, 167, 168 and the number will be shown on each sheet in the case of compartments which extends to several sheets. Numbering should begin at one end of the division and be carried on continuously to the other ends. In case of protected forests, a prefix of "p" is given in the compartment number, so that when it is converted to RF, it can simply retain the compartment number and thereby it does not require further numbering.

3. *Formation of compartments*

The forest divisions have been organised into different RF and PF blocks. These blocks are again divided into different compartments. Compartments are the unit of management and administration. They should have well defined boundary lines. Sometime it becomes necessary to form small compartment by dividing the forest block. However, in many states smaller blocks of compartments are formed by adding freshly acquired revenue areas into the already existing forest areas. Therefore, in such situation necessary precaution should be taken to choose the boundary of the compartment properly.

Procedure : The compartment boundaries are selected in the office before stock mapping work. This provisional boundary should have a conventional sign. This conventional sign is a line and two dots (–··–··–··–). It may become necessary sometime to revise them during the course of actual stock mapping in the forests. In comparatively flat terrain any permanent features can be used as a compartment boundary. Generally, in such situations, roads, main rivers, nalas (small rivulet) are found.

4. *Compartment Area Calculation and its Scrutinization*

Area of the compartment is calculated compartment-wise by the method of plannimetering. During the formative stage of the compartment, the total area of the block would be plannimetered and compared with the notified area of the block. During the process of plannimetering, some eror may take place. This leads to difference between notified area and plannimetered area.

Likewise, during revision of working plan, the stock maps are enlarged to a bigger scale, thereby, fresh maps are required to be prepared. In such case, the maps are developed on the basis of the topographic sheets produced by Survey of India. Such process of enlargement or demarcation of compartment boundaries on the map sheet is always susceptible to errors about area. Such errors are primarily due to changing topographical features like river or nala boundaries, shrinkage of paper during enlargement or drwing, mechanical error in the process of enlargement, human error in drawing and plannimetering. In such case, all such errors upto 3 to

5 per cent of the total area of the block should be ignored and adjusted by redistribution wherever possible. In those cases, where the difference is more than 5 per cent the area should be resurveyed suitably. In hilly and mountainous area, it has been found expedient to form compartments by gradual subdivisions until area of the right size is obtained. A list of features, which constitute the boundaries of block or compartment, is presented in an order of priority.

1. Well-defined ranges and ridges.
2. Main rivers and large nalas.
3. Permanently maintained roads.
4. Permanently maintained fire lines.
5. Kaccha forest road.
6. Small nalas.
7. Well defined cart tracks and paths.

It is also imporatant that all hill ranges and ridges should be used as compartment boundries. Artificial boundaries are generally avoided. However, in unavoidable situation, for a partial part of the boundary, artificial signs should be created. While doing this exercise, existing compartment boundaries should never be disturbed.

5. *Stock Mapping*

A good and accurate stock mapping exercise is a *sine qua non* of the entire working plan exercise. Therefore, every sincere effort should be made to ensure the accuracy of such maps. The time frame decided to complete the plan is a constraint for the working plan officer who gets about one to two years to interact with the forests. Therefore, he should keep on changing his route to traverse the forest every time whenever he visits the forests. In other words, while visiting the forests, he should avail himself of a fresh opportunity to come across a new patch or block of forests. This will help him in developing a balanced and comprehensive opinion about the crop and the factors behind its diminution or development. This should also be borne in mind that while visiting the forests the officer should take note of different features of the forest crop like the distribution forest types, composition, site quality differentiation, inpact of forest operation of last plan on the forest crop.

As a whole, stock mapping is a prolonged exercise. It is carried out by different parties usually 3 to 4 rangers, 5 to 10 subordinate staff and a draftsman. The party must have adequate suppy of maps of 1:15,000 scale including one clean uncut set required for preparing a fair copy in camp daily by the draftsman.

The working plan officer should spend a good part of his time in checking the stockmapping done by rangers. Apart from this, the working plan officer should himself do stock-mapping of few important compartments, at least 10 to 20 per cent of the total number of compartments.

When the stock mapping work is completely left to range officer, there is a risk of underestimation or overestimation of crop. In a good site, there is a possibility of underestimation of site quality and in an inferior site there is a possibility of overestimation of crop quality. This happens primarily because of unprofessionalism on the part of stock mappers. It is always advisable to measure the site quality of the crop by taking the height of the trees with the help of altimeter. The temptation of the stock-mappers to measure the height of the trees by ocular methods should be ruthlessly curbed.

The stock mapping work should be carried out by forest rangers. Generally, a draftsman is attached to a group of stock-mappers. A trace map of 1:15,000 scale should be provided to the party with the topographical sheet of Survey of India in 1:50000 scale. The fair copy of the stock map of the compartment in which the mapping work is completed in the day should be prepared in the same day by the draftsman at the base camp itself. The working plan officer should spend a considerable portion of his time not only in checking the stock mapping done by the forest rangers, but also in stock-mapping certain number of important compartments.

On return from day's work, each stock-mapper should mark in pencil on a fair copy of the trace map of the compartment, all the details that he has stock mapped during the day. This work should be done everyday and in no case be allowed to accumulate for next day. The draftsman should then ink it and colour it and get it checked by the stock mapper who will sign under the compartment number to show that he accepts the fair copy as correct. The working plan officer should also check and sign in it in token of his acceptance.

It is necessary for working plan officer to go through the various inspection notes of senior officer recorded either separately or in the compartment history files. Apart from this, the divisional notebook should also be thoroughly gone through. This helps to acquaint himself with the views expressed by different inspecting officer during the course of their visit in the past. Such views can be taken into account at the time of evolving future proposals.

6. *Programme of Work*

The working plan officer must draw detailed plan for the tours of stock mappers carefully and lay down for each camp the compartments to be examined. The accomplishment of different field work depends largely on careful planning and organisation of the work. The working plan officer should prepare the day's schedule of field work for each stock-mapper and see that he is provided with appropriate map and field compartment description and inventorying forms. One or two local men who have a fair knowledge of the forest routes should accompany each stock-mapper. The working plan officer should use his discretion while deciding the number of local men required for accomapanying the stock-mapping crew. On completion of the day's field work, maps will be fair copied and signed and field compartment descriptive sheets of inventory formats will be typed, checked and signed.

The rate of progress of the work must be constantly monitored by working plan officer so that the timetable laid down for the season will be adhered to strictly. The progress is likely to be somewhat slow in the beginning, until the stock-mapper have got themselves accustomed to the work. In a hilly terrain progress is a usually tardier than the flat areas. However, a commanding hilltop often permits large areas to be mapped after a cursory examination.

Before beginning field work, a comprehensive tour programme along with a timetable should be drawn up showing all prospective camps for the season and the number of compartments to be undertaken for stock-mapping. This would help to monitor and evaluate field works. Likewise, a detailed programme with a timetable for carrying out other field works like socio-economic survey, stem and stump analysis, bio-diversity study, etc. should be drawn personally by the working plan officer.

Therefore, the working plan officer's main work consists of:

1. Checking the stock-mapping and other field work done by the range officer.
2. Stock-mapping some of the important compartments himself.
3. A careful examination of forests from the point of

regeneration, resource use pattern and enthropogenic pressures on the forests.

4. Apart from the aforesaid works, the working plan officer is required to visualize constraints and problems and to generate in his mind ideas of how the different types of forests are to be managed in the future.

Monitoring and Evaluation of Field as well as Office Work

Monitoring and evaluation of progress of field and office work is one of the important management functions of the working plan officer. It is dealt with in four parts.

I. Collection of basic information regarding the forest division.
II. Collection of progress of field works.
III. Assessment of office works.

The details of information and particulars of field and office works are as follows:

I. Basic Information

The following are the basic information required before undertaking monitoring and evaluation:

1. Date of commencement of Plan preparation:
2. Total area in sq. km (Geographical):
3. Total forests Areas:
   I. RF No. Compt. Forest Areas.
   II. PF No. of Compt. Forest Areas.
   III. Other No. of Compt. Forest Area.

   Total compt.
   Total forest area
4. Total No. of enumeration plots (Resource Survey):
5. No. of villages/households identified for socio-economic survey:
6. Other relevant information:

II. Particulars of field works

1. Stock mapping: No. of compartments R.F., P.F. and others.
2. Writing of compartment history. R.F., P.F. and others.

3. Editing and typing of compartment history.
4. Colouring of stock-maps (Field maps).
5. Stem and stumps analysis.
6. Resource survey (No. of plots completed).
7. Socio-economic survey No. of villages/households surveyed.
8. Any other field work:

    1. Survey work.
    2. Identification of plant species.
    3. Soil survey.
    4. Bio-diversity analysis.
    5. Plantation Growth study.
    6. Other study/surveys.

III. Particulars of office works:

1. Checking of Area Register and reconciliation of forest areas.
2. Preparation of appendices Part I.
3. Enlargement of topo-sheets (If any):

    I. Preparation of master set (No. of sets).
    II. Preparation of division set (No. of sets).
    III. Preparation of Range set (No. of sets).

4. Master set (Field) R.F., P.F. and other.
5. Plannimetering of stock mapped compartments.
6. Compilation of data of resource survey.
7. Compilation of data of Socio-economic survey.
8. Compilation of data of other special surveys or studies.
9. Drafting of Part I (Whole) chapters.
10. Lying out of working circle/felling series.
11. Calculation of working circle and felling series areas.
12. Preparation of appendices of Part II.
13. Drafting of Part II (Whole) chapters (Including working rules).
14. Preparation of other appendices and reports.
15. Peparation of management map/other required maps.

IV. Check listing for important records:

The working plan officer requires a number of records and maps for the preparation of the working plan. A list of important records or maps is presented below:

1. Area Register
2. Compartment histories.
3. Stock maps, topo-sheets, indicative and management maps of the division.
4. Plantation journal of all plantations done during the period of old working plans.
5. Coupe control forms with regard to all coupes, which were worked during the last plan period.
6. Other information.
7. Nursery Jouranal.
8. Divisional note book.
9. Deviation statement.
10. Details of changes done in the working plan prescription, if any.
11. Survey and demarcation report, if any, fresh area or lost forest block was surveyed and demarcated afresh.

PART - III

FIELD WORKS
PRINCIPLES, ORGANIZATION AND METHODOLOGY

9

Field Works and Their Organization

Introduction

Collection of primary data is *sin qua non* of any planning exercise. The working plan is no exception to it. These primary data generally pertains to status of tree resources, stock mapping, area of forest under different quality and density classes, change in composition, consumption of forest produce in the neighbourhood of forest areas, growth data of important species, regeneration status in the forests and different uses of forest resources. Secondary data with regard to subjects like forest areas in different ranges, growth data, harvesting of forest produces in the last plan, revenue and expenditure in the last plan period, demographic and other socio-economic conditions of the forest divisions are also necessary. Both the primary and secondary data are harmonized to understand the prevailing environment and resource position of the division. This helps in evolving appropriate management practices for the division. Analysis of the environment and the resource status as well as evolving of management strategy constitutes the main fundamentals required for formulating a working plan.

We gather the secondary data from different sources. However, the primary data have to be collected from the respective forest areas. This is accomplished by conducting field works in the forests.

Type of Field Works

All works or exercises conducted in the field or forest areas to generate primary data are called field works. Working plan exercise involves many field works. Few of important field works are enumerated below:

1. Inventory preparation or Resource survey.
2. Stock-mapping and compartment history writing.
3. Socio-economic survey.
4. Regeneration survey.

5. Stem and stump analysis.
6. Bio-diversity analysis or study.
7. Plantation evaluation study.
8. Micro planning.
9. Special studies like

- Identification of forest areas for Joint Forest Management.
- Assessment of performance in Joint Forest Management practices.
- Prospect of wildlife development and conservation.
- Soil resource survey.
- Identification of suitable areas for plantation of different species.

Organization and Management

Field works are the most important tasks of the working plan exercise. These tasks should be carried out with proper attention and care for quality and technical standard. This necessitates proper organization and management of all field works. A senior forest officer generally mans the working plan division. The division generally consists of 4 to 5 range officers and few foresters and forest guards. Apart from this, it has one or two draftsmen and few ministerial staff. All they constitute the staff of the division.

Management of field work involves the four important functions:

I. Planning (deciding what is to be done)
II. Organizing (deciding how it will be done and who will do)
III. Leading (influencing and behaving)
IV. Controlling (being certain that plans are carried out)

I. *Planning*: Planning involves determination of goals and means of achieving them efficiently and effectively.

II. *Organizing*: Organizing is defined as the management function of assigning duties, grouping tasks, establishing authority and allocating resources to carry out a specific task. After a plan is prepared, the organizing function examines the activities and resources required to implement plan. The resources may be categorized as human resource, physical resources, information resources and financial resources.

III. *Leading*: Leading is defined as a management function of influencing, motivating and directing human resources towards the achievement of the goal that is accomplishment of different field tasks. Basically, in the context of working plan exercise, it involves providing leadership to different field crew or the teams, motivating the staff of the team and improving the work culture and communication process within the team, among different teams and in the working plan organization.

IV. *Controlling*: Controlling is defined as the management function of monitoring the performance of the field teams. It entails establishment of performance standards, time limit and taking of correcting actions. Here, the divisional forest officer must determine outputs which are critical to successful performance of field job.

We can define the function of organizing with respect to field works as:

1. The identification and classification of required field activities.
2. Grouping of activities necessary to attain objectives.
3. The assignment of each job to a forest ranger.
4. Coordinating and controlling vertically by the working plan officer.

V. *Management* : Management of field works is an arduous task. It has different constraints and problems. Proper management will yield good results. Otherwise, poor performance will follow. This will lead to generation of a plan with poor standard. A hypothetical plan for field works planning is presented in Fig. 9.1. Following are some of the practical points pertaining to management of field works:

1. An institution to do the job of preparing working plan should be set up. In most of the states, working plan offices are permanent set-up and in many states, they are set-up as and when required.

2. To carry out the task of field and office work, a cohesive working plan team is required. It is, because, successful preparation of working plan is primarily a result of a team work. Each and every member of the team is equally important. Duties and accountability of each member in the team should be well defined. This should be done in the early days of planning.

3. The working plan officer should properly coordinate the field works, data collection, and plan formulation activities.

4. To accomplish the various field and office works, it is necessary to have a timetable for different operations.

5. There should be a mechanism and clear procedure to monitor and evaluate progress at specified intervals.

6. Field work team and methodology, for different pieces of job should be decided immediately after planning. After the field work teams are organized, each of them should be asked to prepare a working outline explaining survey or field study criteria, methodology and procedures. The outline generally encompasses the following items :

- Objective of the team.
- Composition of the team and coordination mechanism.
- Methodology in field.
- Time schedule for field works, primary data analysis and reporting.
- Survey tools, mode of transportation, and other requirement.
- Team performance report format.

7. Data collection and analysis: Data collection marks the beginning of field exercise. It is a time consuming process. However, in order to accomplish the job within a optimum timeframe, it is necessary to do careful planning and to take proper decision at the beginning of the field work. Adequate attention should be given to what data are really needed for future management of forests and whether they are relevant to the main objective or not. General guidelines for data collection are as follows:

- Data to be collected should be accurate and useful for the final analysis.
- Survey forms, tables, guidelines to be used in the field work should be easily understood. They should be field tested before it is used. The forms and tables must produce objective and not subjective results.
- The survey or field work should be designed to identify problems and their location, extent and the area which will be useful for deciding on management practices.
- All the field works should be so arranged that they could be carried out orderly within the allowable time limit.
- Sampling size and techniques should be decided before embarking on field works.

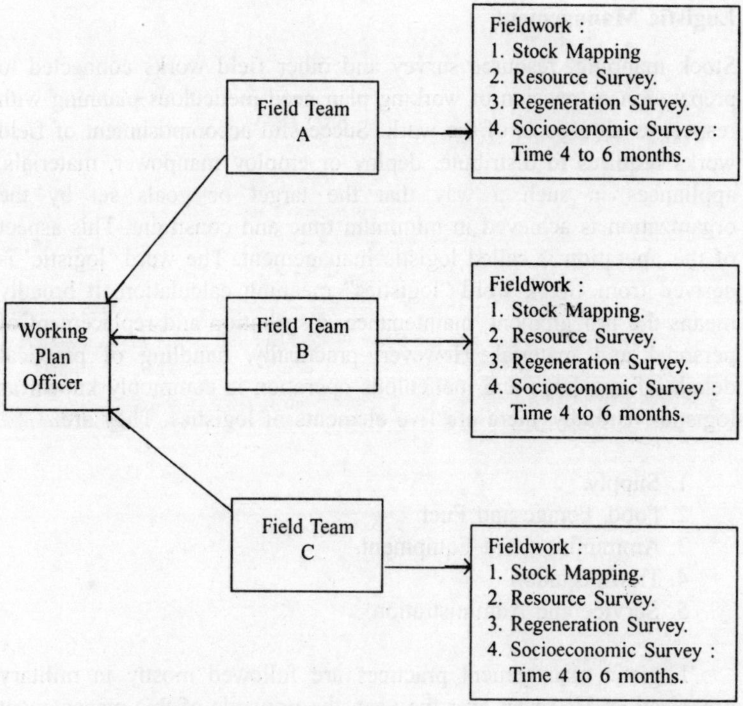

Fig. 9.1 : A Hypothetical scheme for field work planning.

Progress Monitoring

Timely completion of field works is possible if there is firm control over time and progress. Therefore, after each field work team develops its own work schedule, the Working Plan Officer should establish an overall timetable for successful execution of the field works. Reasonable time should be allotted to different jobs. This will ensure the completion of field work within the fixed timeframe. Therefore, a timeframe can be added to each major field activities.

Apart from this, work should be properly scheduled to avoid conflicts. Important target dates for completion of field works, data presentation, submission of collected formats and reports, should be thoroughly discussed and decided. This helps to ensure effective control. Again, it is also necessary that the Working Plan Officer should have periodic and routine supervision and monitoring of progress.

Logistic Management

Stock mapping, resource survey and other field works connected to preparation or revision of working plan need meticulous planning with respect to all aspects of the work. Successful accomplishment of field works requires to distribute, deploy or employ manpower, materials, appliances in such a way that the target or goals set by the organization is achieved in minimum time and constraint. This aspect of the operation is called logistic management. The word 'logistic' is derived from Greek word "logistics" meaning calculation. It broadly means the procurement, maintenance, distribution and replacement of personal and material. However, practically, handling of practical details of any large and meticulous operation is commonly known as logistics. Broadly, there are five elements of logistics. They are:

1. Supply.
2. Food, Forage and Fuel.
3. Ammunition and Equipment.
4. Transportation.
5. Service and Administration.

Logistic management practices are followed mostly in military organization. However, ever the year, the principle of this management has been applied to other civilian operation and exercises. In this context, the stock mapping, resource survey and other field tasks amply fits into this nature of the job. All such forestry exercises need careful planning, precision and enormous physical endurance against the adverse factors of the forests. Therefore, successful accomplishment of field jobs necessitates a well-drawn logistic management.

There are three types of logistics based on supply line. They are self-contained, local supply, and supply from base. In working plan field operation, the types of logistics are either of self-contained or of local supply. It is because of the fact that the crew team is required to stay in the base camp while carrying out field works at least for a fortnight.

10

Compartment History Writing

Introduction

Compartment is the territorial unit of a forest. It is also the unit of management. By definition, compartment is a territorial unit of a forest permanently defined for purposes of administration, description and record. Compartment history is a written account or chronicle of past events pertaining to the compartment. In other words, compartment history deals with a record of what has happened to the compartment since the inception of management. Therefore, compartment history is a record of all events affecting the forestry of an individual compartment.

Organization of Compartment History

Compartment history details are kept in five formats. These formats are:

1. Format I — General description of the compartment.
2. Format II — Record of plantation and changes in growing stock.
3. Format III — Register of operation and out-turn.
4. Format IV — Record of injuries.
5. Format V — Record of Observation.

The working plan revision necessitates updating of information related to general description of the compartment. The information in the other formats are dealt and maintained by the concerned territorial forest division. All such information is supplied to working plan officer during the course of revision.

General Description of the Compartment— The Format

The Format I of the compartment history contains the description of the compartment and changes brought in over a long period on account of silvicultural management. This format is used during field work to update data that are ultimately used for analysis of resource status and strength and weakness of the compartment. Therefore, we shall discuss about the format-I only. Generally, the format-I of the compartment history is organized and described with respect to the following aspects:

1. General Description of the compartment.
2. Information regarding wildlife.
3. Information regarding forest resources.
4. Information regarding results of stock mapping.
5. Information regarding effects of past management.
6. Information regarding results of resource survey.
7. Information regarding proposed management.

General Description of the Compartment

This part deals with the geographical, physical, geological, biotic and administrative features of the compartment. These are: the name of the division, compartment number, total area in hectares, name of range, name of forest block, legal status (RF/PF), details of gazette notification, details of toposheet number, name of stock mapper, latitude, longitude, border details, name of permanent features, height of compartment (MSL), gradient, topography, direction of slope, geological features, soil types, soil structure, soil composition, soil depth, soil consistency, states of humus, status of soil erosion, status of illicit felling and grazing, fire incidence, other biotic pressure, natural events like draught, flood, top drying, mass mortality in the crop, and classification of forests according to Champion and Seth's classification.

Information Regarding Wildlife

This chapter deals with the details of wildlife habitat, wildlife and the constrains in the wildlife management. These details are as follows:

1. Name of wildlife.
2. Description of wildlife habitat.
3. Description of water bodies.
4. Observation of stock mapper during their visit to the compartment. This observation should be enumerated and supported with evidences like animal excreta, rubbing in the

tree trunk, pug mark, browsing pattern in the vegetation and other visible features.

5. Description of naturally salty areas.
6. Description of animal shelter place.
7. Other relevant features with regard to wildlife.

Information Regarding Forest Resources

This Chapter of Compartment History deals with the details regarding total forest resources of the compartment. This is sub-divided into the following categories:

(a) General Description of the compartment

1. Crop composition and types of forests.
2. Crown density.
3. Stratification.
4. Age structures of the crop.
5. Site-qualities of the compartment.
6. Species in the understories like bamboo, climbers and shrubs.

(b) Scattered species.
(c) Rare species.
(d) Species of minor forest products.
(e) Grass species and their classification on the basis of edibility and availability.
(f) Infestation of lantana or other weeds.
(g) Status of root stock, if the forest is degraded one.
(h) Status of productivity.
(i) Other details pertaining to forest resources of the compartment.

Information Regarding Results of Stock-Mapping

This Chapter of the Compartment History deals with the information obtained on account of stock-mapping exercises done in the compartment. It deals with the following aspects:

1. Total area of the compartment.
2. Area unworkable:

(a) Blank areas.
(b) Understocked areas.
(c) Encroachment areas.
(d) Planted up areas : Details of plantation like: year of plantation, plantation area, type of species planted, number of plants planted, survival percentage, growth status of surviving plants and general comment on the plantation, should be given against this point.

3. Workable area:

The details includes:

(a) Workable area, forest type and site qualitywise.
(b) Total natural bamboo area (density and site qualitywise).
(c) Status of natural regeneration.
(d) Details of required management practices for the under-stocked area: The requirement may be in the form of artificial regeneration (plantation), cut-back operation, soil and water conservation, protection and measures having combination of any of the prescriptions.

Information Regarding Effects of Past-Management

This chapter pertains to the history of past-management and their effects on the crop. The points are as given below:

1. Name of working circle.
2. Name of felling cycle.
3. Details of coupes laid in the compartment, year of working and forest produce removed.
4. Effects of past management:

(a) With respect to growing stock.
(b) With respect to density.
(c) With respect to regeneration status.
(d) With respect to age structure, ratio of valuable to non-valuable species.
(e) With respect to improvement in site conditions.
(f) With respect to improvement in the condition of the crop.
(g) With respect to any other objectives of management.

Information Regarding Resource Survey

This Chapter deals with the information or data generated on account of resource survey or inventory prepared for the division. If any of the sample plots has fallen in the compartment, then the details of the data in the tabulated form and also, with some interpretation should be given in the part of the compartment history. The details of the chapter are as follows:

1. Number of sample plot.
2. Grid number.
3. Latitude and longitude.
4. Site quality of the sample plot.
5. Number of trees found in four quadrants of the sample plot. The trees can be again classified into different category based on dominance and also on the basis of species like Teak, Sal, Saja etc.
6. Number of trees in different girth-classes or diametric classes wise.
7. Condition of crop: The crop condition may be categorized on the basis of utility. These categories may be timber, half-timber and fuel species.
8. Volume of wood may be calculated using local volume table. This may again be done with respect to important species.
9. Status of regeneration: Here the regeneration may be enumerated in terms of (i) recruit, (ii) half-established and (iii) established. Again, it is also useful to record data on the basis of importnat species including non-timber tree species.
10. Details of bamboo resources in the sample plot: Average number of culms, total number of clumps in the sample plot, site quality, density, status of regeneration, composition of one-year old, two-year old and three-year old bamboo in the clump, everage height of bamboos, and condition of bamboo.

Information Regarding Proposed Management

This Chapter deals with the management practices proposed for the development of the crop standing on the compartment as well as maintenance and improvement of the site if it is degraded. This consists of the following points :

1. Name of proposed working circle to which the compartment is allotted and the reason for it.
2. Proposed felling series or treatment series.
3. Details of proposed coupe.
4. Details of types of plantation and techniques to be adopted (in brief). This may be explained with plantations like: teak, bamboo, mixed, fuel-wood, and also treatment like restoration of degraded forests (RDF).
5. Status of forest boundaries and pillars.
6. Other relevant details.

In the end of the compartment history, the name of the stock mapper should be given. Again, he should put his signature with the date. In addition to this, the name of base camp should also be mentioned.

11
Stock Mapping

Introduction

Stock map is a map showing distribution of different forests and stand attributes, which have a bearing on management. It comprises information about the crop with respect to the following aspects of the crop in symbolic notation:

1. Composition,
2. Age-class distribution,
3. Site quality,
4. Density,
5. Distribution of species, and
6. Regeneration status.

In other words, stock map is a graphic representation of the nature and composition of the forests. It provides a better general idea of condition and contents of the forest. The stock maps are generally prepared by working plan rangers. However, working plan officer also stock-maps a certain number of compartment in every forest block and ranges. As already said, stock mapping is a process of representing features of forests on a sheet of topographic map. It is done on a trace map of the compartment. Generally, mapping and entering of details of symbolic features on the map is done with the help of pencil. This is inked and coloured in the same day when the stock mapper comes back to his base camp.

The topographic maps are of the scale 1:50,000. Sometime, it is also prepared on bigger scales. It provides a clear picture of crop condition of a forest compartment. In other words, stock map is a map showing the distribution of the different forest or stand type which have a bearing on management with information about their composition, age classes, etc. (BCFT).

Features of Stock Map

Stock maps are prepared or revised in every working plan revision. They are generally prepared by forest Rangers. This task is never assigned to any lower staff, because, stock mapping involves a lot of technicality and the lower staff are not generally trained to possess the competence. However, the working plan officers are required to stock map few compartments as well as to inspect the stock mapping works done by the range officers. Inspection of stock mapping provides an opportunity to estimate forest resources with its constraints and advantages.

The main features which are shown on the topographic map to make it a stock map are:

1. Type of forests or crop composition.
2. Age classes of crop.
3. Density.
4. Site quality.
5. Distribution of dominant species.
6. Regeneration status.

Now we shall discuss different forest features and attributes.

Type of forests or Forest Composition

Forest area with tree cover and density constitutes the criterion for classifying into forest types. This type of classification is called as management types (Table 11.1). Generally, the following management types of forests are found.

1. Sal forest: A stand in which the percentage of sal trees is 20 per cent or more.
2. Teak forest: A stand in which the percentage of teak trees is 20 per cent or more.
3. Miscellaneous forest: A stand in which the percentage of miscellaneous species is 80 per cent or more. Sometime miscellaneous forest are reclassified into such categories like:

— Miscellaneous with sal: In such stand the percentage of sal is less than 20% but more than 10 per cent.

— Miscellaneous with teak: In such stand the percentage of teak is less than 20 per cent but more than 10 per cent.

4. Bamboo forests: Forests having more than 500 culms per hectare are called as bamboo forests. Apart from this, sometime the forest having more than 31 clumps per hectare also constitutes the criterion for qualifying as a bamboo forests.

Table 11.1 : Forest Type Index

Types of forests	Attributes	Index
Teak forests	Population of teak is 20% or more	Blue line
Sal forest	Population of sal is 20 % or more	Green line
Miscellaneous. forests	Neither teak nor sal	Red line
Salai (*Boswellia serrata*)	Population of salai is 40% or more	Violet line
Khair (*Acacia catechu*) forest	Population of khair is 40% or more	Burnt signs

Age Classes of the Crop

The age classes of the crop are those intervals into which the range of age of trees growing in a forest is divided for the purpose of classification or use. In practice, there are mainly three age classes in the crop. They are as follows:

1. Young aged: Crop between 21 cm. to 60 cm. girth.
2. Middle aged: Crop between 60 to 120 cm. girth.
3. Old aged: Crop above 120 cm. girth.

While writing about the age classes, it is necessary to see the crop age for an area of minimum two to five hectares. Apart from this, in a selection forest, it is imperative to look for trees of the dominant age classes standing in the forests (Table 11.2). Number of individuals standing in a forests should be the criterion of age class classification.

Table 11.2 : Age Index Chart

S.No.	Age class	Average girth Range	Colour	Index	Girth class
1.	Matured Crop	Above 120 cm	Faint Red	-	> 120
2.	Middle aged Crop	60cm to 120cm	Faint Red	M_3	100cm to 120cm
				M_2	80cm to 100cm
				M_1	60cm to 80cm
3.	Young aged Crop	Less than 60 cm	Faint Green	Y_3	40cm to 60cm
				Y_2	20cm to 40cm
				Y_1	Less than 20cm

Density

In a stock map exercise, the canopy density is measured. Canopy density is a measure of relative completeness of canopy and is expressed as a decimal co-efficient taking closed canopy as unity. Measurement of density of a stand needs lot of practical experience. It is difficult to determine relative completeness of canopy expressed as coefficient. However, it is never impossible. The foresters are required to master over the skill of density determination. Generally, standing in a point in a forest, one is required to observe the tree cover, basically of dominant trees, immediate above his head and to assess the sky space covered and classify in the said density classes. This exercise has to be repeated in many sites in a compartment and averaging of the value pertaining to canopy for a big patch of forest is done. The minimum unit area for density measurement is 2 hectares. The following classification of canopy density is in vogue in Madhya Pradesh.

Density in tree crop is shown by two systems:

1. Decimal point system.
2. Description system.

1. *Decimal point system*: This is the system, which is generally used. The classification under it is as fellows:

(i) Closed: When density is 1.0.

(ii) Fully stocked area: When density is between 0.8 to 1.0.
(iii) Dense or well stocked: When the density is between 0.7 and 1.0.
(iv) Thin or moderately stocked: When the density is between 0.5 to 0.7.
(v) Under-stocked: When the density is 0.2 to 0.5.
(vi) Blank or open: When the density is under 0.2.

Sometime, for the congested crop, the decimal point used is 1.1. Generally, density of the crop is shown on the stock map like 0.5, 0.6, 0.7, and 0.4. Generally, those areas having density more than 0.5 should show type of forest, site quality, important and valuable species, status of regeneration of dominant and other important species, names of the scattered species and their age. The symbol for density is presented in Table 11.4.

Description System:

- Very dense requiring thinning = D
- Dense requiring early thinning = d
- Medium but likely to require thinning in next 10 years = M
- Medium but not likely to require thinning in next 10 years = m
- Open and not fit for working = X

Density Classes for Bamboo Forests

Different density classes of bamboo forests are as follows:

1. ***Dense*** : When the bamboo forests consist of not less than 1100 number of culms per hectare of forest.
2 ***Medium*** : When the bamboo forests consists of culms ranging from 900 to 1100 number per hectare.
3. ***Low*** : When the bamboo forests consist of culms ranging from 500 to 900 numbers per hectare.

In a bamboo forest with tree growth, it is practically difficult to determine tree density. However, the correct procedure is to visualize and assess the degree of completeness of dominant tree canopy, excluding the bamboo growth. The density of bamboo forests is determined on the basis of culms or clumps per hectare (Table 11.3). Different stock maps are prepared for both tree forest and the bamboo forests. Therefore, in a forest with bamboo growth, stock mapping of tree and bamboo resources is done on two different topo map sheets.

Table 11.3 : Bamboo Density Index Chart

S.No.	Basis of density classification	Density class	Colour
1.	500 to 900 Bamboo per hectare.	Low	Faint yellow
2.	900 to 1100 Bamboo per hectare.	Medium	Faint Rose
3.	Above 1100 Bamboo per hectare.	Dense	Faint green

Table 11.4 : Tree Density Indication Chart

S.No.	Percentage of species in the population	Indication
1.	Types of forest where the one species ranges 20 to 50 %	<
2.	50 to 80%	>
3.	More than 80%	>>

Note: These signs are to be marked on the stock-map. The colour for these signs will be green for Sal, Blue for teak and red for miscellaneous forest. Normally, no sign will be marked if the crop is confined to an area less than 20 hectares.

Site Quality

It is a measure of the relative productive capacity of a site for a particular species. As we know, site is a complex of physical and biological factors of an area. Site determines the forest growth. Every site responds differently to different species.

Site quality is measured on the basis of top height of the crop on a particular age. There are four site quality classes for teak. (According to All India Teak Yield Table) These are as follows:

1. Site Quality I Top height: 36.6 to 30.5 mt.
2. Site Quality II Top height: 30.5 to 24.4 mt.
3. Site Quality III Top height: 24.4 to 18.3 mt.
4. Site Quality IV Top height: 18.3 to 12.2 mt.

The top height is attained at the age of 50 years. In case, the crop is young, the corresponding top height for that age has to be determined. In case of sal, there are again four-site qualities prevalent in forestry practice. They are based on the top height of the crop at 80 years. These site qualities are as follows:

1. Site Quality I Top height 36.6 to 30.5mt.
2. Site Quality II Top height 30.5 to 24.4 mt.
3. Site Quality III Top height 24.4 to 18.3.mt.
4. Site Quality IV Top height 18.3 to 12.2 mt.

From the description of site quality and its classification given above, it is clear that the site quality has a sufficiently long range of top heights between each of its sub-division against a particular age. In other words, the range of variation of top height in a same site quality is so larger that sometime it is difficult to measure the productive capacity of a site for a particular species for the purpose of comparison with other sites. In order to describe the quality more correctly and precisely, the method of fractional quality has been evolved. Fractional quality is the site quality expressed as a decimal sub-division of the height range of half a quality class, the figure running 0.0 to 2.0 within the whole.

In Madhya Pradesh there is a slightly different site quality classification in vogue. There are seven site qualities with respect to teak and miscellaneous forests. They are as follows:

1. Site Quality I Top height over 30 mt.
2. Site Quality II Top height 25 to 30 mt.
3. Site Quality III Top height 20 to 25 mt.
4. Site Quality IV Top height 15 to 20 mt.
5. Site Quality V Top height 12 to 15 mt.
6. Site Quality VI Top height 9 to 12 mt.
7. Site Quality VII Top height up to 9 mt.

The symbol or notation for representing site quality is given in Table 11.5. The site quality for bamboo has been categorized into three qualities in Madhy Pradesh. The notation has been given in Table 11.6. The basis of classification is as follows:

Bamboo	Site Quality I	Height of tallest culm over 9 mt.
Bamboo	Site Quality II	Height of tallest culm 6 to 9 mt.
Bamboo	Site Quality III	Height of tallest culm up to 6 mt.

However, the classification for site quality for bamboo prevalent in other parts of the country is somewhat different. These classifications are given below:

— *Quality class I* : Average culm height 6mt or more for *Dendrocalamus strictus* and 14 metres or more for *Bambusa arundinancea*.

— *Quality class II* : Average culm height 4 mt or more but less than 6 metres for *Dendrocalamus strictus* and 10 mt or more but less than 14 metres for *Bambusa arundinacea*.

— *Quality class III* : Average culm height 2 mt or more but less than 4 metres for *Dendrocalamus strictus* and 2 metres and more but less than 10 metres for *Bambusa arundinacea*.

The height attained by the dominant trees of a crop depends primarily on the site quality. This is called top height and is used as a measure of site quality. Top height is standard for young and middle-aged crop and it is the height corresponding to the mean diameter of 250 biggest trees per hectare as read from the height diameter curve. When the number of trees fall below 125 per hectare, top height is called crop height.

For the determination of site quality, it is necessary to measure the top height of the crop. Nevertheless, calculation of the top height of the crop involves a lengthy procedure. Therefore, it is not practicable to measure top height for determining site quality of the crop standing in a compartment. So, in place of top height of the crop, the emerge height of dominant trees occurring in an area not less than 2 hectares is measured. This estimated height would be checked by measuring a few dominant trees again. Measurement of height of trees should be done with the help of Hagga Altimeter or other prescribed instruments.

Distribution of Dominant Species

A stock mapper generally cris-crosses the compartment. He is required to look for the pattern of distribution of different species in the compartment. Commonly, in a miscellaneous forests, certain species are found abundant in certain patches and some other species in other areas. Such pattern of distribution of species should be depicted on the stock map with the appropriate symbolic notation. A list of notation indicating different species is presented in Table 11.9. In teak and sal forests, it is also necessary to assess the dominance of teak or sal in different patches with their appropriate percentage signs (Table 11.4).

Table 11.5 : Site Quality Index Chart

S. No.	Site Quality	Height of tree	Index	Symbol
1.	Ist Quality	Above 30 mt	One single vertical lines	||
2.	IInd Quality	25 mt. to 30 mt.	Two horizontal line	==
3.	IIIrd Quality	20 mt. to 25 mt.	One oblique line	//
4.	IVa Quality	15 mt to 20 mt.	Two oblique lines (NE-SW-SE-NW) The distance between two lines is 1 cm.	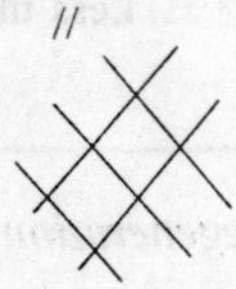
5.	IVb Quality	12 mt to 15 mt.	Both vertical and horizontal lines forming a square box of size 1 cm on all sides	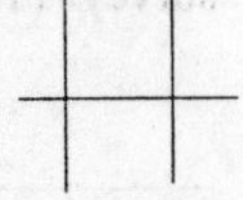
6.	Va Site Quality	9 mt to 12 mt	One vertical broken line (N-S) (length of such broken line is 1cm)	
7.	Vb Site Quality	less than 9 mt	One oblique broken line. (Each broken line should have 1cm. in length)	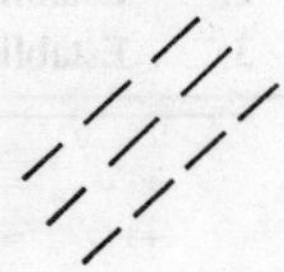
8.	Understocked	Density (0.2 to 0.4)	U.S.1: Suitable for plantation. U.S.2: Suitable for R.D.F. U.S.3: Stony areas	Yellow colour
9.	Blank	Density .2 and less.	In the mid of the centre, the classification as shown on the above understocked category is given.	Bl

Table 11.6 : Bamboo Quality Index Chart

S.No.	Basis of quality classification (Height of culm)	Site Quality	Sign	Colour
1.	9 mt and above	I	===	Black
2.	6 mt to 9 mt	II	||	Black
3.	Less than 6 mt	III		Black

Regeneration Status

Regeneration of main species can also be shown on the stock map. More about regeneration is illustrated in the Chapter 'Regeneration survey' (Table 11.7).

Table 11.7 : Regeneration Index Chart

S.No.	Regeneration of Species	Colour	Sign/Symbol
1.	Established teak regeneration :	Blue	+++
2.	Established sal :	Green	+++
3.	Established Misc. regeneration	Red	+++

+ = For un-established

++ = For half-established

+++ = For established

The status of regeneration will be marked on the stock map for an approximate area of 20 hectares.

Sample Plot/Preservation Plot/Experimental Plot

Sometime sample plots/preservation plots/experimental plots are found in the compartment. The basic purpose of laying out such plots is to conduct forest research. Such plots are laid and maintained by research wing of the forest department. Symbols for sample plots, preservation plots and experimental plots are shown in the map as presented in Table 11.8.

Table 11.8 : Sample Plot/Preservation Plot/Experimental Plot Index Chart

1. Sample Plot	SP No.
2. Preservation Plot	PP
3. Experimental Plot	EP

Table 11.9 : Species Index Chart

Common Name	Botanical Name	Index	Block Colour
1. Haldu	*Adina Cordifolia*	H	Black
2. Salai	*Boswellia Serrata*	Sa	Orange
3. Kusum	*Schleichera Oleosa*	Km	Violet
4. Prija	*Pterocarpus Marsupium*	Bj	Black
5. Tinsa	*Ougeinia Ojeinensis*	T	Black
6. Saja	*Terminalia Tomentosa*	Sj	Black
7. Shisham	*Dalbergia Latifolia*	Sr	Black
8. Harra	*Terminalia Chebula*	H	Black
9. Garari	*Cleistanthus Collinus*	Gr	Black
10. Dhawada	*Angeissus Latifolia*	D	Black
11. Bamboo	*Dendrocalamus Strictus*	B	Black
12. Amla	*Emblica Officinalis*	A	Black
13. Arjun	*Terminalia Arjuna*	Ar	Black
14. Achor	*Buchanania Lanzan*	Ac	Black
15. Mahua	*Madhuka Indica*	M	Black
16. Ber	*Ziziphus Maratiana*	Br	Black
17. Pales	*Butea Monosperma*	Ps	Black
18. Teak	*Tectona Grandis*	T	Blue
19. Sal	*Shorea Robusta*	S	Green
20. Semal	*Bombax Malacaricum*	Si	Black
21. Khair	*Acacia Catechu*	K	Brunt Siena

Materials Required for the Crew for Stock Mapping

An officer with a rank of forest ranger always heads the crew or the stock mapping party. The party should have the following materials before initiating the stock mapping exercise.

1. Toposheet : Toposheet (in the scale of 1:50,000) of the area for which stock mapping is being carried out.

2. Trace maps (in the scale of 1:15000) of the compartment with the outer boundary lines drawn and the pillers shown and few blank tracing papers.
3. Tape: 30 mt. 2 in number
   20 mt. 1 in number
   2 mt. 6 in number
4. Axe 1 or 2 in number
7. Abneys level one in number
8. Haga Altimeter one in number
9. Pocket compass one in number
10. Colour pensil, (one packet), chalk (1 packet), Black paint (Small tin. 1/2 litre), white paint (1/2 litre), Brush and scale (one)
11. Prismatic compass (one in number)
12. Ranging Rod, field book, list of conversion sign, list of flora
13. Herbarium press.

Methodology

Stock mapping is an intensive exercise of observation, assessment and determination of certain attributes of the standing crop and the forests as a whole. To do this job, the stock mapper needs a lot of not only practical experience and exposure to different phenomenon and operations in the forests but also adequate skill to translate the observation and assessment into symbolic presentation in the map. For stock mapping exercise, the compartment is taken as the unit area. The main steps in stock mapping exercise are as follows:

1. Prepare a detailed plan for stock mapping for different compartments of a forest block or area. The plan should have a list of compartments and time schedule. It should also have a route chart drawn meticulously to cover different compartments in a forest block.
2. Draw the outline of a compartment in a tracing paper before going to the field. However, when trace map is provided, then there is no need to trace the toposheet. The trace map should be on a 1:15000 scale. The contours and other natural topographical features should be exhibited on the trace-map.
3. Enter into the compartment and take a known point in the forests. With the help of pocket compass, orient the Survey of India map (1:50,000) on the spot. Then chose a direction

Stock Map of a Compartment No. P747

Area : 318,250 Hact.
Range: Chowki
Division : Rajnandgaon Forest Division
Sheet No : 64D/10NE, D/10NW
Scale : 1:25,000

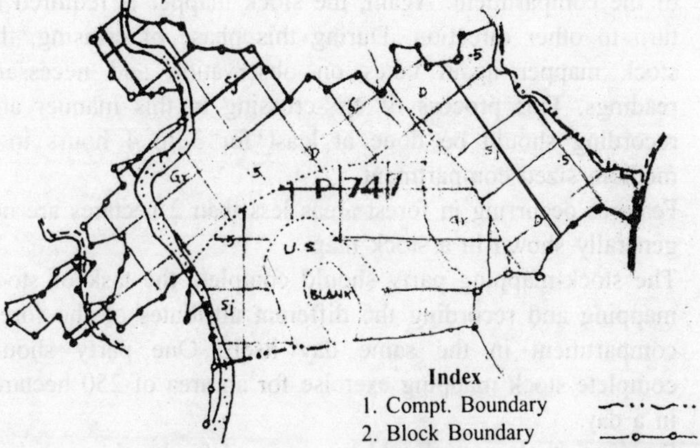

Index

1. Compt. Boundary
2. Block Boundary
3. Road
4. River/Nala
5. Block Pillar

Forest Crop (Quality class)

6. Mixed IV A ....
7. Mixed IV B ....

Age Class

8. Mainly Young ....
9. Mainly Middle ....
10. Under Stpck ....
11. Blank .... BLANK

Fig. 11.1

and proceed. While cruising along the forest, it is necessary to look at different features/aspects of the forest like crop age, type of forest, distribution of species, crop condition, crop composition, regeneration status etc. Besides, it is also essential to measure the heights of few tallest trees with the help of Abneys level or Hagga Altimeter. All such attributes are simultaneously recorded by the stock-mapper in his note-book. By this way, the stock mapper reaches the boundary of the compartment. Again, the stock mapper is required to turn to other direction. During this phase of cruising, the stock mapper again takes on observation and necessary readings. This process of cris-crossing in this manner and recording should be done at least for 3 to 4 hours in a medium sized compartment.

4. Features occurring in forest areas less than 2 hectares are not generally shown in a stock map.
5. The stock-mapping party should complete the task of stock mapping and recording the different attributes of the forest compartment in the same day itself. One party should complete stock mapping exercise for an area of 250 hectares in a day.
6. Separate stock map should be prepared for bamboo crop, in addition to the general stock map for tree crop. A stock map of a compartment is given in Fig. 11.1.

12
Socio-Economic Survey

Introduction

Indian forests are facing tremendous anthopogenic pressures. This has resulted in deforestation and degradation of valuable forest resources. Over the years, alienation of local people from the forest management has made them insensitive to the primary need of forest conservation. It is now an established fact that unless there is stabilization of population and participatory management in forest sector, the sustainable forest development will remain a dream only. In an effort to draw an effective strategy for sustainable forest development, it is required to have sufficient database with regard to availability, consumption pattern and demand of forest produces in the vicinity of forests. Today, there is a dearth of data, especially with regard to socio-economic aspects of forests. Therefore, considered in this context, the socio-economic survey of the villages in the vicinity of forest assumes utmost relevance.

Socio-economic survey is an exercise to collect socio-economic data from the rural areas adjoining the forests and to analyze and interpret in relation to the forest resources. It involves surveying of selected villages by means of a variety of methods such as Participatory Rural Appraisal (PRA), questionnaire survey, interviewing, directly observing, measuring or combining any or all of them. Sample survey method is generally the most accepted method of data collection. It is used both on large and small scale.

The generally conducted socio-economic survey involves selection of representative study units by using statistical sampling techniques such as multi-visit household survey and single-visit household survey. Apart from this, other methods like Rapid Rural Approach (RRA), case studies, community interviewing are also used to conduct survey. However, the two important methods are used to undertake socio-economic survey with respect to forestry related issues. They are as follows:

1. Sample Survey.
2. Questionnaire Survey.

Sample Survey

The sample survey is generally applied to village level studies. It covers socio-economic activities pertaining to households, individuals, or organizations. The data collection may be:

A. Through filling up of a questionnaire.
B. Through interviews.
C. Through observation of respondent's consumptive pattern or phenomenon.
D. Through direct estimation of consumption of forest produces.

Questionnaire Survey

This is also a commonly used method of survey. It involves preparing a written questionnaire for the respondents and eliciting information on it. A questionnaire is group of question designed to elicit information on a subject from a respondent.

Objects of Socio-economic Survey

The following are the objects of the survey:

1. To assess the local demand of forest produce like timber, fuelwood, bamboo, grass and other minor forest produce.
2. To assess the source of supply of forest produce presently used.
3. To assess the degree of dependence of local people on the forest with respect to forest produce.
4. To identify those forest produces which have higher demand than their supply and to develop strategy to conserve and develop those resources on a sustainable basis and also to determine the impact of over use on the sustainability of the forests.
5. To conduct site suitability survey in order to recommend different useful species outside forest areas, i.e. in the agro-forestry and social-forestry sectors.
6. To identify the villages which have higher degree of participation and involvement in the forestry and allied activities and commitment towards the conservation of forest.

7. To study the socio-economic condition of people living in the area of the division and their pattern of energy use.

Steps in the Socio-economic Survey

Socio-economic survey comprises many distinct but integrated tasks (Fig. 12.1). The following are the tasks/steps in the socio-economic survey exercise.

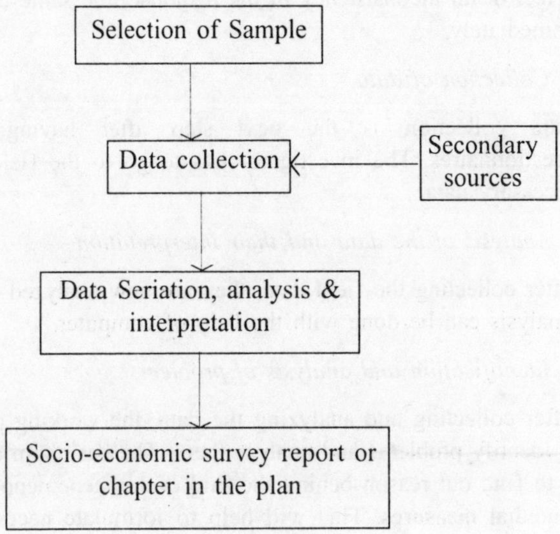

Fig 12.1 : Tasks in Socio-economic Survey.

1. *Review of existing report and data*

Existing reports on general socio-economic conditions of the rural areas of the country should be collected and reviewed before beginning detailed studies. The existing reports and data will provide working plan officer basic information, which may be valuable for the preparation and design of survey proposals, related forms and questionnaires.

2. *Selection of survey methodology and technique*

Mostly, random, stratified and cluster-sampling methods are selected as the sampling techniques for socio-economic surveys. Details regarding the methods are given separately.

3. *Design and use of questionnaire*

Designing of questionnaires requires clarity in thinking and objectives. Questionnaires should be concise and constructed in a logical order. Ambiguous questions should be avoided. Questionnaires should include a double checking system so that if the enumerator finds a defect or an inconsistency in the response, the same can be corrected immediately.

4. *Collection of data*

Data collection is the next step after having designed the questionnaires. The investigator has to go to the field to collect the necessary data.

5. *Analysis of the data and their interpretation*

After collecting the field data, they are now analyzed and interpreted. Analysis can be done with the help of computer.

6. *Identification and analysis of problem*

After collecting and analyzing the data, the working plan officer has to identify problems and analyze them. The basic aim of interpretation is to find out reason behind the said event/phenomenon and to evolve remedial measures. This will help to formulate necessary regulation or practices for the sustainable management of the natural resources.

Problems should be classified into two categories. They are :

1. Problem which requires long term solution.
2. Problem which can be solved in a relatively short time.

7. *Preparation of report*

After a close and thorough examination of the above problems, the Working Plan Officer prepares a report discussing the problem and evolves pragmatic solution to the problem. The remedial measures suggested would be integrated into the management practices recommended in the plan. Therefore, finding appropriate and pragmatic solution to the problem is the most difficult challenge the Working Plan Officer faces in the whole exercise.

In the socio-economic survey, the method of stratified random sampling is generally used in Madhya Pradesh. This method consists of selecting a large number of sample areas (villages) from different strata on a random basis and carrying out collection of field-data from few randomly selected families from such sampled villages. This technique enables the selection of a sample of attributes out of a given population. It also helps in maintaining the representative character of the sample.

Sources of Data

Data regarding socio-economic aspects can be generated from two sources. One is primary source and the other is the secondary source. Fresh field studies generate primary data. So, it constitutes the primary source. The already available documents in the form of reports and records generate secondary data. So they constitute secondary source. Both types of data are necessary to make the analysis and to draw inferences with regard to the aforesaid objectives. The primary data are to be generated from the field studies and exercises. The secondary data may be collected from district statistical book or census report or Government records. The secondary data required in connection with the socio-economic survey are as follows:

(a) Name of the village and topo-sheet of the survey of India (scale 1:50000).
(b) Number of households in the villages.
(c) Population data villagewise.
(d) Literacy percentage.
(e) Animal stock of the district.
(f) Land use classification of the district.
(g) Agricultural land data, especially irrigated and un-irrigated or rain-fed.
(h) Agricultural crop data.
(i) Communication infrastructure of the district.
(j) Educational institutions.
(k) Primary health facility of the district.

The Methodology of Selection of Villages and Households

The methodology of selection of villages for carrying out socio-economic survey is illustrated below:

1. The area of the forest division is divided into two or three

strata on the basis of their geographical proximity to forest area.

It has been found that consumption of forest produce in the households increases with the increased accessibility and availability of the said produces. The villages, whichever lying in the proximity of the forests has high consumption rate than the villages away from forest. Because of this character of consumption of forest produce among the villages, the proximity of village to the nearby forest is taken as the criterion for stratification of the total area. Generally, the villages are categorized into three strata. They are:

Strata I :	The villages adjoining the forest, i.e. situated within a radius of less than 1 km.
Strata II :	The villages situated within range of 1 to 5 kilometre distance from the boundary of the forest.
Strata III :	The villages situated beyond the distance of 5 kilometre from the outer periphery of the forests.

When the district has relatively less forest area, then an additional strata may be added to categorize villages situated beyond 10 kilometre distance from the forest. This may help in getting better representative data from each stratum.

2. List out the name of villages in the area of the division or district. Categorize the name of village on the basis of strata to which they fall. Villages are selected from the strata on the basis of random sampling method.

3. Prepare a list of selected villages in each strata and the number of households living in the village.

4. Prepare a list of households in the sampled villages. Every household will be numbered. In this preliminary survey, a list containing the names of the family heads and their occupation will be categorized into nine categories on the basis of given criteria.

5. Now the households are selected again by a method of random sampling primarily basing on the economic criterion of land-holding and occupation. Again, from each category, village households are selected on a random sampling basis. During this exercise it should be ensured that each category has at least five representatives in the final lists of selected families in the entire area.

6. In total about 5 to 10% of families are to be selected to do the household socio-economic survey. Villages are selected taking the help of the table of random numbers. Schedules of questionnaires are

prepared to gather data pertaining to socio-economic conditions, energy consumption pattern, dependence on forest resources, and land use pattern and any other issues for which the survey is being conducted.

7. Various staffs like Range Officers and Foresters can be trained in the field of socio-economic investigation. These trained persons will now be called as investigators or enumerators. Such enumerators with the help of schedules collect data by the primary method of personal investigation. Generally, the enumerator approaches the selected households and obtain the necessary information by direct method.

It has been seen that land holding and economic status of the village people determines the consumption pattern. Therefore, the economic criterion has been adopted to classify the households in the villages into seven categories. These categories are as follow:

(a) Big farmer (Household having 5 hectare of land or more).
(b) Medium farmer (Household having 2 to 5 hectare of land).
(c) Small farmer (Household having less than 2 hectares).
(d) Land-less farmers.
(e) Families involved in animal rearing.
(f) Village artisan.
(g) Traders.
(h) Service holders.
(i) Other occupation holder.

Analysis of Data

Data collected during the socio-economic survey will be analyzed in combination with relevant secondary data. The general method of analysis is as follows:

1. Data will be serialized and tabulated according to different nine categories in a same village. Data will be averaged per family in the same category in every village.

2. Based on the aforesaid method, family size, number of males and females, demand and supply of forest produce will be estimated for each family under different category in one village. Dividing the aggregate by the population size of the sample families in a village the average consumption of forest resource per family will be estimated.

3. This method of calculation will be applied to all the sample villages. Such village data will be arranged again on the basis of

strata. Now, on every stratum, there will be two sets of data with respect to each of nine categories. One set is total consumption for the family and the other one is for the per capita consumption.

4. Per capita consumption in each category under different strata will be multiplied by their respective size of population to find out the total consumption of different forest resources. Finally, the consumption in different nine categories may be estimated for the division or the district. This method of estimation is basically applied to rural areas. Different survey designs, however, can be prepared for the urban areas also.

5. The consumption pattern of forest produce in urban areas differs from the rural areas. Therefore, there is a need to develop a separate methodology of estimating consumption pattern in urban areas. The consumption of timber and fuelwood in households needs to be surveyed in the same manner as done for the rural areas. The income criteria may vary in urban areas. There may be no need of stratification for urban areas. In addition to this method, information related to total supply of fuelwood, timber and other forest produce can be collected from secondary sources. With respect to those resources, there is a dearth of information. A suitable organization may be formed to gather the data. Processing and analysis of data are generally done in the same manner as done for rural areas. In the manner described above, in addition to the total consumption, supply and demand of different forest produce will also be estimated.

13

Soil Survey and Mapping

Introduction

The soils of India are varied because of diverse geological formation. In a homogeneous climatic region, the productivity capacity of a site is directly related to characteristics of the soil and land. The soil in conjunction with climate and topography determines the nature, composition and distribution of forests.

Forest Soil Survey provides an inventory of different kinds of soil and soil types occurring in different parts the forest division. The primary objective of the survey is to classify soils into well-defined types to map different kinds of soil. The data collected during soil survey are used to correlate soil conditions with the character and growth of the existing forest vegetation and in deciding the suitability of specific blocks or areas for specific association of trees. It also helps in evaluating the potentiality of an area for specific management practices and in determining the ecological status. A careful correlation of soil characteristics with vegetation enables the delineation of natural homogeneous area exhibiting homogenous soil properties.

A salient feature of the forest ecosystem is the development of the distinct forest floor on account of continuous litter fall. This characteristic contributes to unique soil properties. The nature and content of organic debris returned to the forest floor and varied with vegetation affects the physio-chemical properties of soil. (Benerjee, *et al.* 1996). Forest Soil Survey aims at furnishing comprehensive information about soils and an inventory of soil resources of the area. The exercise consists of:

1. Studying and recording important characteristics of the soils in the forests and in the laboratory.
2. Classifying the charaçteristics of the soils into well defined units and locating its extent and boundaries on a map.

Forest soil survey helps in generating relevant soil data and in expanding our knowledge and understanding of the soils with regard to their genesis, development, classification and nomenclature. The knowledge and understanding of the forest soils is applied to improve the existing forest crop as well as to undertake artificial plantation. In other words, forest soil survey helps in evolving a rational land use plan for a forest division.

Types of Survey

Soil surveys are classified into three types. They are:

1. Reconnaissance surveys.
2. Detailed surveys.
3. Detailed reconnaissance survey.

The basis of categorization of such types of survey is as follows:

1. The method and intensity of the survey.
2. The scale of the base map.
3. The resulting details.
4. The precision of mapping.

It is increasingly necessary to understand soils and supporting vegetation for establishing a relationship with growth, stand composition and other factors of optimum forest production. Therefore, soil study helps in better management of forests. It is the first step towards sustained forest management. It defines the area, which are under forestry use and delineates its suitability for planting of different indigenous tree species.

Reconnaissance Survey

In this type of soil survey the extensive or major soil series (taxonomic or classification units) are identified, defined and named. Here, soil survey is done on 1:50,000 scale toposheets of Survey of India. In reconnaissance mapping, the series occurring together in a regularly repeating geographic pattern are grouped into an "association" and such "associations" are delienated on the map. Such areas are delienated by a common boundary. In the course of field survey, monoliths of representative profiles, considered as the "typical profile" or "norm" are collected and kept in record. Reconnaissance survey serves the purpose of evaluating the correlation between soil and vegetation in an area.

Detailed Survey

In this type of soil survey, the boundaries between mapping units are plotted on base maps. Observations are made with respect to prominent spots or sites. The type and phases are mapped to the extent details are possible in order to show all boundaries. The maps used as base maps are generally of the scale of 20cm : 1km. Such type of soil survey is done for agricultural purposes.

In detailed survey, on account of being an elaborate exercise, maps with large scale are used. In practice, forest stock maps of 1:15,000 or 1:25,000 are used as base maps. Such large-scale helps in delineating the phases of soil types, slopes, erosion, etc. within a series are used. Since, under such survey on account of extent of details required, the soil profile examination and auger-bore observations are made at closer intervals and the mapping units at the phase level are delineated.

The detailed soil map provides information required for understanding the soil and land problems of a forest division. This helps in evolving suitable strategy for soil and water conservation, for allocation of areas to different working circles, and tackling problem areas with suitable plantation and soil amendment measures.

The National Bureau of Soil Survey and Land-use Planning (NBSS and LUP) and state soil survey units have taken up Soil Map of India Project. The soil resources and land-use maps of West Bengal, Madhya Pradesh, Haryana, Pondicherry, Punjab, Bihar, Orissa, Tamil Nadu and Gujarat have been prepared in 1:250000 scale and released for users and planners. Such maps can be used, developed and applied to forestry.

Detailed Reconnaissance Survey

This is a combination of the reconnaissance and detailed survey methods. Here, it satisfies the objectives of both these survey partially. The method is to make a reconnaissance survey over an area with a purpose of classifying and grouping of the soils into soil series or associations. The areas, which require further study, are taken up for detailed survey. In this way, this method economizes time, money and manpower. This method is generally used for soil-vegetation survey. This type of survey requires two broad types of base maps for survey and mapping. The first is topographic maps and the second is arial photographs. On account of limited availability of arial photographs the method using topographic maps are generally in vogue for soil mapping.

Method of Survey

Here the, geology, soil and vegetation of the survey area are examined, and thoroughly reconnoitered. All sort of variation in the distribution patterns of soil and vegetation are identified. For each variation, a representative site is selected. A profile pit in the selected site is excavated. Site characteristics, profile and morphological features are recorded on the specified format. All the aspects of forest vegetation at the selected site are studied along with the soil. This provides a complete picture about soil and vegetation of the representative area of the variations identified. Where the variation is too large, more representatives' sites are selected. The norm is to select about one representative site for a chunk of 200 to 300 hectares of forests. Now, a traversing plan is worked out and spots are taken at suitable intervals. Field-tests for other soil attributes are carried out and vegetational attributes are recorded. The survey is conducted in such a fashion that all possible variations found in the area are covered. After having completed the survey, delineation of soil vegetation is done.

Soil Examination in the Field

A complete appraisal of soil involves three aspects. One is surface feature, second is surface characteristics and the third is characteristics of depth. A soil profile exhibits characteristic succession of layers or horizons and the description of all those that provide a complete picture of the soil. The following are the steps in soil examinations:

1. Preparation of Profile pit
2. Examination of Profile pit
3. Designation for horizons and layers.

Details Description of the Procedure

1. *Preparation of the Profile Pit*

Soil develops distinct layers at varying depths below the land surface. A vertical section of the soil to expose tha layering is called a profile. In the selected site, the profile pit is dug. The pit is always rectangular in shpae. It has a dimension of 1.8 mt x 1.8 mt x 0.75 mt. The long side or the vertical side is examined and recorded. Then the whole pit is left to dry. Now a portion of the profile pit is pricked with the finger or small knife-blade so as to bring out the characteristics of the structural elements.

2. *Examinations of the Profile Pit*

Now the characteristics of the structural elements in the removed portion are examined. The arrangement of the individual soil particles with respect to each other into a pattern is called soil structure. In other words, soil structure symbolizes the arrangement of small, medium and large pores into a structural pattern. Soil structure is generally classified according to size of aggregates and pore sized. The other characteristics examined are soil depth and horizon thickness. Horizon thickness is generally considered as subdivision of soil depth. Apart from this, particle size analysis and soil texture estimate are also undertaken.

The determination of the soil separates present in the sample is called particle-size analysis. A complete particle-size analysis involves sample preparation followed by two kinds of procedures. The sand separates are determined by sieving and silt and slay separates are determined by their rates of settling in water.

Soil surveyors generally estimate texture of soil by feeling the soil between thumb and fingers. The technique used can be divided into two parts. First the amount of clay is estimated by how hard the sample is when dry, how much water it can absorb and its degree of stickness and plasticity when nearly saturated with water. Plasticity is evaluated by attempting to squeeze the soil into a thin ribbon as it is pressed with a rolling motion between thumb and forefinger. A moist clay soil can be formed into a long, durable ribbon or wire. A clay loam will form a ribbon or a wire but not as long or durable as that formed by a clay. The second step is to decide if the sample is sandy or silty. The sample should be wetter for this test than for the clay test. Sand gives the wet soil a gritty feel; silt gives it a smooth, floury feel. Thus a soil that forms a moderately good ribbon would be called a silty clay loam if it had a very smooth feel, a sandy clay loam, if it had a very gritty feel or simply a clay loam if the smooth and gritty materials were approximately equel. Other characteristics, which are measured, are soil porosity, soil consistence and soil colour.

3. *Designation for the Horizon and Layers*

This step involves designation for horizons and layers. The upper layer is usually higher in organic and darker in colour than the layers below. This upper layer is called a horizon or top soil. The middle part of the profile usually contains more clay and has a brighter colour. This layer is called the B-horizon or subsoil. The A and B-horions together are referred as the solum or the true soil. The C-horizon, commonly referred to as the soil parent material, occurs beneath the solum and extends downwards to dedrock. It may be thick, thin or even absent. Both the horizons A and B are normally measured by a metric scale and thickness recorded.

Soil Mapping

Soil Maps are prepared to show what kinds of soil occur in each part of the area mapped. The most comprehensive soil survey programme is generally carried out by agricultural department of States and National Bureau of Soil Survey and Land Use Planning, a unit functioning under Indian Council of Agricultural Research. Soil maps prepared by this institute for States are on a scale of 1:250000. The state agricultural department prepares soil maps on different scales.

Each area delineated on a soil map is designated as soil mapping unit. Most soil mapping units shown on detailed soil maps are phases of soil series. Soil phases are made by dividing soil series into more specific units that are significant for practical use.

Soil maps contains information that is useful for many different purposes, but this information must be interpreted before it can be used. Different, uses require different interpretations of the same soil map. Therefore, based on the objective, suitability maps can be prepared from soil maps. A suitability map can be prepared for any type of land. Such maps are made by first evaluating all soil mapping units in the area according to suitability for the specified use. Symbols representing the suitability evaluation of each soil are placed on a map. The map used is usually a copy or a tracing of the soil map.

Preparing Soil Resource Map

A detailed soil survey is a costly affair. Apart this, such maps prepared on the basis of detailed soil survey adopt inferior technology and old methods. Generally no uniform technology and sound criteria are used to generate such soil map on a scale, which can be used for optimising forest land use.

The National Bureau of Soil Survey and Land Use Planning (NBSS&LUP), Nagpur, an institution under Indian Council of Agricultural Research, has developed soil resource map for many states. The technology adopted for mapping soil resource was three-tier approach, i.e. satellite immage interpretation, field survey and lab analysis and cartography. The soil resource map prepared by NBSS & LUP indicated the kinds, extent and distribution of soils on the landscapes. It specifies different characteristics of soils of different sites. It unfolds the erosion status of ths soil also.

These soil resource maps can be used to generate a division specific soil resource map. Such division specific soil resource map based on modern technology, saves a lot of cost as well as time. Suh maps are highly useful for forest management planning in the division. Some of the uses of such maps are illustrated below:

- To generate forest soil resource map indicating the kinds, extent and distribution of different soils in a forest division.
- To delineate problematic/degraded/eroded soils for their amelioration.
- To generate several thematic maps as per the requirement of forest management practices.
- To promote forest research on mapped and defined soils for generating soil-based technology and its analogous transfer to other areas having comparable soil site characteristics.
- To generate database for storage and retrieval as and when desired for future interpretation and use.
- To predict forest productivity based on soil site suitability models.
- To monitor the soil health and to forewarn the emerging constraints.
- To preapre perspective land use plan for the area, including degraded forest land, in order to optimize land use for sustainable forest management.

The procedure for preparing soil resource map is described in the Chapter Technique for Optimizing Forest Land Use.

14
Natural Regeneration Survey

Introduction

Regeneration symbolizes renewal of a forest crop by natural or artificial means. It also refers to the crop so obtained. Renewal of a forest crop by self-sown seed or by coppice or root suckers is called natural regeneration. For sustainable development of forest resources, it is necessary that forest crop continues with natural regeneration. For natural regeneration to take place successfully, the following ground conditions are to be met:

— Adequate seed supply.
— Adequate light condition, i.e. obtained via the regulation of the density and composition of various strata or canopies.
— Proper soil composition and degree of development of undergrowth to protect the seedlings from desiccation and biotic influences without undue competition for light, moisture and nutrients and also to improve the soil structure and nutrient status.
— Efficient protection from damage from all kinds.
— Adequate drainage, infiltration and aeration in the soil mass.
— Effective soil and moisture conservation.

In a situation where the forest ground conditions are compatible and congenial to natural regeneration, renewal of forest is a continuous process. But when such conditions are not fulfilled, then the regeneration process is affected. As a result, no regeneration or inadequate regeneration occurs. Apart from this, the natural regeneration so obtained, commonly in the form of recruits, half-established and established or pole crop, appears on the forest floor in different varieties of species and in different sizes. Therefore, by observing the forest floor, it becomes difficult to make an estimate of regeneration status correctly. Besides, ocular method is also susceptible to various errors. Hence, there is a need to develop a scientific method to determine the regeneration status in the forests. In the absence of hundred per cent enumeration, reasonably accurate assessment of natural regeneration can be made by regeneration survey. Regeneration survey, by definition, is a survey for the assessment of half-established and established regeneration generally by the method of sample enumeration. This helps to draw a pragmatic regeneration strategy.

Methods of regeneration surveys being practiced in forest departments are explained below:

1. Method of regeneration survey along with inventory.
2. Traditional method of regeneration survey.
3. Method of regeneration survey in Assisted Natural Regeneration (ANR) areas.

Method of Regeneration Survey along with Inventory

Natural regeneration survey is a *sine qua non* for forest management. It is a significant component in the process of revision or preparation of working plan. However, during the preparation of plan, inventory of tree resources is prepared. Therefore, it is always a practice and most economical to combine a regeneration survey with the inventory exercise. In the regeneration survey, a low intensity sub-sampling is adequate on a pattern of sampling made to suit the system of partial enumeration for the forests. The intensity of sampling could be increased if there is any particular reason to do so. But the cost associated with increased sampling intensity should be considered thoroughly before deciding the method.

The most common method of sampling consists of strip surveys in selected compartments after stratification of forest. In some areas, line-plot surveys are in vogue. Sub-sampling of strips or line-plots is an obvious choice for the regeneration surveys. In case of strip sampling, whether random sub-sample of strips or systematic sub-sample of strips, the component quadrant will be treated as the recording unit. Line-plot survey for tree enumeration are carried out by taking plots systematically along straight lines spaced at equal distances with the first plot and first line with random starts. A systematic sub-sample of the plots is used to locate quadrants for the regeneration survey of forests.

In regeneration survey of large tracts, only the most promising categories of regeneration like poles, saplings, established plants, half-established, recruits etc. would be recorded. These categories should be defined for each species. The data should be recorded separately for each recording unit, which is generally a quadrant of 2m × 2m size. This small quadrant can be called regeneration plot. These plots can be located on the center of two diagonals of sample plots and on mid points of each arm. A layout of the plot is given in Fig. 14.1. Thus, generally, five regeneration plots, in each main plot could be surveyed for assessing regeneration status of plant species. However, in order to increase the sampling intensity of regeneration survey, the number of regeneration plot can be increased by locating additional four regeneration plots in the diagonal line as shown in Fig 14.2. Now, after laying out the required number of regeneration plots inside the main sample plot area, recording of data in the prescribed format (Table 14.1) is begun.

Table 14.1 : Regeneration Survey Format

Working plan Unit : Range :
Compartment No. : Plot No. :

Species	Regeneration	No. of seedlings per plot					Total
		Central	North	South	West	East	
Sal\Teak	Established (E) Un-established (UE) Recruit (R)						
Bija	(E) (Ue) (R)						
Saja	(E) (Ue) (R)						
Lendia	(E) (Ue) (R)						
Dhaora	(E) (Ue) (R)						
Tendu	(E) (Ue) (R)						
Salai	(E) (Ue) (R)						
Misc.	(E) (Ue) (R)						

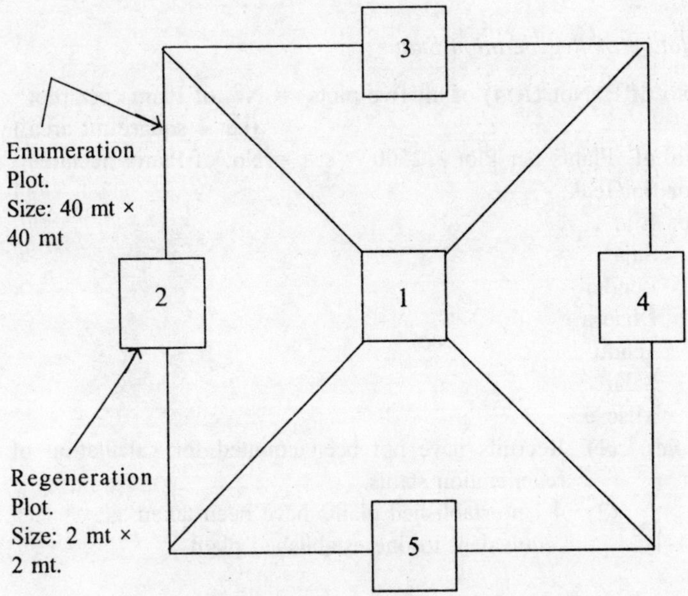

Fig. 14.1 : Enumeration and Regeneration Plot.
(With five regeneration plots.)

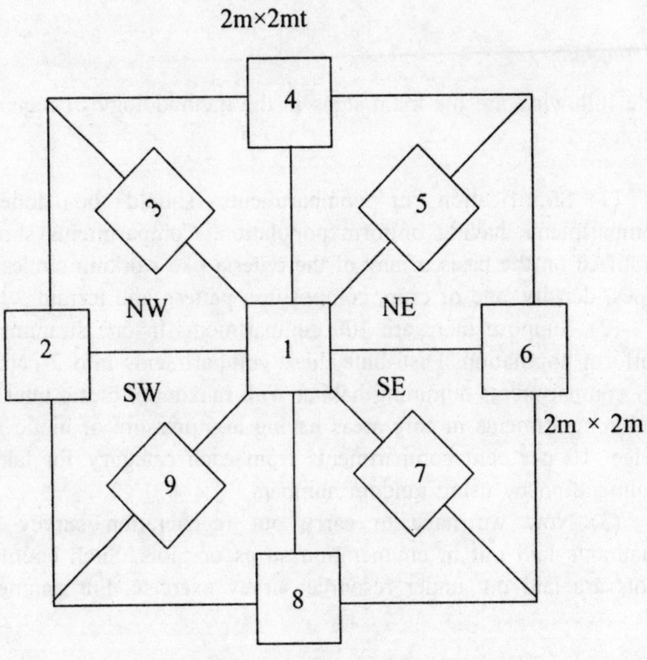

Fig. 14.2 : Enumeration with Nine Regeneration Plots.

Calculation of Regeneration Status

(No of E+Not Ue/4) of all five plots = No. of Plants per plot (i.e. 4 square mt area.)

No. of Plants per Plot × 2500 = No. of Plants/hectare.

For Sal/Teak: -

For Bija

Saja

Lendia

Dhaora

Tendu

Salai

Misc.

Notes: (1) Recruits have not been counted for calculation of regeneration status.

(2) 4 Un-established plants have been taken as equivalent to one established plant.

Date:

Signature

Name

Designation

Methodology

The following are the main steps in the methodology of regeneration survey:

(1) Stratification of compartments should be done with compartments having uniform population. Compartments should be stratified on the basis of any of the criteria like working circles, forest types, density, age of crop, composition pattern and terrain.

(2) Suppose there are 100 compartments in one stratum having uniform population. Distribute these compartments into 2 categories, i.e. compartments adjoining habitat with maximum biotic interference and compartments in core areas having less pressure of biotic factors. Select 10 per cent compartments from each category for taking up enumeration by using random numbers.

(3) Now we have to carry out regeneration survey in the quadrants laid out in enumeration strips or plots. Such enumeration plots are laid out under resource survey exercise. For enumeration,

40m × 40m plots are generally laid out. However, small plot can also be preferred on the basis of required precision. In both cases small plots having size of 2m × 2m would be adequate for regeneration survey. These plots would be located on the centre of two diagonals of sample plots, and on mid-points of each arm. Thus, five regeneration plots in each main plot could be located and suitably surveyed for regeneration status of established and half-established and the data recorded in the format enclosed herewith. As already said, in case where more regeneration plots are required to be laid, such additional regeneration plots can be located on the diagonal lines.

Traditional Method of Regeneration Survey

Regeneration survey can be done along with the inventory or independently. In flat terrain, the regeneration survey may be carried out in linear strips or line plot survey. In mountainous terrain, the survey can be carried out by topographical units. In the past, in case of forests managed under shelter-wood system the intensity of sampling for regeneration survey used to be kept generally from 2 to 4 per cent. For this purpose, a base line was drawn on the base of the map and parallel survey lines were drawn at right angles to the base line 200 mt or 100 mt apart, after selecting the position of the first line by random sample. Regeneration surveys were used to be done in plots having size of 2mt × 2mt square on either side of the survey line. In this method of survey, it was sufficient to record the occurrence of the most promising category of regeneration only. In dense forests, attention was generally concentrated on established saplings and poles below the lowest diameter limit of tree enumeration and on woody shoots, which were on their way to establishment. It was only when these classes were absent then only the lower classes of regeneration were recorded (Fig. 14.3). Category of regeneration and their symbol is used in method given in Table 14.2.

The regeneration categories as shown in Table 14.2 are being allotted different values shown in Table 14.3. The values are computed for every hectare of forests to find out the status of regeneration. Values calculated for these rectangles are shown on the regeneration stock map as given in Fig. 14.4 Regeneration status of the forest is computed and assessed on the basis of certain predecided weightage or value. The symbol, value, and indices are presented in Tables 14.3 and 14.4.

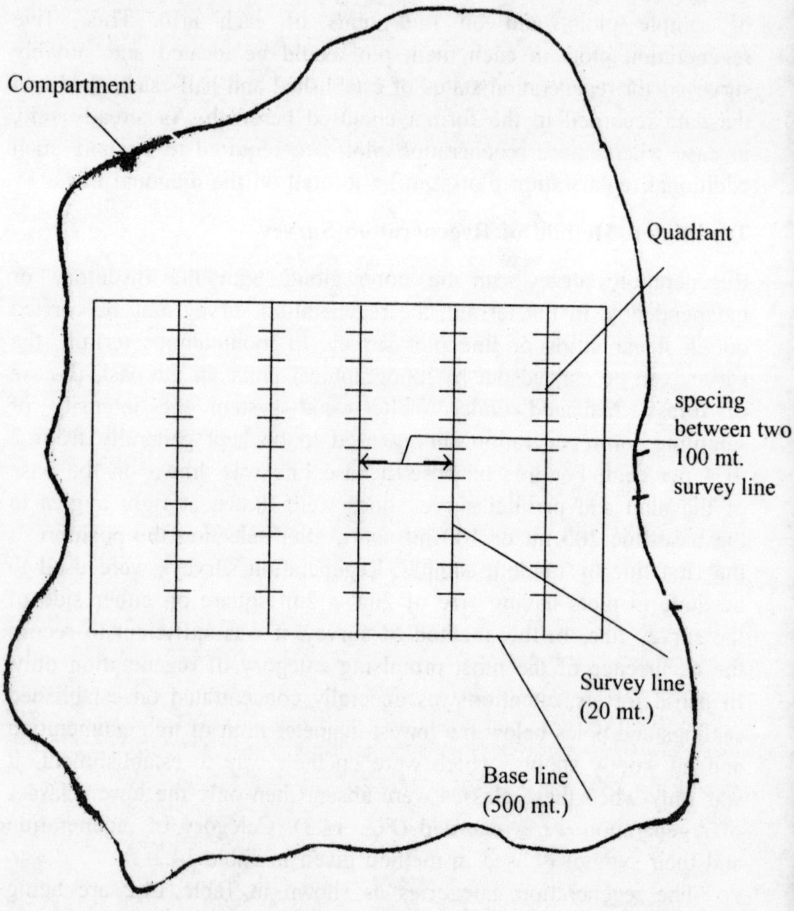

Each survey line : 10 Quadrant on each side
One unit (1 hact.) = 500 mt × 20 mt.
= 100 quadrant

Fig. 14.3 : Layout of Regeneration Survey.

Table 14.2 : Category of Regeneration and Their Symbol

S.No.	Category of Regeneration	Symbol	Explanation
1.	Established regeneration whose height should be 2.5 m or more and d.b.h. 10 cm	E	Symbol indicated that at least one such plant is present which is sufficient to stock the quadrant.
2.	Woody shoot, which is not established but large and vigorous enough to become established early.	W	In the absence of 'e', 'w' indicates that there is at least one woody shoot, which is sufficient to stock the quadrant.
3.	Woody shoots which has been browsed.	W+	—
4.	Whippy unestablished seedling whose height is more than 50 cm	U+	In the absence of 'e' and 'w', 'u+' indicates that there are more than one unestablished whippy seedlings more than 50cm height in the quadrant.
5.	Whippy unestablished seedlings whose height is more than 50 cm	U	In the absence of 'e', 'w' and 'u+', 'u' indicates that there is one unestablished whippy seedling more than 50 cm in height in the quadrant.
6.	Sub-whippy unestablished seedling whose height is less than 50 cm	S+	In the absence of other symbols (i.e. e, w, u+, u), s+ indicates that there are more than one sub-whippy seedling in the quadrant.
7.	Sub-whippy unestablished seedlings whose height is less than 50cm	S	In the absence of other symbols, 's' indicates that there is one sub-whippy seedling in the quadrant.
8.	Recruit (Current year's seedling)	R	In the absence of other symbols, 'r' indicates that there are only current year's seedlings in the quadrant.
9.	Blank	O	'O' indicates that there is no regeneration in the quadrant.

Table 14.3 : Value of Different Categories of Regeneration

S.No.	Regeneration categories	Symbol	Value
1.	Established	E	5.00
2.	Woody shoots	W	4.00
3.	Woody shoots browsed	W+	4.00
4.	Whippy unestablished shoots (More than one)	U+	2.00
5.	Whippy unestablished shoots (Only one)	U	1.00
6.	Sub-whippy shoots (More than one)	S+	0.50
7.	Sub-whippy shoots (Only one)	S	0.25
8.	Recruit	R	0.00

Table 14.4 : Regeneration Estimation Index

Total value of regeneration categories in one unit of forest (i.e. one hectare)	Description of the condition of regeneration	Colour in the map
81 to 100	Excellent	Green
61 to 80	Good	Blue
41 to 60	Moderate	Yellow
21 to 40	Fair	Red
0 to 20	Deficient	Blank (no colour)

Method of Regeneration Survey in ANR Areas

Off late, there has been a concerted effort to *prop* up natural regeneration in different parts of forests. Most of the foreign aided projects focus on revival of natural regeneration activity. By minimizing the external factors that impair natural regeneration, there is a possibility of rejuvenating the forest system towards sustained natural regeneration. In forestry parlance, such programme encompassing all activities, which make the forest system congenial to revival of natural regeneration, is called Assisted Natural Regeneration (ANR). While taking up ANR Operations in forests after main felling, there is a need to carry out benchmark analysis of regeneration status in the compartment. Again, after having implemented the ANR programme, it is also necessary to undertake periodic natural regeneration survey to evaluate the progress in the regeneration front. Therefore, in the state of M.P., a model has been developed to conduct regeneration survey in the compartment before

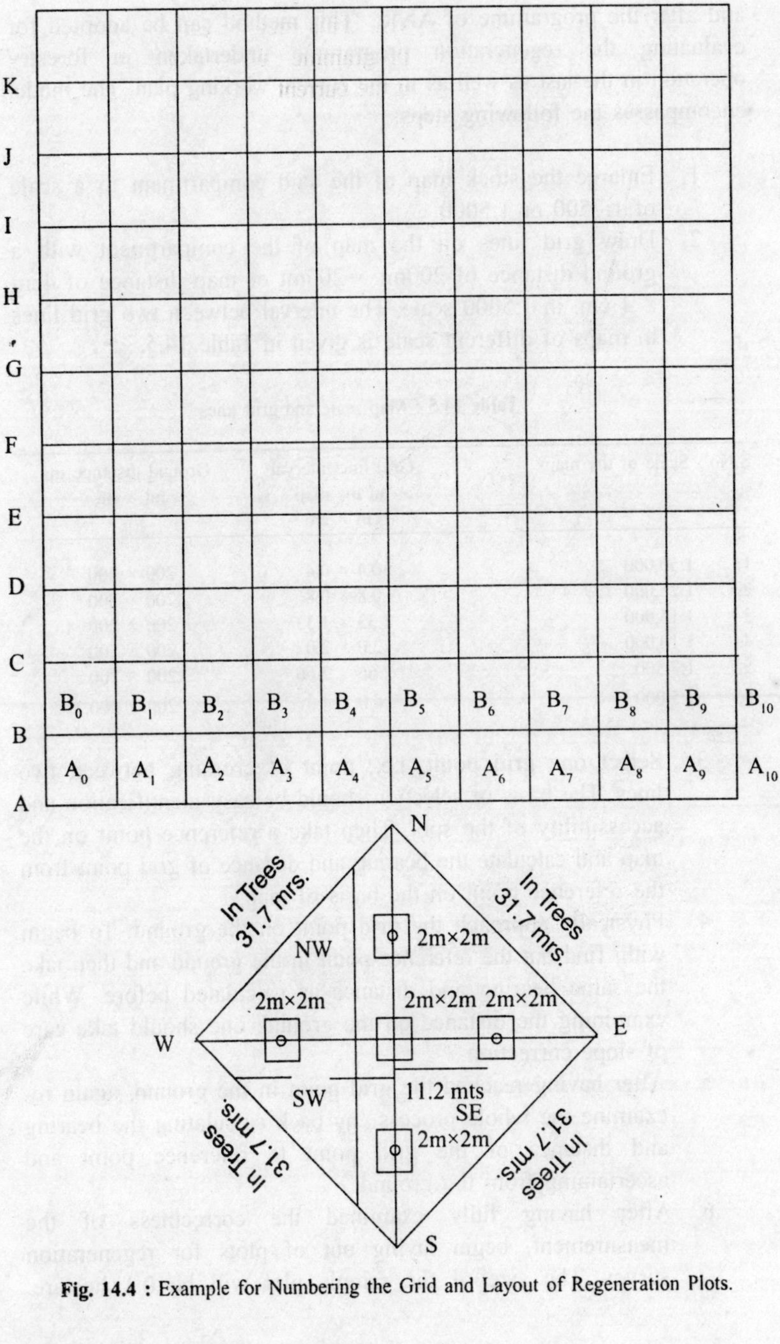

Fig. 14.4 : Example for Numbering the Grid and Layout of Regeneration Plots.

and after the programme of ANR. This method can be adopted for evaluating the regeneration programme undertaken in forestry operation in the last as well as in the current working plan. The model encompasses the following steps:

1. Enlarge the stock map of the said compartment to a scale of 1:7500 or 1:5000.
2. Draw grid lines on the map of the compartment with a ground distance of 200mt × 200mt or map distance of 4cm × 4 cm. in 1:5000 scale. The interval between two grid lines in maps of different scale is given in Table 14.5.

Table 14.5 : Map scale and grid lines

S.No.	Scale of the map	Grid line interval on the map cm × cm	Ground distance in mt × mt
1.	1:50,000	0.4 × 0.4	200 × 200
2.	1:25,000	0.8 × 0.8	200 × 200
3.	1:15,000	1.33 × 1.33	200 × 200
4.	1:10,000	2.0 × 2.0	200 × 200
5.	1:7,500	2.66 × 2.66	200 × 200
6.	1:5,000	4.0 × 4.0	200 × 200

3. Select one grid point, i.e., point of crossing between two lines. The basis of selection should be easy identification and accessibility of the spot. Then take a reference point on the map and calculate the bearing and distance of grid point from the reference point on the basis of map.
4. Physically approach the grid point on the ground. To begin with, find out the reference point in the ground and then take the same bearing and distance as calculated before. While examining the distance on the ground, one should take care of slope correction.
5. After having reached the grid point in the ground, again re-examine the whole process by back calculating the bearing and distance of the grid point to reference point and ascertaining from the ground.
6. After having fully examined the correctness of the measurement, begin laying out of plots for regeneration survey. The size of the sample plot will be 0.1 hectare.

Regeneration plot will be of the size of 2 mt × 2 mt. Such plots will be laid out at a distance of 11.2 mt from the central point of the sample plot. In addition to such four regeneration plots, there will be one regeneration plot in the centre. Therefore, there will be five regeneration plots in the same sample plot. (Fig. 14.1)

7. Estimate regeneration of tree species in all the five regeneration plots. The format for counting or recording regeneration is provided in Fig. 14.5. The regeneration of tree speies has been classified into the following categories :

Comptt. No. : Range :
Grid No. : Division :

R.F./P.F.

Northern		Eastern		Central		Southern		Western	

Fig. 14.5 : Format for counting or recording regeneration.

1. Recruit (Plant of one year old and contains two to four leaves).
2. Unestablished (Includes all whippy and sub-whippy plants having collar girth of thumb size and height upto 1.25 mt).
3. Established (Plants having collar girth more than thumb size and height more than 1.25 mt).

8. Repeat the same process for undertaking regeneration survey in other grid points also.

15

Study of Bio-diversity : Principles and Practices

Introduction

Off late, bio-diversity conservation has become an important strategy to achieve sustainable development of forest resources. In 1992, the World Resources Institute (WRI), the World Conservation Union (WCU) and the United Nations Environment Programme (UNEP) released a global Bio-diversity Strategy. The strategy lays out a comprehensive plan of action to conserve diversity.

The fundamental goal of bio-diversity conservation is to support sustainable development by protecting and using biological resources in ways that do not diminish the existing variety of genus and species or destroy important habitat or ecosystem. Thus, bio-diversity conservation encompasses the following three basic policy concerns:

(i) To minimise the loss of genetic diversity
(ii) To minimise the loss of species diversity.
(iii) To manage biodiversity within and along with biological communities so that human needs are met without reducing potential contributions to future generations.

The third concern, involving bio-diversity at community and ecosystem levels is distinct from the first two for different reasons. First, many different units or shades of diversity are conspicuous. Such units of diversity are pattern of habitat in a community, the relative abundance of species, the age structure of population, the pattern of communities, trophic structure and dynamics. It is said that communities are transient associations of species. All communities can be defined on the basis of existing species composition and structure over shorter time periods. Communities are ephemeral and sensitive environmental change. Therefore, management of bio-diversity symbolises concern and action for bio-diversity conservation at the species level. Two basic reasons are attributed to this.

(i) Ensuring maintenance of the species comprising the community.

(ii) Helping to maintain the life supporting system and ecological benefits obtained from the ecosystems.

Therefore, there is a need to understand community structure and bio-diversity of ecosystem. Conservation does not purport to preserve community's attributes *in toto*. Rather, it is not possible to conserve all the ecological attributes of a community. However, in order to ensure maintenance of the species within a community and to maintain the life-supporting system of the ecosystem, there is an imperative need to conserve the basic structure and diversity of community attributes and functions. This will prevent loss of species. To sum up, sustainable forest management for diversity aims at maintaining the benefits that man obtains from ecosystems, by managing the basic community structure and diversity in the ecosystems, to meet human needs, support species and genetic diversity and enable the system to adopt to changing conditions.

Community Structure and Bio-Diversity

Forests constitute an ecosystem. As we know, an ecosystem comprises of interactions of organisms and their environmental factors involving flow of energy and cycling of minerals. Naturally, it has structural characteristics depending upon the spatial distribution of plants and animals. Its functional aspects include tapping of energy by green plants and its passage into herbaceous and carnivorous animals, which are either parasite or saprophyte. The solar radiation originally received in the system gets dissipated out of it by the respiratory activity or the participatory organism at various stages so that, for maintenance of the ecosystem energy must be continuously added and transferred from organism within the community. The amount of energy gained by the system and the mode and quantities of it transferred, affect the community structure of the system such as density, dominance, abundance, frequency and diversity of species. The concept of community structure is important and relevant in the practice of forest ecology, because with the disappearance of community, the organism also vanishes. Man's welfare depends ultimately on the nature of communities and ecosystems upon which he superimposes his culture (P.G. Odum, 1971).

While energy flows through the ecosystem, minerals are absorbed from the soil by the plants and then passed on to the organisms in the form of elaborated compounds. The use and reuse of the minerals by the organism keep them circulating within the ecosystem. However, climatic, edaphic and biotic factors impose limitation on the rates and amounts involved in soil circulation. Therefore, community structure, function, and factors of the environment become triangular base of interaction of the ecosystem. This triangular base constitutes the scope for ecological study of the ecosystem.

Plants growing together have mutual relationship among themselves and with the environment. Such a group of plants in one area constitutes a stand. Several such similar stands represent a community. By definition, a biotic community constitutes an assemblage of population living in a prescribed area or physical habitat; it is an organized unit to the extent that it has characteristics additional to its individuals and population components and functions as a unit through coupled metabolism transformations. Communities not only have a definite functional utility with characteristic trophic structures and patterns of energy flow but they also have compositional unity in that there is a certain probability that certain species will occur together (Odum, 1971). In other words, community is the part of the ecosystem, in which transformation, accumulation and flow of energy are involved. The functioning of this system is intimately related with the components of the community. The components vary in quality as well as in quantity and impart a structure to the community.

Communities can also be classified into two categories. One is major community and the other is minor community. Major communities are those which are of sufficient size and completeness of organization that they are relatively independent; that is, they need only to receive sun energy from outside and are relatively independently of inputs and outputs from adjacent communities. Minor communities are those which are more or less dependent on neighbouring aggregations. In a forest, both major as well as minor communities are found. Therefore, the structure and diversity of plant community in a forest can be studied by taking consideration of numbers of characters, which are described below:

1. Density, abundance and dominance.
2. Frequency.

3. Importance Value Index (IVI).
4. Comparison of two communities or stands.
5. Species diversity.

Density, Abundance and Dominance

Density represents the number of individuals per unit area. The term abundance and density represent the numerical strength of species in a community. The former, if considered along with frequency, gives an idea of the distribution pattern of species. The density and frequency taken together are of immense importance in determining community structure.

Characters of community structure are determined by adopting sampling method. There are two main sampling methods which are used for studying community structure in a forest ecosystem. They are quadrat and transact method.

A. Quadrat Method

It is the name given to the sampling unit. The sampling unit is an area of definite size. In shape, it is generally rectangular or square or circular. Because of the fact that the assemblage of plants in a community is largely heterogeneous, there is a need to determine the size and number of quadrats necessary for adequate sampling.

1. *Procedure to determine size of quadrat*

(1) Field procedure to lay out plots:

1. Lay out an 'L' shaped structure in the field and demarcate the point of joining of two arms of 'L' shaped structure as point 'O'.
2. Take 'O' as the southwest corner of the quadrat and demarcate an area of 10 cm×10 cm.
3. Record the species present with in this area.
4. Now enlarge the area to 20 cm × 20 cm, taking 'O' as the same southwest corner point.
5. Record the additional species found in that area.
6. Again enlarge the area to 30 cm × 30 cm and record the additional species.
7. Likewise, go on increasing the area and record the additional species till the quadrat size is reached at about 2 mt × 2 mt.
8. The fashion in which the area is enlarged is shown in Fig. 15.1.

9. Summarize the data in suitable table.

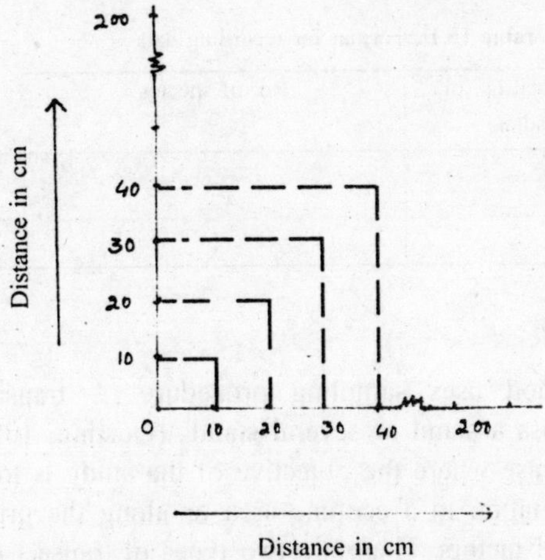

Fig. 15.1 : Determination of Requisite Size of Quadart.

Procedure for analysis of data

The data is analyzed in the following manner :

1. Take a graph paper and draw 'X' and 'Y' axis on it.
2. Plot the number of species on the Y-axis and the area on the X-axis.
3. The point at which the curve starts flattening up represents the minimal area of the quadrat required for sampling the field. This curve is known as 'species area curve'. This is shown in Fig. 15.2. In other words, the point 'b' indicates the area required to take for laying out sampling units.
4. After having ascertained the requisite size of quadrat, lay 30 to 50 quadrats of that size in a stand under study.
5. Record the species occurring in each quadrat and tabulate the data in the format given in Table 15.1.
6. Take a graph sheet and plot the number of quadrats on the X-axis and number of species on Y-axis. Here also the point at which the curve starts flattening indicates the number of

quadrats required for adequate sampling of all the stands of the same community.

Table 15.1 : Format for recording data

Serial no. of quadrats	No. of species

B. Transact method

The transact method uses sampling procedure. A transact is a sampling strip across a stand or several stands (Oosting, 1958). It is applied in such a case where the objective of the study is to analyze the change in vegetation in a ecotone area or along the gradient of some environmental factors. There are two types of transact methods. One is Line transact method and the other is Belt transact method. The size and numbers of transacts are determined by extent of the area to be taken for study.

In line transact method, a measuring tape is run across the vegetation to be analyzed and plants coming in touch with the measuring tape are recorded. It is useful for grassland ecosystem. From the data generated from this field work, various characteristics like density, dominance can be found out. In belt transect method, a belt or strip of definite width is studied across the vegetation. This strip is divided into a number of small sections and plant species and number of individuals of each species coming in each section are recorded. By mapping the sections on a graph paper, a useful picture of the vegetation is emerged.

Lying out of quadrat in the field

After determining the requisite number of quadrats, it is time to lay the quadrats of desired size on the field. Then record the number of individuals of each species in each quadrat as per the format given in Table 15.2.

Table 15.2 : Format for recording of species

Plant species	No. of individuals in quadrant number 1, 2, 3, ...10	Total No. individuals	No. of quadrats of occurrence	Total No. quadrats studied	Abundance	Density
1	2	3	4	5	6	7

Calculate abundance for each species by the following formula:

$$* \text{ Abundance} = \frac{\text{Total No. of individuals}}{\text{No. of quadrats of occurrence}}$$

$$* \text{ Density} = \frac{\text{Total No. of individuals}}{\text{Total No. of quadrats studied}}$$

Note: Abundance is described as the number of individuals per quadrat of occurrence. Density is described as the number of individuals per quadrat.

*** Index of Dominance**

Communities, at least major ones, have producers, macro-consumers, and micro-consumers. Within these group species or species groups which largely control the energy flow and strongly affect the environment of all other species are known as ecological dominants. The degree to which dominance is concentrated in one, several, or many species can be expressed by an appropriate index of dominance that sums each species importance in relation to the community as a whole (Odum, Eugene, 1971).

Simpson (1949) has suggested the following formula to estimate the index of dominance

$$C = \Sigma \, (ni/N)^2$$
$$ni$$

Whereas, *ni* is the importance value of the species in terms of number of individuals; biomass, production or so forth over an unit area;

N is the total of corresponding importance values of all the component species in the same area and period; and

Σ (sigma) indicates summation.

In a mixed crop where all most all the constituent species share almost the same number of individuals or biomass almost equally, the dominance value will be very low. Whereas, in an almost pure crop dominated by one or two species, the dominance will be high showing values more than 0.5 (Fig. 15.2).

Frequency

Frequency as described by Raunkiaer (1934) indicates the number of sampling units in which a given species occurs. It expresses the distribution or dispersion of various species in a community. From this, the percentage frequency is calculated as follows :

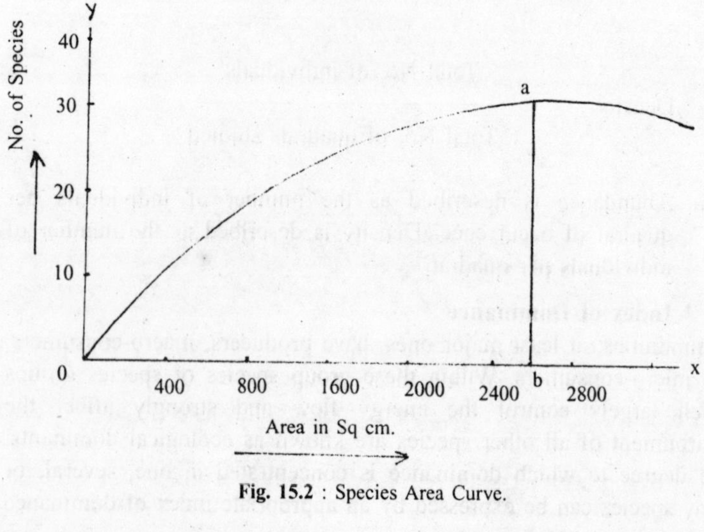

Fig. 15.2 : Species Area Curve.

$$\% \text{ Frequency} = \frac{\text{No. of sampling units in which the species occurred}}{\text{Total No. of units studied}} \times 100$$

Procedure:

1. Lay out the required number of quadrats of appropriate size in the field.
2. Record presence or absence of each species by + or – sign respectively in each of the quadrats in the way shown in the format (Table 15.3),

Table 15.3 : Format for recording presence or absence of each species

S.No.	Plant species	Quadrats 1,2,3,4,...10	Total No. of quadrats of occurrence	Total No. of quadrats studied	% frequency	Frequency class
1	2	3	4	5	6	7

3. From the data collected in the above table, prepare a frequency diagram for the field under study as described below.
4. Divide the species into 5 classes based on percentage frequncy.
   * Class A 1 to 20 %.
   * Class B 21 to 40 %.
   * Class C 41 to 60 %.
   * Class D 61 to 80 %.
   * Class E 81 to 100 %.
5. Take a graph sheet and represent the per cent of the total number of species coming into each class on Y-axis against the five frequency classes on X-axis (Fig. 15.3).

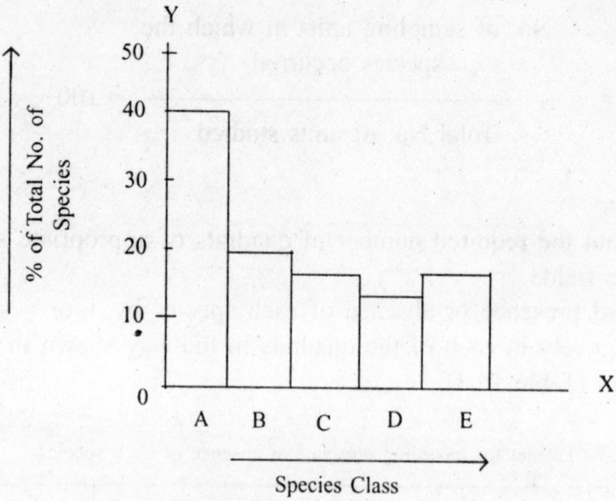

Fig. 15.3 : Raunkiaer's Normal Frequency Diagram.

Raunkiaer's law of frequency class is A > B > C >/< D < E. Frequency diagram represents the homogeneity or heterogeneity of a community as floristic uniformity varies directly with the value of classes A and E. When classes B, C, and D are relatively high, the stand is heterogeneous. In other words, higher the class E, greater is the homogeneity.

Importance Value Index (IVI)

In order to express the dominance and ecological success of any species, with a single value, the concept of importance value index has been developed. This index utilizes three characteristics. These are (1) relative frequency, (2) relative density, and (3) relative dominance.

Procedure to Calculate the Index

I. Evaluate frequency, density and basal area of all species growing at the study site through methods already described and calculate the following value. (Phillips 1959)

$$\text{Relative dominance} = \frac{\text{Total basal area of the species}}{\text{Total basal area of all species}} \times 100.$$

$$\text{Relative density} = \frac{\text{Number of individuals of the species}}{\text{Number of individuals of all species}} \times 100$$

$$\text{Relative frequency} = \frac{\text{Number of occurrence of species}}{\text{Number of occurrence of all species}} \times 100$$

II. Add all the above three values to get Importance Value Index (IVI).
III. Calculate the index for all the species and arrange them in decreasing order of values.
IV. Compare the values of IVI of important species from different sites, differing in intensity of biotic disturbance and discuss.
V. While calculating IVI in a forest ecosystem, the herbaceous species are not taken into consideration. The structure of ground flora community of a forest is studied through the method described above.

Comparison of Stands

Sometime it becomes necessary to determine whether two communities or stands studied belong to the same community or how far two different communities resemble each other. This is determined by using the concept of *community coefficient*. In this concept any quantitative character of all the species present in the two stands is taken up and an index is calculated.

Procedure

I. Determine the frequency of all species growing in the two stands or communities and tabulate the data as shown in the format (Table 15.4).
II. Underline the values for the species which are common to both the stands.
III. Total up the lower of each pair (A and B) of frequency percentage of these common species and put the sum as "W".

Table 15.4 : Format for frequency percentage of species present in both the stands/communities

Species	Stand 'A'	Stand 'B'
1 2 Etc.		
Total	a	b

IV. Total up all the values in the columns 'A' and 'B' separately and call it as 'a' and 'b' respectively.

V. Calculate the community coefficient by using the following formula (Kulczynski, 1937) :

$$\text{Community coefficient} = \frac{2W}{a + b} \times 100$$

Note: Higher the coefficient, more similar will be the communities.

Species Diversity

Biodiversity is a synonym for the variety and variability of life forms on this earth. In other words, all variety of living organisms like plant, animals and microbes that exist on the earth constitute the bio diversity.

Broadly, bio-diversity is categorised into three kinds. They are:

1. *Ecosystem diversity*: It constitutes all different kinds of ecosystems that exist on earth.
2. *Species diversity*: It constitutes all different species of plants and animals in a given area that is flora and fauna of an area.
3. *Genetic diversity*: It constitutes genetic or other variations within a species such as ecotype, etc.

Bio-diversity is a precious resource, which should be conserved for our future generation. It is because of the fact that conservation of bio-diversity holds the key to survival and progress of humankind as well as civilisation. Today, the bio-diversity resources are under tremendous pressure on account of uncontrolled increase in human as well as livestock population, changing life style pattern and environment degrading trade practices. In the context of rapidly social and economic transformation in the society, the conservation and sustainable utilisation of biological resources have assumed paramount importance. A list indicating estimated numbers of species found in India is provided in Table 15.5.

Table 15.5 : Estimated number of species in India

S.No.	Taxon	Number of spp	Percentage
1.	Bacteria	850	0.8
2.	Fungi	23000	21.2
3.	Algae	2500	2.3
4.	Pteridophyta	1022	0.9
5.	Gymnosperm	64	0.1
6.	Angiosperm	15000	13.9
7.	Insect	53430	49.3
8.	Molluscs	5050	4.7
9.	Pisces	2546	2.4
10.	Amphibia	204	0.2
11.	Reptilia	446	0.4
12.	Aves	1228	1.1
13.	Mammalia	372	0.8
	TOTAL	108276	100

Source: GOI, 1994

While the common species or dominants largely account for the energy flow in each trophic group, it is the large number of rare species that primarily determines the species diversity of the whole communities. Ratio between the number of species and "importance value" (number, biomass, productivity and so on) of individuals are called species diversity indices. Species diversity tends to be low in physically controlled ecosystem and high in biologically controlled ecosystem. It is directly related to stability (Odum, Eugine 1971). Higher diversity means longer food chains and more cases of symbiosis (mutualism, parasitism, commensalism, and so forth) and greater possibilities for negative feedback control, which reduces oscillations and hence increases stability.

Two broad approaches are used to analyse species diversity in different situations. These are as follows:

1. Comparisons based on the shapes, patterns, or equation of species abundance curves.
2. Comparisons based on diversity indices, which are ratios or other mathematical expression of species-importance relationships.

The first approach has already been discussed in the previous paragraphs. The methods relating to second approach is taken up for illustration. Different bio-diversity indices used under the second approach are as follows:

Species Diversity Index

Different kinds of indices for species diversity have been suggested by number of ecologists.

1. Odum, Cantlon and Kornicker (1960) have taken diversity (d) as the number of species (S) for one thousand individuals.

$$d = \frac{S}{1000}$$

2. Margalef (1958) has suggested the following formula:

$$d = \frac{S - 1}{\text{Log } N}$$

3. Menhinick (1964) has given the following formula:

$$d = \frac{S}{\sqrt{N}}$$

where d = diversity index.
S = Number of species.
N = Number of individuals, etc.

Evenness Index (Pielou 1966)

$$e = \frac{\overline{H}}{\text{Log } S}$$

where $\bar{H}$ = Shannon index (see below)
S = Number of species.

Shannon index of general diversity (H) (Shannon and Weaver (1949)

$$\bar{H} = -\ \Sigma\ (ni/N) \log (ni/N)$$

Or

$$-\ \Sigma\ pi \log pi$$

where ni = Importance value for each species
N = Total importance values
Pi = Importance probability for each species = ni/N.

Note: (In d, e, $\bar{H}$ natural logarithms (log e) are usually employed but log e is often used to calculate $\bar{H}$, so as to obtain 'bits per individual'.

Evenness or equitability Index shows the apportionment of individuals among the species. For example, two systems or habitat containing 10 species and 100 individuals has the same S/N index, but these could have widely different evenness indices depending on the apportionment of the 100 individuals among the 10 species. The 'e' index is an example for a convenient expression for this component. The widely used "Shannon function" or $\bar{H}$ index combines the variety and evenness components as one overall index of diversity.

Similarity Index

Sorensen (1948) has given a simple formula to determine the index of similarity (s) between two samples of vegetation.

$$S = \frac{2C}{A+C}$$

where A = Number of species in sample A.
B = Number of species in sample B.
C = Number of species common to both samples.

Such indices can be used to compare one community or group of populations with another, provided it is first determined that the S is a linear function of the log or square root of N. If there are 20 species on site A and 15 species on site B and 10 species are common for both the samples, then the similarity index can be computed in the following way:

$$S = 2C/(A+B) = 20/(20+15) = 20/35 = 0.57$$

Dissimilarity index

$$D = 1 - S$$

whereas D = Dissimilarity index
S = Similarity index

Taking the example given above, the dissimilarity index will be

$$D = 1 - S$$
$$= 1 - 0.57$$
$$= 0.43$$

Procedure for field work

Now, with the enunciation of New Forest Policy (1998) and the promulgation of bio-diversity convention under the aegis of United Nation's Commission on Sustainable Development, Conservation of Biological Diversity has assumed primacy in forest management. Therefore, apart from analysis of community structure, bio-diversity study is also an important component of present day working plan exercise.

The procedure explained in regard to analysis of community structure applies to study of bio-diversity also. Apart from this, while doing the inventory of tree resources, it is required to lay sample plots in the field. In an area which is selected for bio-diversity study, all the data regarding tree species and shrubs collected from the sample plots can be taken up for this analysis. Alternatively, to reduce the work, one quadrant instead of all four quadrants out of the said sample plot may be taken up for enumeration for all types of plant species. If the size of the plot is 0.1 hectare, then the size of a quadrant would be about 0.025 hectare. This size would satisfy the principles of sampling. However, if the number of plots falls short of the required statistical norm, additional plots may be laid in the study area to increase the sampling per cent.

In a working plan exercise, many kinds of diversity studies are undertaken. The first one is ecosystem diversity. Under it, it is required to name different existing ecosystems or forest types based on Champion and Seth classification. Each ecosystem should be properly analysed with their characteristics and demarcated with distinct boundaries. The second one is species diversity study. Details of species diversity indicators are provided in the coming paragraph. Next, the most important is the study of community structure. In the exercise, all relevant community attributes like dominance, abundance, density, frequency can be taken up. All such studies help the planner to estimate status of community structure and bio-diversity of the forests for which a management plan is being prepared. This also help to develop a base line ecological data for monitoring and evaluation of ongoing forestry plans/projects.

Bio-Diversity Indicators

Indicators are key statistical series. They are used to help policy makers, public and resource managers to assess conditions, trends and the achievement of specific goals. More specifically, in one hand, indicators enable policy-makers to evaluate and ascertain whether existing policies are having the desired effects on bio-diversity and to help clarify where problems exist in the current policy framework. In the other hand, it provides decision-makers with useful information on the status of and trends in bio-diversity and help them to evolve the suitable action plan. Apart from this, indicators are also used for forecasting and projecting trends of bio-diversity in the context of given level of management. What is most important is the fact that indicators help to determine priorities for action.

Indicators should be based on data that are timely, accurate and of known quality and standard. They are most valuable in the context of a specific management objective. If the objective is to minimise species extinction then a usual indicator would be changed in the number of species over time. If the objective is to minimise the loss of "species diversity" then the determination of the number of species along with species richness and distinctiveness may be required. There are two types of indicator. One set of indicator is required to assess pursuit of the broadly stated bio-diversity conservation concerns and objectives and the second set of indicator measures management of species and habitat. Noss (1990) has presented a set of indicators, which are of enormous value at the level of the management of diversity at communities and species level.

There are thirteen indicators, which deal with mild species and genetic diversity of natural forests. These indicators are used by policy makers, planners and decision-makers in the pursuit of goals of sustainable management of forests.

There are thirteen indicators developed till today to assess species and genetic diversity. These indicators are explained below:

1. *Species richness* (number of species per unit area and number of species per habitat type) : Species richness is one of the most important indicators. However, it is only a partial measure of species diversity. It indicates the status and trends of change. Generally, species richness is measured with respect to better-known texa, particularly plants and vertebrates. Out of plants, again, identifiable species are enumerated. Therefore, lack of adequate data regarding identification leads to the measurement of few groups of species in the field, leaving aside a major part of the texa. However, such approach has the following merits:

— Plants and vertebrates tend to be of considerable direct importance to humanity.
— Areas with high species richness in these groups tend to have high species richness in other groups.
— The diversity of plants in a region does not fluctuate significantly over short time intervals.

Measurement of species richness should be done for specific types of habitats such as dry and moist deciduous tropical forests, wetland, temperate forests, and tropical evergreen forests.

2. *Species (populations) threatened with extinction (number or per cent):* The number of species threatened with extinction is one of the most common indicators with respect to the status of bio-diversity.
3. *Species (population) threatened with extirpation (Number or per cent):* The threat of local or national extirpation of a species assumes significance in the context of sustainable management of forest resources. Therefore, this is one of the important indicators.

4. *Endemic species (Number or per cent):* The measurement of endemic species is one of the important indicators which portrays the status of bio-diversity.
5. *Endemic species threatened with extinction (Number or percent):* Levels of endemism are useful indicators because they reveal ecosystems or community that are distinct biologically. It is appropriate to express levels of endemism on per unit area basis. Myers (1988, 1990) used such a measure in his analysis of potential bio-diversity hot spots by examining the percentage of endemic species in relation to the percentage of land area involved. Conservation actions in sites having a high number of endemic species relative to the land area are likely to be most efficient in conserving species diversity.
6. *Species Risk Index :* A species risk index combines information on endemic species within a community. It provides sufficient insights into the status of species, even though there is an absence of detailed list of threatened species. The index is calculated by multiplying the number of endemic species (per unit area) in a community by the percentage of the natural community that has been lost. Thus, an ecological community with many endemics and with a high proportion of its area lost would be ranked at high risk, while a community with few endemic or an area which witnessed little loss would be ranked at low risk.

Generally, this index is applied to bio-geographic units for which endemism data concerning plants, birds and mammal's are available. To calculate the index, the number of endemic species is converted to a per unit area basis, using a species-area curve, determined by the exponent z, which varies from region to region because of differences in the number of habitats present. Empirical studies of the species area relationship have found that the z value typically varies from 0.15 to 0.35. Therefore, a mid-part of 0.25 was chosen for calculations for the species Risk index (Conner and McCoy, 1979; Simberloff, 1986). The species area curve is expressed as:

$$S = CA^{z}$$

When S is the number of endemic species of the area of the bio-geographical unit, C is a constant and z is the exponent. Solving for

C and then substituting this value into the same equation gives:

$$S2 = (S1\ A2)^z / A^z$$

Where *S*2 is the number of expected per standard area, and *A*2 is the standard area (100,000 sq. km.). Multiplying this by the percentage of the bio-geographical unit that has witnessed a loss provides the species Risk Index Value.

7. *Species (population) with stable or increasing populations (number or percent):* This is also an important indicator of species bio-diversity. It helps to indicate the changes taking place in the community and to evaluate the impact of conservation efforts.
8. *Species (population) with decreasing populations (number or percent) :* There is a red list of 'Threatened Species', documenting native species at risk of extirpation or extinction. Analysis in the context of that list helps in better monitoring of conservation efforts.

Population monitoring is practicable only when only a few indicator species are tracked. Such key stone species, having particular significance to ecosystems function cater to the habitat requirements of smaller ranging species and ensures their protection. Therefore, data on the population trends of indicator species assumes significance from the point of management.

9. *Threatened species in protected areas (number or per cent):* This is also an important indicator of bio-diversity in the protected areas.
10. *Endemic species in protected areas (number or per cent):* This is another important indicator of bio-diversity in the protected areas.
11. *Threatened species maintained in exsitu collections. (Number or per cent) :* This is also an important indicator of bio-diversity in those areas where the objectives is to monitor the changes in the *exsitu* conservation areas.
12. *Threatened species with viable (reproducing) exsitu populations (number or per cent) :* This constitutes the important indicator used in the conservation of bio-diversity in protected areas and *exsitu* conservation centres.

Enumeration of known threatened or endemic species in protected areas provides a useful baseline indicator for the potential conservation of these areas.

Bio-diversity conservation involves the class linkage of *exsitu* and *insitu* conservation. Generally, species threatened with extinction in the wild should be maintained *exsitu* as an insurance until the wild population is stabilised.

13. *Species used by local residents (number or per cent):* The dependence of local residents on bio-diversity is the most important factor that influences the prospects for its conservation. People use a variety of species for food, fuelwood, building materials and medicines and for economic benefits by selling minor forest produces in the market. Therefore, there is always a necessity to ensure that the benefits derived from the forests on account of diversity remains sustainable without impairing the resource base. Such an indicator can be used to project the dependence of local communities on bio-diversity.

All these thirteen indicators help in the accomplishment of objectives of bio-diversity management. They provide the minimum framework required for monitoring the effectiveness of specific conservation practices. Above all, data regarding the estimate of status of species diversity helps to draw pragmatic strategy for sustainable management of bio-diversity in a forest area.

Bio-diversity Survey and Conservation Practices

Conservation and sustainable use of bio-diversity is fundamental to ecologically sustainable development. Bio-diversity is part of our lives and livelihoods and constitutes resources upon which families, communities, nations, and future generations depend. Every country has the responsibility to restore and sustainably conserve the biological diversity within its jurisdiction. Biological diversity is fundamental to human needs. An environment rich in biological diversity offers the broadest aray of options for sustainable economic activity, for sustaining human welfare and for adopting for change. Loss of bio-diversity has serious economic and social costs for any society.

Considered in this contest, biodiversity conservation assumes stupendous importance. To begin with, it is necessary to design and plan bio-diversity survey and conservation strategy. Such survey and strategy should be an integral part of the forest management planning process. The working plan officer, while doing his survey and field work, can also undertake the task of collecting all information regarding bio-diversity related practices.

Bio-diversity related practices encompass all facets and components of human life, which depends on bioresources. Important steps in bio-diversity conservation planning are enumerated below:

- Identifying bio-diversity components.
- Enumerating bio-diversity components.
- Mapping bio-diversity.
- Preparing bio-diversity database.
- Documenting the components of bio-diversity.

The aforesaid steps which encompass data collection, processing and documentation can be integrated and developed into an action strategy. This may be called "Biodiversity Action Strategy". Such biodivesity action strategy specific to a district can be developed. Normally, such strategy sonsists of thrust areas as (1) Survey of Bio-diversity and Preparation of Database, (2) *In situ* Conservation, (3) *Ex situ* Conservation, (4) Sustainable Utilization, (5) Indigeneous Knowledge System, Practices and Benefit Sharing, (6) People's Participation. Different thrust areas and action points under them, have been enumerated below.

Bio-diversity Conservation Action Strategy

Thrust Areas of Bio-diversity Conservation and Action points

Thrust Area I : Survey of Biodiversity and Preparation of Database:

Action Points

- Preparation of data-format for collection of information regarding biocultural diversity.
- Survey, Identification and inventorization of Components (By involving agencies like Universities/colleges, Concerned Departments, NGO's)
- Preparing a directory of taxonomist

- Directory of resource patches which need protection.
- Enlisting of endangered species of flora and fauna.
- Identification of keystone, umbrella, endemic species which need to be conserved and developing models for their conservation.
- Listing of sacred grooves, sacred ponds etc.
- Location of sacred groves or ponds.
- Name of god /goddess of sacred groves/ponds.
- Festival, priest and social/cultural relevance of the components of bio-diversity
- Listing of taboos and do's and don'ts
- Survey of land races, traditional cultivators of crop, wild relatives of crop plant and breed of domesticated animals.
- To prepare a Biocultural map of the division and superimposing over the forest map (Scale 1:2,50000).

Thrust Areas II : In-situ Conservation

Action Points

- Expanding PA Network to cover all bio-geographical zones of the country.
- Explore options for expanding/strengthening the network of protected areas through identification of buffer zone, corridors etc.
- Formation of management plan for all PA areas.
- Develop and implement eco-development projects adjacent to the projected areas.
- Strengthen research programme to develop baseline data on biological and managerial parameter.
- Devise effective management and conservation technique for forest preservation plots to ensure conservation of representative area of different forest types.
- Identify Hotspots of agro-diversity under different agro-ecozones and cropping systems and strengthen research programmes for their conservation.
- Provide incentive for *in situ* conservation wherever there is treat of such conservation being endangered by other economically renumerative cultivars by people.

Thrust Area III : Ex-situ Conservation

Action Points

- Strengthen the Network of *ex situ* centres of zoos, botanical gardens, herbaria, equiria, etc.
- Study feasibility of establishment of private zoological park, botanical garden and other *ex situ* facilities.
- Exhibition of bicultural diversity:
- Bio-diversity Park (Including selection of themes for the park and of media, e.g. audio-visual, painting artifacts, and virtual model)

Thrust Area IV : Sustainable Utilization

Action Points

- Initiate capacity building for participatory decision-making to ensure sustainable utilization.
- Enlarge farming of medicinal plants.
- Develop suitable mechanism for protection of knowledge, and related rights including Intellectual Property Rights (IPR). How Protection of IPR can be done?
  1. Who are the beneficiaries of such IPR.
  2. How knowledge will be fixed in different forms like textural, festival and electronical form.
  3. Endemicity of such knowledge.
  4. Other issues related to IPR.

Thrust Area V : Indigenous Knowledge Systems, Practices, Innovations and Benefit sharing

Action Points

- Document traditional knowledge on conservation and upgrade Use of Biodiversity (including knowledge of bioactive principles)
- Name of bioactive principles.
- Name of local people using bioactive principles.
- Methods and protection of documentation.
- Preparation of a directory of resource persons of the district.

- Prepare a directory of knowledge spokesman, e.g. medicine man, village Vaid etc.
- Revise sustainable traditional and other folk uses of components of biodiversity.
- Create a system for documenting benefits sharing and practice of wider use.
- Create people's awareness about the need to conserve, protect and gainfully use those knowledge systems.

Thrust Area VI : People's Participation

Action Points

- Strengthen Participatory Rural Appraisal (PRA) technique to promote biomass production.
- Strengthen JFM/Self-help Group in preservation and conservation of bio-diversity including cultural diversity.
- Recognize and integrate traditional knowledge and practices into bio-diversity conservation and management of common property resources.
- Devise ways and means to meet the subsistence needs of local people by different activities.
- Preserve and strengthen traditional religions, ethical and cultural methods of conservation e.g. Sacred grooves, worshipping plants and animals.

16
Stump and Stem Analysis

Introduction

Trees exhibits growth generally expressed in terms of volume. There are two types of growth in a tree. One is height growth and the other is radial or diametric growth. Variation occurs in growth patterns of the tree. Such variations are conspicuous in tropical trees.

The tree has two main regions of growth. They are the buds and the cambium layers of meristematic tissue extending throughout between the bark and the wood. As a result of the activity of the growth regions, a tree increases in size in two ways. The first one is known as primary or apical or height growth. This is confined to the buds. This results into elongation of stem. The second one is known as secondary or radial or cambial growth. These types of growth occur due to the physiological activity of the cambium. The cambiums are group of actively dividing cells, which cut of new cells both externally and internally during the growing seasons. The cells, which are cut off on the inner side, constitute the wood and those cut off on the outside from the bark. As a rule, relatively more cells are cut off in the inside to form the cambium cells. This activity, over the years, leads to increase in diametric size of the tree.

When a tree is cut on the stump or on any point of the log, wood shows a number of concentric rings. These rings may be visible or not. Sometime, they are very conspicuous. The interval between two such consecutive growth marks represents the wood laid down by the tree during a growing season and is called as "growth ring" or "annual ring". In temperate forests, on account of distinct growth season, growth rings provide a reliable index to the age of the tree. In tropical forests, because of variation and irregularity in growth season, rings are either absent or inconspicuous, leaving aside a few species. However, in conifers like chir (*Pinus roxburghi*), deodar (*Cedrus deodara*) and some of the broad leafed species like teak (*Tectona grandis*), khair (*Acacia catechu*) and toona (*Cedrela toona*), growth rings give correct indication about the rate of growth and the age of the tree.

The growth of trees with distinct annual rings are determined by the following methods:

1. Stump Analysis.
2. Stem Analysis.
3. Increment Boring.

Stump Analysis

Stump Analysis is an exercise, which gives ideas about diametric increment and age of the tree. In other words, it is an analysis of a stump-section by measuring annual rings in order to estimate the age of the tree and its past rate of diametric growth. There are two main objectives of the exercise. They are:

1. To determine the progress of diametric increment on stump over the past years.
2. To find out age-diameter relationship to decide the rotation age corresponding to exploitable diameter.

Field Works

The field work for carrying out this exercise involves the following activities.

1. Old stumps of average representative trees should be selected. In the case of non-availability of old stumps, trees are selected, felled and analysed.
2. The trees should belong to the rotation size or beyond that size.
3. The trees should belong to the site quality for which the analysis is being carried out.
4. Trees with abnormalities or marked malformation should be avoided.
5. In case of a standing tree, the breast-height diameter is measured and recorded.
6. After completing this selection process, the tree is felled.
7. The stumps are again cross-cut by a saw, so that the cross-section of the stump will present smooth and horizontal surface.

Measurement and Recording

1. Stump height and under-bark girth of each stump are then measured and recorded in the scheduled form. Average radius under bark is computed from the girth of each stump and recorded in the scheduled form.

2. The surface of the stump is moistened with water to make the rings conspicuous. Four radii are marked on the stump by pencil. On each radius, decades (Ten rings) are counted. Counting is to be done from the central point of the pith towards outer end. A pin is inserted in the wood to mark the completion of 10 rings. By counting the rings in this fashion, there will be some rings short of a decade on the outer periphery. Fixing of pins is done at decade point along the four radii independently. Now it is examined to see weather four pins of the same decade along the four radii lie on the same continuous ring. If necessary, correction is made.

Now the radius of each decade is measured from the pith towards outside. Such measurements are recorded in the scheduled form

1. Stump analysis Form No. 1 (Radii at successive decades).
2. Stump analysis Form No. 2 (Seedling Height data).
3. Stump analysis Form No. 3 (Taper data).

Growth Curves

Out of the information collected in the three forms indicated above, the following graphs are drawn:

1. Seedling height curve is drawn form data collected in form No 2. This curve shows the average age at various heights. From the seedling height curve the corresponding age of the stump is read.
2. A curve for conversion of stump diameter under-bark to diameter at breast height (d.b.h.) is drawn by using data from the form No. 3. This provides curves for each height classes. In the same graph, an additional curve is required to be interpolated corresponding to the average stump height. Now, the diameter at breast height (d.b.h.) over-bark corresponding to stump diameter under bark (u.b.) at different height range like 10 cm., 15 cm., and 20 cm. and 50 cm. are read from this new curve and data tabulated in a different format.

3. Based on the data read from the curves mentioned above, a curve showing diameter at breast height (over bark) against total age is drawn finally and necessary correction is done. This curve is known as age-diameter curve and it provides diameter at breast height (over bark) corresponding to the age of the tree and the age at which a tree assumes a particular diameter.

Stem Analysis

Stem analysis is defined as the analysis of a complete stem by measuring annual rings on a number of cross-sections at different heights in order to determine its past rate of growth. (Chaturvedi, A. N. *et al.*, 1994).

The principal objective of this exercise is to determine the following relationship of the tree with respect to the whole span of its life.

1. Age — diameter
2. Age — height.
3. Age — volume.

In addition, it aims to assess the average rate of diameter, height and volume increment in the said tree.

Field Work

Field work of stem analysis consists of the following steps:

1. Trees selected for the analysis should have shape and bole development typical of the representative crop. Trees should also be selected from the rotation age or exploitable diameter class.
2. Diameter at breast height is measured and recorded.
3. The tree is felled and the following measurements are recorded.

   (a) The height of the tree and its crown width.
   (b) The height of the first green branch and the lowest point of the full crown to determine the clear bole and crown length.
   (c) The height of the points in the trunk having 20 cm. diameter over-bark.

4. Now the felled tree trunk is divided into sections from the base point or ground level point. (Not from the stump level). The base section is kept at 2.74 cm. long so that the breast height shall correspond to its mid-point. The other sections, from the base towards the crown, will be 3mt long except the upper most tip section. This upper most section will be treated as a separate section if it exceeds 1.5 mt. in length but, in case of its length being less than 1.5 mt, it should be included in the previous section.
5. The mid-points of these sections will be marked on the tree at height ranging from 1.37 mt, 4.24, 7.24 mt, 10.24 mt, and so on (Fig. 16.1).
6. The diameter over bark is measured by a calliper and recorded.
7. At these mid-points, the ring of bark is removed and mid-diameter (under bark) of these sections are measured and recorded.
8. Now, after recording the diameter over bark and under bark at mid-points of different sections, the tree trunk is cut at those mid-points by a cross-cut saw.
9. Average radius is calculated with respect to each section. Two radii being angularly apart and equal to average radius are identified on each section and marked with pencil.
10. Rings are now counted on base sections. Counting is carried out from the central point of pith towards the outer side along the marked lines. At the completion of each decade, a pin is inserted. However, this process may lead to an incomplete decade at the circumference. This incomplete decade is left off.
11. Now the counting of rings is taken up for the remaining sections. Here the counting is done from outer and proceeded inward towards pith. Number of rings corresponding to the outermost incomplete decade on the 1.37 mt. base section will be marked off first. The remaining rings are marked of in decade and the total number recorded. However, the radii in each section are measured from the pith towards the circumference. Measurements of radii should be taken up to the nearest millimetre. First, the radial measurement for the outermost full decade along with the marked radii on the base section, i.e. 1.37mt section piece is done. Then the

successive radial measurement to other decades on the same section is done. The corresponding data regarding radial measurement are posted in the Form No. 1 of stem analysis exercise under appropriate decade headings. The measurements of incomplete decade on the circumference are not recorded. It is a healthy practice to take all the reading along the same radius.

12. Few seedlings are cut and rings there in and the height of seedlings are measured and recorded in Form 2 of stem analysis of exercise.

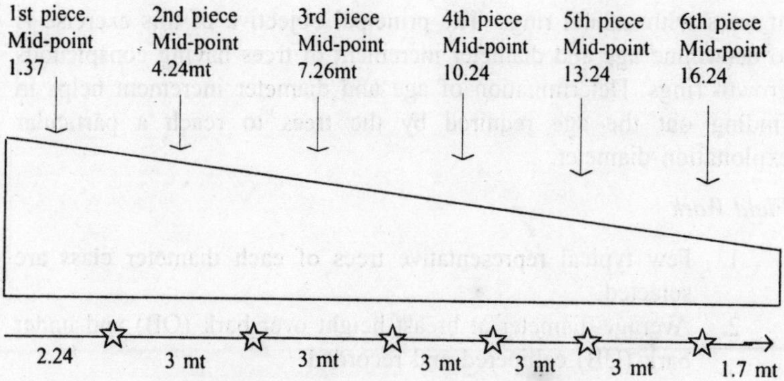

Fig. 16.1 : Division of tree trunk into sections.

The relevant data generated from this stem analysis exercise are posted in the following tables:

1. Stem Analysis Form No. 1 — Radial Measurement.
2. Stem Analysis Form No. 2 — Seedling height data.
3. Stem Analysis Form No. 3(a) — Age computation.
4. Stem Analysis Form No. 3(b) — Mean Diameter computation for decades.
5. Stem Analysis Form No. 3(c) — Mean tree volume computation for decades.
6. Stem Analysis Form No. 4 — Computation for bark thickness.

Growth Curves

The data are used to draw certain graphs, which indicate the rate of past growth of the tree. The curves are as follows:

1. Age-height curve.
2. Age-diameter curve.
3. Diameter (under Bark) / Bark thickness (Twice) curve.
4. Age-height of seedling curve.
5. Total Age-diameter at breast height (OB) curve.
6. Age-volume curve.

Increment Boring

Increment boring is defined as the boring of a tree stem with Pressler's increment borer or any other borer to determine increment of trees with annual rings. The principal objective of this exercise is to determine age and diameter increment of trees having conspicuous growth rings. Determination of age and diameter increment helps in finding out the age required by the trees to reach a particular exploitation diameter.

Field Work

1. Few typical representative trees of each diameter class are selected.
2. Average-diameter at breast height over bark (OB) and under bark (UB) callipered and recorded.
3. Now the tree is bored at two ends of the diameter at right angle to the axis of the tree trunk up to a depth of 5 cm. This facilitates to remove chunk of wood having 15 to 20 rings on each radial core.
4. The length of the outermost five rings of one end of each of the two diameters are noted and aggregated.

The data collected are tabulated in diameterwise in the scheduled form. Based on this analysis and interpretation of data is done and curves are drawn. The curves generally drawn are as follows:

1. Diameter increment curve.
2. Growth curve.
3. Age-diameter at breast height (over bark) curve.

17

Forest Inventory : Principles, Planning and Methodology

Introduction

The main objective of forest inventory is to collect qualitative and quantitative information about the forest resources within the prescribed precision limits. The data generated are used for planning forest management practices and preparing reports on potentiality of forest resources. Forestry inventory helps in describing the forest area, estimating volume of standing tress and their growth, primarily timber within a specified area. In the good old days, forest inventory was synonymous with timber estimate. Nevertheless, today connotation of forest inventory has changed. Now, timber estimate is a part of the inventory. Generally, a forest inventory describes the quantity and quality of forest trees and many of the characteristics of the land upon which the trees are growing. While assessing the forest resources over an extensive forest tract, instead of total enumeration being impractical and prohibitively expensive, invariably sampling is resorted to.

Over the years, the scope of forest inventory has widened with the increasing use of forests for multifarious purposes. When such other values of forests are increasingly important, their relationship with the forests and the land upon which it is situated is also taken up for study and analysis. Such information like forest composition, bio-diversity pattern, forest productivity, utility pattern of forest products in forest areas, etc. are some of the subjects taken for study.

Objective of the inventory or Resource Survey

The objectives of the inventory as a part of the preparation of management plan or working plan in general are enumerated below:

- To estimate the growing stock within a precision limit of 5 per cent at 95 per cent probability level.
- To estimate total growing stock for each major forest type

and management sub-type within a precision limit of 10 per cent at 95 per cent probability level.

- To estimate the species and stem distribution by different girth classes of important species occurring in the plan area.
- To estimate volume of growing stock by categorizing the forests into different types like stocked and understocked ares, young, middle aged or mature crop etc.

Purpose and size of the study determines the scale and complexity of a forest inventory. Always the inventory exercise is taken up for a specific purpose. If the purpose is to evolve a management plan for the forest, then it is necessary to gather information relating to resource status in different forest types and in different forest areas.

Table 17.1 : Relative emphasis of the elements of a forest inventory

S.No.	Type of inventory	Forest areas		Volume or other estimation	Growth estimation	Other information
		Area estimate	Topographic description			
1.	National Forest inventory	B	B	B	B	B
2.	Working plan Resource surveys	B	B	A	A	A
3.	Feasibility study for forest industries	B	B	A	A	B
4.	Land-use study	A	A	A	C	C
5.	Monitoring of forestry works	C	C	A	A	A
6.	Bio-diversity study	A	A	A	C	A

Source : FAO (1971).
A = Very Important, required in detail.
B = General estimate.
C = Little emphasis or can be eliminated.

Inventories provide detailed information about the resources of the forests. Such information are used for evolving management practices. Volume estimates for different site qualities and for different type of forests (Sal, Teak, Pine, Misc.) are prepared on the basis of information. Growth estimates for different species and topographical attributes of the areas are also prepared. Based on all such information, detailed forest maps are prepared.

Attributes of Resource Survey

The attributes of resource survey conducted for working plan exercise are as follows:

1. *Classification of forest areas*

It is breaking down of general categories of forests into a series of more detailed classifications. Forest areas are divided into (1) different site quality area, (2) different areas with varying densities, or (3) areas with different management types of forest like Sal forest, Miscellaneous forest, or Teak forests. These types of forests may further be classified into sub-categories. Apart from this, there is a need to make estimates of the areas in each class or sub-class or sub-categories.

2. *Measures of quantity of standing volume*

Quantitative expression of the forest growth or growing stock is the second attribute of the inventory. Presently, use of volume in the unit of cubic metre is in vogue. However, the use of weight as a measure of forest production is steadily increasing in the field of pulp and paper.

3. *Growth*

Determination of present or potential growth is an important prerequisite for preparing forest management plans. This is done basing on the growth data generated from "research plots" or post production records and correlating it with the present.

4. *Yield*

It is necessary to estimate the quantity of timber cut from a forest. Estimation of timber yield of a forest should be included in any inventory. The inventory provides information for sustained yield management planning. This should also include timber removal from the forests illegally by the villagers. This is undoubtedly a difficult task but now-a-days, a socio economic survey is being conducted in the course of working plan exercise. This helps to estimate the wood consumption pattern and demand and supply position of forest produces in the villages near forests.

5. Periodicity of Resource Surveys

Resource surveys should be done at specific period to assess the change taking place in forests. The period may be five years, ten years or fifteen years or twenty years. Generally, the differences between the two inventories indicate the change with respect to such attributes like forest composition, growing stock, diversity and productivity.

Factors Influencing Survey Costs

Cost involved in resource survey is an important consideration. There are many factors, which affect the cost of a resource survey. Few of important factors are enumerated below:

1. Type of information required.
2. Standard of accuracy.
3. Size of area to be surveyed.
4. Minimum size of unit area with the forests for which independent estimates of timber quality. Bamboo stock or any they characteristic are to be undertaken.

While deciding different attributes of resource survey, it is necessary to examine the aforesaid factors. This helps in making proper planning about resource survey.

Planning for Resource Survey

Management of forest resources requires, *inter alia*, knowledge of the area, estimate of quantity and quality of tree resources and their growth pattern and past yield. Gathering such information efficiently and presenting it in utilizable form requires detailed planning.

Planning a forest inventory or resource survey involves the consideration of a complex series of problems. In one hand, it is necessary to decide precisely what information is required. In the other hand, methods must be selected which will yield the desired information with optimum time, money and effort.

Check-list of Items Considered for Planning

In such survey exercise, there are a number of items, which must always be considered. These items are shown below in the form of checklist:

1. Decisions on type of information required.
2. Time and funds available.
3. Forest classification.
4. Inventory sampling design.
5. Maps.
6. Quantity relationship.
7. Personnel and training.
8. Logistical support.
9. Field measurement procedures.
10. Calculation and compilation.

The planning involves an array of activities. Some of such activities are given below :

I. *Compilation of preliminary information and reconnaissance*

This is the first stage in the planning. This initial stage constitutes a compilation of and study of all available information concerning the forest area. This includes area records, old-working plans, past survey reports, old maps, compartment histories. Then follows reconnaissance trips to the forests. This help in understanding the forest types and composition and geographical attributes.

2. *Decision on subjects or aspects about which information required*

It is imperative to identify the information required for revision or preparation of working plan. This is a critical part of the first phase. Unless one does not understand his requirement, it is not possible to proceed with the exercise. Apart from this, all people involved in such exercise should know before proceeding to next stage of planning all important aspects. Few of such important aspects are as follows:

1. Descriptions of stand and their parameter.
2. Unit of measurement for timber, trees and bamboo resources.
3. The desired limit of accuracy for the estimates.
4. Allowable sampling error and probability level in the planning process. This is determined by considering the objective behind the exercise and precision required.
5. Degree of precision should be sufficient for the use of the information.
6. Preparation of formats and outline of tables that will appear in the final report. All such tables should include necessary

titles, column, headings, classes, range limits, measurement units.

7. Kind and form of final data. This aspect needs to be considered in the planning stage itself. This helps in designing proper inventory procedures to gather required information.
8. Areas to be covered and forest classification.
9. Size classes and limits of their parameter.
10. Maps.
11. Growth and yield data.

3. *Inventory Design and Procedures*

After having decided the quality and quantity of information required, a procedure must be determined to conduct the resources survey. Amalgamation of knowledge and experience of specialists are necessary to evolve suitable methodology for inventory. Such methodology gives satisfactory results. In the course of inventory design there must be sufficient flexibilities to meet the variations in the matter of forests attributes, topographic characters and socio-economic conditions.

Volume is the most frequently used parameter in estimation of wood quantity in a forest. Volume estimates can be based on the volumes of individual trees or on measurement of average stands. The volume of timber in the forest can also be estimated from field measurements of few representative sites of the entire stand rather than by the summation of the volumes of individual trees. Hence, stand characteristics such as diameter or girth or basal area and average height can be measured and used to express stand volume.

In a small forest area of five hectares, all the trees can be measured and a complete inventory of the entire forest can be prepared. The accuracy of such an estimate will be maximum. However, for larger areas the census or 100 per cent survey is not possible because of the magnitude of area, small size of working force, the length of time required and the exorbitant cost involved. Therefore, inventories are prepared for such large area by using some type of statistical method, *i.e.*, sampling technique. Basically, sampling technique involves.

1. Selection of small sample portions of the forest, which are, considered to represent the forests as a whole.

2. Measurement of quantity of wood on the samples.
3. Application of this estimate to the whole forest area.

The total forest areas are described statistically, as "universe" or population and each sample portion is called as "a sampling unit." Collection of all sampling units is referred to as the "sample". Therefore, a series of plots of definite size. (For example, the plot size may be kept as 0.1 hectare or 0.16 hectare) may be laid out in the forests and trees present in the plot are measured to estimate the volume of wood. Each such sample plot is called the sampling unit, the entire series of plots is called the sample, and the entire forest is called as the population.

Sample of plots does not reflect the true picture of the forests. However, it helps in drawing logical inferences based on the characteristic of the sample plots. Besides, in order to evolve suitable management prescriptions for the forest, it is imperative to have cogent information regarding the forest resources. This necessities that the sample plots should give an unbiased estimate of the average volume of growing stock for the plot and a measure of how much variability has been found. Since, a forest is seldom uniform, individual plots on the entire series do not represent exactly the characteristic of the entire forest. Because of this, the sampling procedure always has a sampling error.

There are two main sources of error in the estimate of the population. They are:

1. Sampling error.
2. Non-sampling error.

Sampling errors are those errors, which result from the fact that only a very small area of the forest or population is enumerated and its results are applied to interpret the whole population. In a forest inventory, it is not possible to compute the true value of sampling error. It is desirable to keep sampling errors within prescribed limits. This is possible only by making proper choice of the size of sampling units, their numbers and distribution in the population.

Non-sampling errors are those errors which arise because of reasons like error in laying of sampling unit, errors in measurement and recording of trees, biases of the enumerators, instrumental errors etc. Such errors cumulatively contribute to total errors in the estimate.

The non-sampling errors are basically man-made. Therefore, such errors can be corrected by taking effective and appropriate measures.

Sampling errors obviously result from the chances of selecting wrong plots. The sampling error is the probable difference between the estimate from the sample and the true value for the population or the forest. As the size of the sample increases, the average sampling error decreases and the reliability of the estimate is increased. Size of the sample increases with the increase in number of plots or size of the plots. Therefore, it is always desirable that the type of sampling technique chosen should indicate the estimate of sampling error. Nevertheless, sampling error in no way related to accuracy with which measurements are taken in the plots. Such error results from mistakes in choice of samples, measurements, their recording and faulty procedure and all such errors are called non-sampling errors. Steps should be taken to remove such errors; otherwise, errors may have an influence on the final estimate. If the non-sampling errors are conspicuous, then the effort of keeping the sampling error at a low level becomes meaningless. Therefore, to find an accurate estimate of growing stock in a forest, it is imperative to keep the sampling error at pre-decided minimum level and to eliminate the non-sampling errors.

The sampling error is estimated on the basis of standard error of the mean. Sometime, percentage of the mean is also taken as the basis for assessing sampling error. The mean value plus or minus the standard error of the mean determines a range. This range is called the confidence interval. Its boundary values are called as confidence limits. To complete the expression of the range, in which the true population value will fall, it is necessary to express the probability of this occurrence. Thus, we may have a sampling error with a probability 0.95, which means that the true population mean will fall within the confidence limits 95 times out of 100.

Precision and Accuracy

These terms are frequently used in the field of sampling. The term 'precision' is indicated by the size of the sampling error and excludes the effects of biases. Accuracy refers to the size of the total error including the effect of biases. In a forest inventory, much emphasis is laid on accuracy of an estimate. We always aim at designing an inventory for the maximum precision that is achieved by reducing bias to a minimum. A well-designed sample yields best results.

Sampling Methodology

Sampling methodology encompasses the following aspects and considerations:

1. The stratification system.
2. Decision on the use of a systematic or random distribution of sampling units.
3. Distribution of sampling units (on maps and on fields) in different strata.
4. Estimation of the required number of sampling units to yield the desired estimate within the accepted limits of accuracy.

Stratification

In order to arrive at an estimate having a predetermined and required precision, we need to take a number of sampling units in a given population. The number of sampling units depends upon the quantum of variability or variance of the characteristics to be measured within that population. If the variance is large, the number of sampling units (or intensity of sampling) required will be relatively high. It the variance is small, the number of units can be greatly reduced. Therefore, it is always prudent to divide the population into strata because within each stratum the variance is relatively small. This leads to satisfactory estimate of conditions within each stratum from a correspondingly smaller number of sample units. Different forest types based on a simple criteria like density, composition, types of forests and management status can be considered as the basis for constituting separate strata for sampling purposes. In such methodology of stratification, the variance with in a given stratum will be small in comparison with the variance in the forest as a whole. Therefore, there is a need of stratification before sampling is commenced. This procedure makes it possible to vary the intensity of sample between different strata in accordance with their relative value as well as in accordance with the quantum of variance within each class. Now-a-days, stratification of forest is possible before sampling, because detailed information pertaining to management status, forest composition, densities and forest distribution are available in old or existing working plan.

A look at the existing working plan would reveal that the following main information/data is generally available for the plan area:

- Total plan area, its break-up with respect to Reserved and Protected Forests and their distribution pattern.
- Total number of compartments.
- Forest types, its distribution pattern, quality and stand structure.
- Break-up of forest area into various site qualities and density classes.
- Management sub-types existing within the plan area and the break-up of plan area into various working circles.
- Distribution pattern of various important species within the plan area in qualitative terms.
- Predominant age class distribution pattern.

Stratification Possibilities

Based on the existing information, the stratification possibilities are available on the following lines:

(a) Based on functional classification

The information regarding the compartment numbers, which fall in different category of forest, such as protection forest, tree forest or production forest, minor forest and others is generally available for Reserved Forests as well as Protected. Such types of functional status of forests can constitute the basis of stratification.

(b) Based on working circles

Working circle symbolizes intensity of management in the forests. Its existance existing within the plan area indicates different intensity of management. Therefore, all such working circles may be grouped into the following three strata (as an example):

Strata	Working Circle
(I)	Conversion working circle
(II)	Coppice With Research (CWR)
(III)	Improvement working circle / Selection-cum-Improvement.

Area and volume contribution by each stratum can be analysed and distribution of sample plot in each stratum could be done on proportional basis depending upon the volume contribution to the total growing stock.

(c) Based on Forest Types

Stratification could also be done according to predominant forest types occurring within the plan area. But if the plan are consists of one forest type spreading over most of the total area, the stratification of the forest types may not serve the purpose. Therefore, in such case, other criteria should be opted for stratification.

Sampling Units

The sampling units of an inventory are usually distributed in one of the two ways:

1. Systematic or non-random sampling.
2. Random sampling.

1. *Systematic Sampling or Non-Random Sampling*

Systematic sampling is that method of sampling in which samples are selected according to the predetermined pattern. In other words, it is a way of measuring forest attributes in sampling units distributed according to a fixed pattern. Systematic sampling is an old method. It was devised to obtain mean and total estimates without any measure of statistical accuracy. Under this method, lines are drawn perpendicular to the pattern of topographic variations. Sampling units are then located on these lines. This method is also called as linear forest survey or line plot survey (Fig. 17.1).

Estimation and measurement of tree attributes can be taken along a strip of predetermined width lying on either side of the survey line. This is called as strip survey. Measurement may be taken in a small sample unit distributed at predetermined places alteratively along the survey line.

Another method can be used in which a rectangular grid is imposed on a map. The intersection of grid lines, i.e. at the corners of the squares determines the location of sampling unit. The field party reaches the plot by reference to the nearest easily recognizable point. The sampling points may be distributed systematically and located at fixed intervals along the line which traverses the forest area (Fig. 17.2).

Systematic surveys designed in the previously mentioned fashion may be on an extensive or intensive scale. The purpose of the study determines the scale. The objective of the study here is to estimate

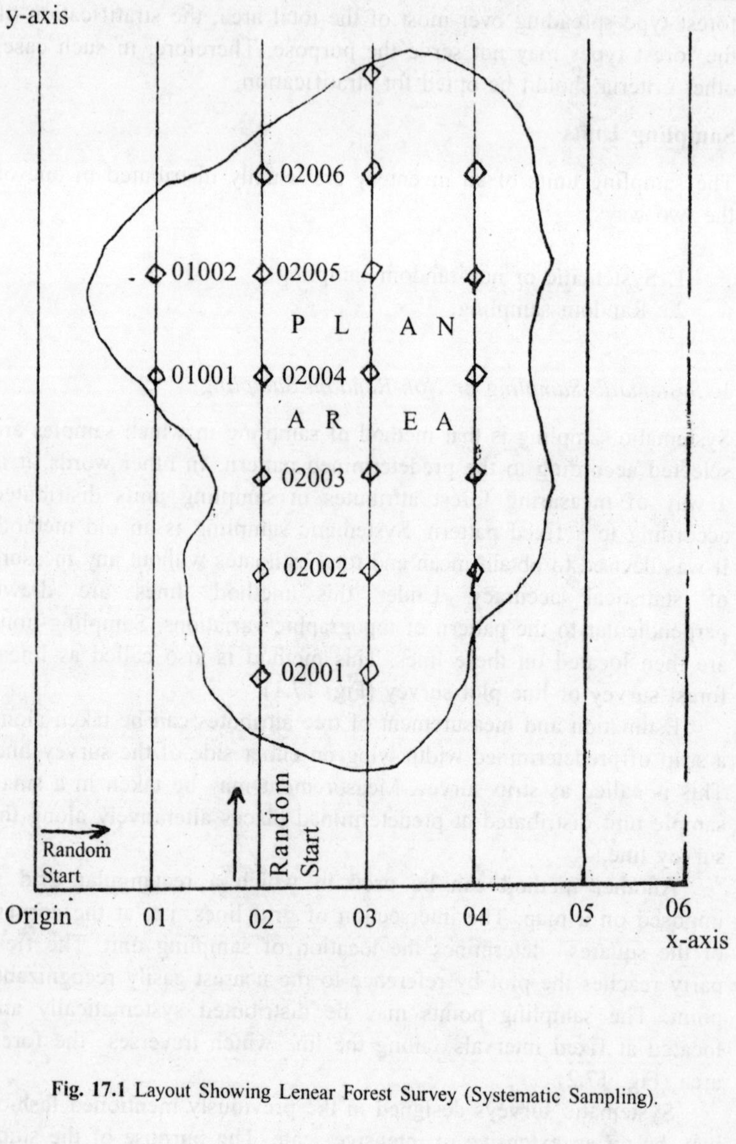

Fig. 17.1 Layout Showing Lenear Forest Survey (Systematic Sampling).

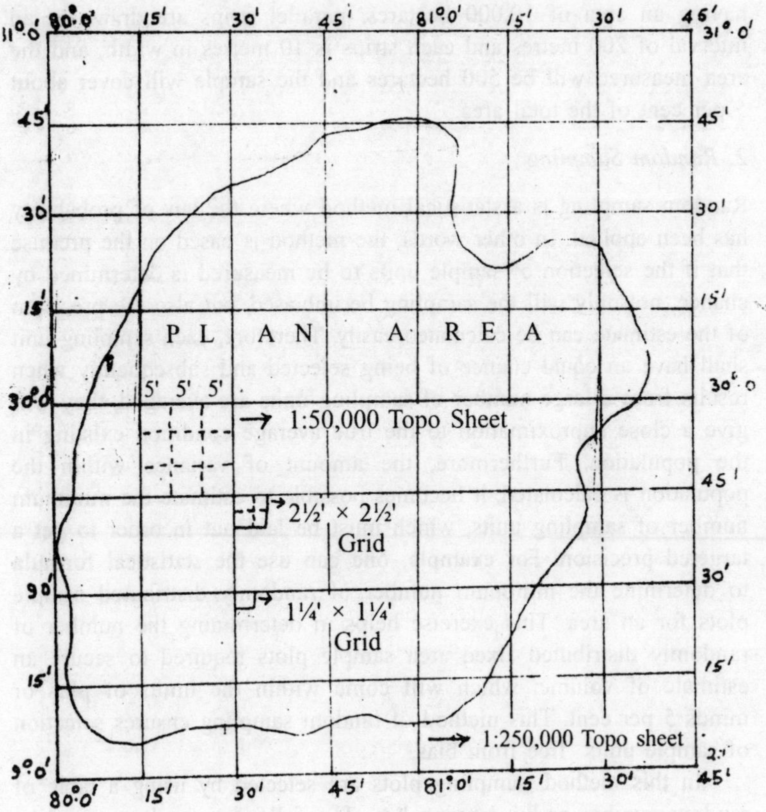

Fig. 17.2 Layout Showing Grid Survey Method (Systematic Sampling).

forest resources and conditions on forest areas. Here, lines are usually spaced at wide interval ranging from 1 to 10 km. When the purpose of the study is to estimate the forest resources and conditions on an intensive basis like inventory of a forest division or district, lines are closely spaced, keeping an interval of 200 to 400 metres. In a forest having an area of 10,000 hectares, parallel strips are drawn at an interval of 200 metres and each strips is 10 metres in width, and the area measured will be 500 hectares and the sample will cover about 5 per cent of the total area.

2. *Random Sampling*

Random sampling is a statistical method where the law of probability has been applied. In other words, the method is based on the premise that if the selection of sample units to be measured is determined by chance, not only will the sampling be unbiased, but also the precision of the estimate can be calculated easily. Therefore, each sampling unit shall have an equal chance of being selected and subsequently when results from a large number of sampling units are averaged, they will give a close approximation to the true average condition existing in the population. Furthermore, the amount of variance within the population is calculated, it becomes possible to estimate the minimum number of sampling units, which must be laid out in order to get a targeted precision. For example, one can use the statistical formula to determine the minimum number of randomly distributed sample plots for an area. This exercise helps in determining the number of randomly distributed fixed area sample plots required to secure an estimate of volume, which will come within the limits of plus or minus 5 per cent. This method of random sampling ensures selection of sample units free from bias.

In this method, sampling plots are selected by using a table of random number or by drawing lots. The following random sampling methods are used in forest inventories with little variations.

1. *Simple Random Sampling :* In this method, sampling units are selected from the whole population by random process. If the size of the sampling unit is 1 hectare and the area of the population is 10,000 hectares, then the whole area is divided into 10,000 sampling units and each of them is given a number 0000 to 9999. Now for a 5% simple random sampling, 500 different numbers are selected by using a table of random numbers. Each sampling unit so determined shall have equal chance of selection. This method is commonly

applied when the population is small and uniform.

2. *Stratified Random Sampling*: In this method, the population is divided into different strata and sampling units are selected from each of them in proportion to their size. Strata are homogenous in nature. Forest attributes like density, site quality, type of forests and management pattern can constitute the basis for dividing the whole forests into different strata. This method of dividing the heterogeneous population into different homogeneous groups or strata is called stratification. This method is widely adopted in forest departments for undertaking an inventory.

3. *Multistage sampling*: In this method, sampling units are drawn from a population in two or three stages. The sampling units become smaller at each successive stage. A forest area of 100 hectares may be divided into 200 sampling units having a size of 0.5 hectare each. If the intensity of sampling is decided as 5 per cent, then 10 sampling units will be selected in the primary or the first stage. For the second stage sampling, previously selected sampling units are again divided into 5 sampling units having a size of 0.1 hectare each. One or more sampling units from the second stage sampling are again selected at random for the third stage. This pattern of dividing the sampling units further can be continued as per the objective of the exercise. Different stages of sampling are used to estimate different attributes of forests. The first stage may be used to determine the crop composition or tree density and the second stage may be used to estimate mean volume of forest growth or growing stock per hectare in different forests (Fig. 17.3).

4. *Multiphase sampling* : In this method, same sampling units are used in more than one phase of sampling. Different phases of sampling lead to different information. Here, there is no need of delineating sampling design repeatedly. Different information collected in the field is used in more than one phase. For example, in the first phase, the number of trees or volume per hectare can be determined by large systematic survey. Again in the second phase, the condition of the crop can be estimated from a small enumeration survey of a much smaller area.

Sampling Design: Some Considerations

The following consideration will have to be kept in view while evolving a suitable sampling design:

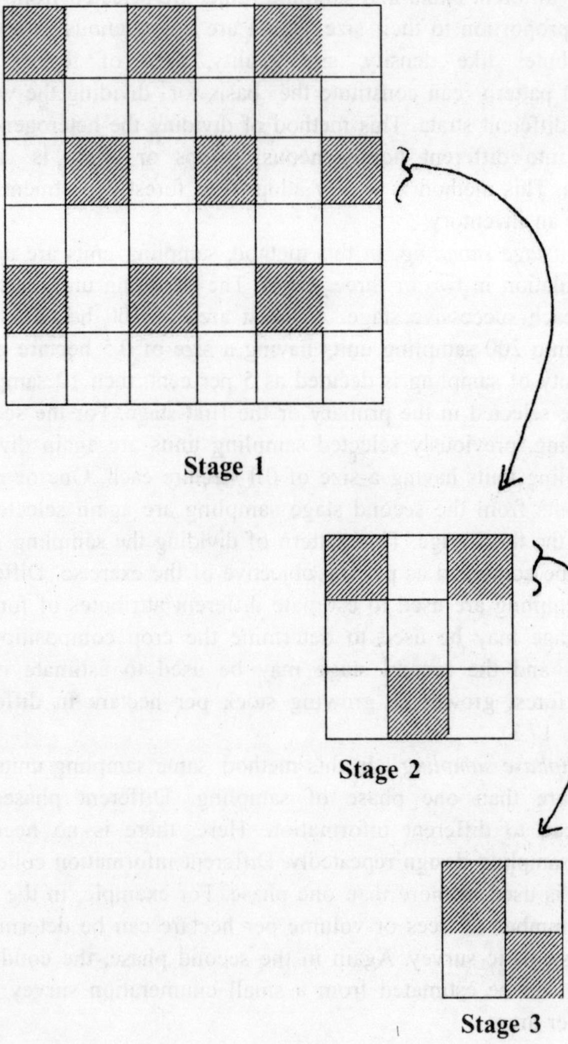

Fig. 17.3 : Multistage Sampling.

(1) The resource survey work constitutes an integral part of the working plan formulation process. The same crew may do stock-mapping work as well as inventory work.

(2) Since all the compartments are not be visited by the crew for the purpose of stock-mapping, no extra time and cost will be involved in case the sampling units are spread over the entire plan area, rather then attempting any two stage sampling design limited to some compartments only.

(3) Some localized stands, which may be of interest for the purposes of management planning, have greater probability of being included in the sample, if mono-stage sampling is adopted instead of two-stage sampling.

(4) The field work of one sample plot must be calculated in one day only.

In view of the above considerations, a mono-stage sampling may be more efficient for working plan resource survey. With mono-stage sampling, a systematic distribution of sampling units may assume more relevance. It is because, of the fact, that such type of systematic distribution facilitates implementation and control of field work. However, sizeable patches of localized species would be better represented under systematic sampling design rather than distributing the sampling randomly. Therefore, laying-out of random sampling units should be preferably done. In general, the common practice in most parts of our country, the sampling design is a systematic stratified sampling by stands of uniform characters whose area information is already available in the working plan.

Choice of Sampling Techniques

Selection of correct sampling technique necessitates proper understanding of the objective of the exercise as well as merits and demerits of different sampling methods. Sampling methods have their own merits and demerits.

Merits of Random Sampling method

1. It eliminates bias in the choice of sampling units.
2. It permits calculation of sampling error.

Merits of non-radom or Systematic Sampling Method

1. It provides estimates that are more accurate for the same number of samples because of better distribution of sample points over the population or forests.
2. Sampling bias is commonly negligible if strip or line plot survey is taken across the forest types.

However, the systematic sampling method has one major demerit. The method does not lead to correct and valid calculation of sampling error. Because of this, it is not possible to assess the reliability of the method.

Intensity of sampling

Intensity of sampling refers to the percentage of the areas of the population or universe to be included in the sample. It depends on the following factors:

1. The object of inventory.
2. The type of forest taken for making inventory.
3. Kind or method of sampling used. Using of advanced sampling technique or method which can reduce the sampling intensity.
4. Degree of accuracy and precision aimed at.
5. Availability of time and money.
6. Adequacy and technical competency of staff.

It is generally an accepted principle that the smaller the area is, the higher is need for the sampling intensity. In other words, for larger area, the required sampling intensity is less. Thus, a small area like plantation or a block/compartment or a range at a certain intensity does not give the same accuracy as obtained when the same intensity is used for a larger forest tract like an area of a forest division. There are two methods of deciding the desired sampling intensity. The first method entails calculation of the number of field plots required for a given probability level and allowable sampling error. This method is generally adopted when an inventory of tree resources is being undertaken for a forest division for the first time. The second method is to opt a fixed sampling intensity or the number of sampling units before commencing field mark. This optimum number is generally decided on the basis of past record and experiences.

It is always desirable to adopt optimal sampling intensity for better and reliable result. Griffith recommended the following sampling intensities for various types of forests.

Table 17.2 : Optimum sampling intensity in different type of forests

S.No.	Type of forests	Sampling intensity in %
1.	Tropical moist evergreen forests	10
2.	Tropical moist deciduous forests	2.5
3.	Sub-tropical Pine forest	5

By using either of both the methods, the intensity of sampling or the number of sampling units are determined. Afterwards, sample units are established in the fields and all necessary measurement of trees is taken. Now, the data collected are processed and sampling errors computed. If the sampling error computed is greater than the acceptable limit, it indicates that the sample was too small. In a case where the intensity of sampling is pre-decided, then a systematic distribution of sampling units are done.

After deciding a broad sampling intensity for a large tract of forest area, it is necessary to allocate proportional number of sample plots to various strata into which the forest has been categorized. This kind of proportional allocation technique can be done in those areas where variability is distinct.

Before deciding the sampling intensity, it is necessary to take a decision about the acceptable size of the sampling error or the acceptable size of standard error of means. This is generally expressed in terms of percentage. Thus, an allowable sampling error of the five per cent means that the largest acceptable error of the mean shall be no more than plus or minus five per cent of the mean itself. Apart from determining the acceptable size of the sampling error, a decision is required to be made about the probability level compatible to the chosen sampling error. A probability of 95 per cent means that an estimate of the means with its standard error (the confidence limits) is desired, which will contain the true mean value of the forest population 95 times out of 100 with only 5 chances in 100 of its occurring outside these units. The commonly followed practice of determining sampling intensity in forestry is the method in which limits of precision are first chosen and then the number of plots estimated to achieve this.

Size of Sampling Plots

Size of sample plot determines relative efficiency about economy of the statistical exercise. Therefore, the size and shape of the plot should be chosen cautiously. The guiding principle for determining the size of the plot is that the area under the plot should be sufficient to include a representative number of trees. Generally, in deciduous forests a plot should be sufficiently large to encompass 20 measurable trees. Such plot can have different shapes like rectangular or square. Square plots are preferred, because it is easy to lay out such plots in a forest. In forest inventory, two categories of plot sizes are in vogue. One is 0.16 hectare and the other is 0.10 hectare. Lay out of both these sizes is shown in Fig. 17.4.

Optimal Size of Sample Plot

The precision of inventory exercise based on sample plots method largely depends on the number of units examined. Apart from this, choice of sampling units affects the sampling error. In tropical and sub-tropical forests, sampling designs based on conventional method of any fixed area units are generally in practice. When such conventional methods are used, the size and shape of the unit should be essentially specified. If the heterogeneity in the tree population is pronounced, larger samples units will be necessary because such plot tends to average out the effect of "clumps and gaps". In relatively homogeneous forest, smaller sampling units may be considered adequate (Roy, 1993).

It has been pointed out by many statisticians that intensive inventories with large sampling units or clusters of sample plots must be considered while inventorying irregular mixed tropical forests. However, within a desired sampling intensity and precision, the objective of inventory can be achieved by taking fewer but large sampling units or by taking more sampling units of small size. However, in both the cases, there are limits. Cost and time associated with laying out plots in the ground are the limiting factors. Laying out of large sized plot in the forests by taking compass bearing, traversing long distance, clearing under growth and enumerating for a longer time involves high variable cost. Similarly, in a case where number of plots are more, in spite of the size being small, location and laying of plots in more points, traversing long distance in the forests involves the variable cost.

(1) Sample Plot of Size 0.1 HA

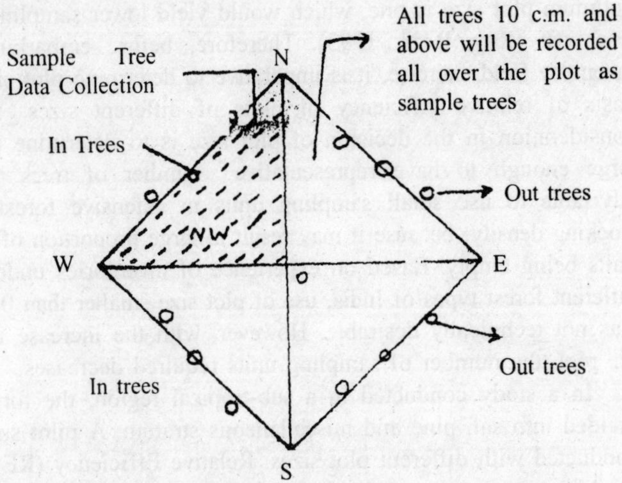

(2) Sample Plot of Size 0.16 HA

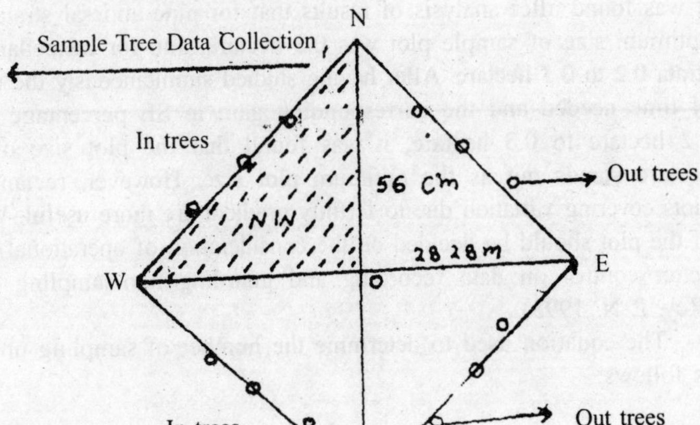

Fig. 17.4 : Lay-out of Sample Plot.

Considered in this perspective, the need for optimization of plot size with regard to desired precision assumes significance. An optimum plot size is one, which would yield lower sampling error at less cost (Roy, P. N. 1993). Therefore, before embarking on an inventory field exercise, it is imperative to determine plot size on the basis of relative efficiency of plots of different sizes. The basic consideration in the decision of plot size is to determine plot size large enough to have representative number of trees. It is not advisable to use small sampling units in extensive forests of low stocking density; because it may result in large proportion of sampling units being empty. Based on experience of inventories undertaken in different forest types of India, use of plot size smaller than 0.1 hectare was not technically desirable. However, with the increase in size of the plot, the number of sampling units required decreases.

In a study conducted in a sub-tropical region, the forests were divided into sal, pine and miscellaneous stratum. A pilot survey was conducted with different plot sizes. Relative Efficiency (RE) of plots of different sizes was calculated. Different statistical parameters of such plots in different strata are given in Tables 17.3, 17.4 and 17.5. It was found after analysis of results that for pine and sal strata, the optimum size of sample plot was 0.2 hectare, and for miscellaneous strata 0.2 to 0.3 hectare. After having studied simultaneously the effect of time needed and the corresponding gain in SE percentage from 0.2 hectare to 0.3 hectare, it was found that the plot size of 0.2 hectare stands out as the optimum plot size. However, rectangular plots covering variation due to fertility gradient are more useful. Width of the plot should be decided on the consideration of operational ease, better control on data recording and guarding non-sampling error (Roy, P. N. 1993).

The equation used to determine the number of sampling units is as follows:

$$N = t\, cv \,/\, (\text{AE } \%)$$

where : cv is coefficient of variation,
N is number of sampling units,
AE is allowable error, and
t is Students t with corresponding degree of freedom.

(*Source*: Roy, P.N., 1993)

Table 17.3 : Pine Stratum

Plot size	Mean volume (cu. m)	SE %	Co-efficient of variation (cv)	Total number of sampling units required
0.1	10.6136	18.70	74.79	224
0.2	20.6322	17.70	70.80	200
0.3	30.7519	17.37	69.47	195

Table 17.4 : Miscellaneous Stratum

Plot size	Mean volume (cu.m)	SE %	Co-efficient of variation (cv)	Total number of sampling units required
0.1	10.3648	26.34	87.35	305
0.2	21.1950	20.45	67.84	184
0.3	35.5591	17.61	58.42	136

Table 17.5 : Sal Stratum

Plot size	Mean volume (cu.m)	SE %	Co-efficient of variation (cv)	Total number of sampling units required
0.1	14568.64	10.24	56.16	126
0.2	28884.73	7.63	41.79	70
0.3	42925.58	7.40	40.55	66

Source : P. N. Roy (1993).

Pilot Survey

The working plan unit has a limited staff, time and budget. Nevertheless, the target with respect to time and cost of preparing or revising the plan is specified. Therefore, a suitable field sampling design has to be conceived to cater to the needs of the working plan at a minimum cost and time, without sacrificing the accuracy of the exercise. In order to accomplish this daunting task, especially in view of the bare minimum available infrastructure in terms of manpower, time and cost, it would be necessary to carry out a pilot survey in the plan area. This exercise shall provide the actual value of co-efficient of variation existing within the forest of the plan area. Apart from this, pilot survey helps in exploring the possibility of stratification, which consequently increase the efficiency of the sampling design.

Pilot survey aims at determining the number of sample plots required minimally for achieving the desired precision in the inventory.

On the basis of available attributes, forest will be stratified for allocation of sampling unit. This survey is a prerequisite to decide the most efficient survey design within the ambit of available time, money and manpower. Now almost all the working plans revision exercise undertakes this survey before going for inventorying.

For the propose of this study about 50 plots of 0.1 hectare each are laid out all over the forest in proportion to their representation. This is due to the existence of wide range of variation in the forest in terms of composition, density, and quality. Stratum areas are demarcated on Survey of India map sheets and are divided into 1×1 grid block. Now, such blocks are numbered stratum-wise. With the help of Fisher and Yate's random number tables, required number of plots are selected at random and allocated to different strata. The point of intersection of diagonals of grid block is the first corner point of the plot. The plot is laid along the fertility gradient. The plot is 10 metre wide and 10 metre long. Total enumeration of trees down to 10 cm. diameter at breast height over bark are carried out for all plots and data are recorded plotwise. In M.P., there is practice of enumerating trees down to 20 cm girth over bark.

On the basis of data collected, the volume per hectare and number of stems per hectare may be worked out and various statistical parameters such as mean, standard deviation and co-efficient of variation will be calculated for each stratum as well as for the whole area. Volume of growing stock for all the plots in different strata is estimated separately by using suitable local volume equation. Apart from this, standard error in percent, the Coefficient of Variation (cv) of volume or basal area and the total number of plots required are calculated. Analysis of variance for each stratum is carried out to ascertain degrees of variation in plot volume/basal area estimate in different plot sizes.

The value of these parameters would enable the working plan officer to know the number of sample plots needed to be laid-out in the field to achieve the objectives of the inventory.

Number of sample plots required = (Co-efficient of variation)^2/(Standard error)^2

Depending upon the number of sample plots required to be laid-out in the field for various strata, the pattern of laying out of sample plots will be decided. As mentioned above, the number of sampling units would depend upon the value of various statistical parameters as found out from the pilot survey carried out in the plan area.

Having decided upon the number of sampling units to be laid-out, the next part of the exercise in pilot survey will be to decide upon the pattern of laying-out of sample units.

Forests in our country have large variation in type and topography varying from flat terrain to hilly and very hilly areas. It may, therefore, be not possible to prescribe one uniform pattern of laying out of sample units for the entire country. However, any of the following method can be adopted.

1. *Adoption of list sampling method*

Having known the number of sampling units for each stratum, it may be possible to select the same number of compartments by adopting the method of list sampling which would give due weightage to the area of each compartment. Having selected the required number of compartments, the sample units may be laid out at the centre of each compartment which may be determined by laying out a cross (×) in each compartment.

2. *By adopting Grid Survey method*

Another possibility could be to layout the sampling unit at the centre of each grid spread over the entire plan area. The size of the grid would be determined according to the number of sampling units required to be laid-out and the total area of each stratum. It is advisable to divide the plan area into grids which may either coincide with the grids already laid-out on Survey of India topo-sheets or constitute part thereof, *e.g.* 5′ × 5′ grid having an area of 80 square km or say 2-1/2′ × 2-1/2′ grid having an area of 20 square km or say 1-1/4′ × 1-1/4′ grid having an area of 5 square km.

3. *Line Plot Survey*

In line plot sampling method, the standard practice is to space cruise line wider apart. The sample plots are laid-out at a close interval on the lines. Cruise lines are laid-out in such a manner that they run at right angle to the topography or river system so that it falls along the fertility gradient. Knowing the number of sampling units to be laid-out and the total area figures, it would be possible to determine the number of cruise lines and the number of sample units within each cruise line and the spacing of sampling. This would determine the distance among units within the cruise lines. The first cruise line and the location of first sample plot on the first cruise line may be determined randomly along X-axis and parallel to Y-axis taking south-west corner of the plan area enclosed within X and Y axis running in East-west and North-South direction as the origin. Subsequent lines may be spaced parallel to the first line and the location of first plot in the first line will be given a random start and the same pattern would be followed in the subsequent lines.

The working plan officer may select any of the above alternatives for his plan area, which he finds suitable looking into the prevailing condition in the area and objective of the exercise.

Shape and Size of Sampling Units for Pilot Survey

Based on the past experience of the sampling unit is always recommended to be square. This facilitates laying-out of plots in the field easily. However, if it is found convenient to lay-out rectangular plots, the working plan officer may consider for doing so. Two different sizes of the sampling units are in vogue in the inventory exercise. However, generally 0.1 ha plot is preferred. The layout of both these sizes are shown in Fig. 17.4. The choice regarding size of sample plots for the pilot survey is generally made by the working plan officer primarily on the basis of the consideration that some 40-50 trees fall within the sample unit and the field work for the said plot is completed in one day.

4. *Quantity Relationship*

The primary objective of a forest inventory is to make an estimate of the quantity of timber or wood in a forest. For this, we generally classify the forests on the basis of either of few attributes like species composition, densities, site qualities, management intensity, etc. Now we shall delve into how the measurements taken in the course of an inventory are transformed into a wood or timber estimate.

A decision on the quantity relationship required to estimate is always made before the commencement of field work. However, over the year, because of repeated exercises, such decisions have been conventional. In practice, the most frequently required expression of quantity in a forest is volume. Other expressions of quantity are number of trees in the entire forests, number of trees per hectare and the basal area. However, volume is the widely used and most suitable indicator in quantity relationship. When, we want to express the volume of trees or growing stock in a forest, a procedure is required to estimate such volume of standing trees in an unit area. Since, it is not possible to measure the volume directly, there is a need to apply a relationship between tree characteristics and the volume.

The procedure followed extensively in India is to estimate the volume of trees utilizing general relationship previously derived from other similar trees or stands. In the field, tree dimension like dbh (diameter at breast height) or gbh (girth at breast height) and height are measured. The forms of trees with respect to different species in different regions of the country have already been recorded by many research institutes. Such institutes and organizations have developed volume equation for different regions. Based on volume equation, regional volume table and local volume table can be easily prepared. Such tables contain co-efficients which when multiplied by the number of trees in the specified size class will give volume of the tree. This approach is called volume table approach. One should take all precautions to use the right kind of local volume table specified for the said district or forest division.

Volume tables are categorized into local, regional, standard volume tables. A local volume table shows volume in the unit of cubic metre, taking into account one parameter of tree dimension, i.e., diameter at breast height. Regional and standard volume tables are based on two parameters of tree, that is, diameter and height. There are different volume tables for different species. It is because of the fact, that trees of different species show differences in their shape and form. Therefore, in order to estimate the volume, it is always judicious to use local volume tables prepared for different species found in a division.

Before applying the existing volume table to estimate the volume, it is necessary to examine the suitability and appealability of that volume table. It will lead to grave mistakes if one uses volume table prepared for some other locality. In other words, applying a wrong volume table shall always result into under-estimation or over-estimation of crop, as things stand.

5. *Estimation of Growing Stock*

Estimation of growing stock is an important step in the process of working plan formulation. This is necessary to assess changes that have taken place in the growing stock over the years. Apart from this, estimation of growing stock is necessary to regulate yield according to the principles of sustained yield. As we already know that growing stock is the sum total of the biomass of trees standing in a forest. In other words, this is the total volume of woody trees in forests. However, according to the prevailing practice in forestry, only those trees, which have a girth more than 20 cm. at breast height (Over Bark), are considered for calculation of growing stock. In other words, those very young pole crop having girth less than 20 cm. are not counted towards volume calculation. This is the practice in Madhya Pradesh and in many other states. However, in some states, trees, having girth more than 30 cm. or diameter more than 10cm. are considered for volume calculation.

I. Tree Forests

The process of estimation of growing stock begins after the compilation of inventory data. Generally, the inventory data are organized according to different strata. Strata are further classified into sub-strata: Then sub-strata can be formed on the basis of any of the following criteria.

- Density -
  - I. Dense forest.
  - II. Medium forests.
  - III. Open forest/blanks/encroachment/plantation/others.
- Site Quality, like I, II, III, IVa, IVb, Va, Vb
- Age of the crop— like young, middle aged, matured age.
- Type of forests, like (I) Teak, (II) Sal, (III) Misc., (IV) Other.

Mean volume per hectare with respect to different sub-categories of forest is computed from findings of resource survey or inventory results. The area concerning to each strata or sub-strata or category are determined after having plannimetered different forest area for which stock mapping has been completed. Area of each category under different types of forest will be multiplied with mean volume per hectare. The three density strata can also be further categorized into different quality classes, if necessary. This procedure is adopted for calculating growing stock of large tract of forests having enormous variation in its attributes. Total volume in different density classes under a single forest type will indicate the growing stock in that type of forests. Likewise, the growing stock for other type of forest will be calculated. In the end, growing stock of different types of forests like the teak forests, miscellaneous forests, etc. will be added to find out total growing stock of the forest division. A format for calculation of growing stock of a forest is given in Table 17.6.

Table 17.6 : Format for calculation of growing stock: Tree forests.

S.No.	Density/ Special Categories of tree forests	Area in hectare	Mean volume per hectare in cu. mt.	Growing stock (Product of columns 2 & 3) in cu. mt.
	(1)	(2)	(3)	(4)
1.	Dense or Well-stocked			
2.	Medium			
3.	Understocked			
4.	Blank			
5.	Encroachment			
6.	Plantation			
7.	Others			
	Total			

II. Bamboo Forest

Estimation of growing stock in bamboo forests is relatively simple. First, densitywise bamboo forests areas are arranged. Under each density class, the area is further divided into different three quality classes. Now, the number of bamboo culms per hectare found in the resource survey is multiplied with their respective area. The product will indicate the number of culms standing in that particular sub-stratum. Now all the bamboo culms are added to find out the growing stock of bamboo resources. (Table 17.7)

Table 17.7 : Format for calculation of growing stock of bamboo

Density	Quality classes	Total area in hectare	Number of culms per hectare	Total culms in numbers (Product of column Nos. 3 & 4)
(1)	(2)	(3)	(4)	(5)
Dense	1st Quality			
	2nd Quality			
	3rd Quality			
Medium	1st Quality			
	2nd Quality			
	3rd Quality			
	Total culms			

Execution of Field Work

The pilot survey determines the number of plots to be laid in different strata. Then, the plots are marked in the toposheet or old stock map. Now, the process of field working begins. The crew leaders is provided with a list of plots to be tackled along with a set of 1:50,000 scale topographical sheet of Survey of India or old stock map on which the plot centre has been duly marked. After deciding the plot number, which is to be surveyed, the crew leader should find a nearest convenient route to reach the plot with minimum traverse by motorcycle or foot. After reaching in a nearby location of the plot, the next job is to search a reference point, which can be read on the map as well as located on the ground.

The reference point selected on a map should be such that it is not a temporary structure, which may disappear within a year or two. Normally, the following features are taken as reference points:

1. Bench mark.
2. Triangulation points.
3. Village tri-junction points.
4. Old bridges and culverts.
5. Crossing of rail track with roads, rivers or streams.
6. Old temple, mosque and church.
7. Junction of rivers or streams and roads.
8. Permanent bends in roads, rivers, streams.
9. Old ponds and wells.
10. Springs.
11. Prominent topographical features in hilly areas such as spurs, knots etc.
12. Mile stone or kilometre stones.
13. Boundary pillars (state, district and forests).

The crew leader may select any of the above features, which is most prominent on the map. The location of reference point and its correct description in the form is very important, so that the very same point can be reached by working plan officer or any other superior officer to verify the plot in future.

Marking of Reference Point

Having reached to the field reference point, the following details will be recorded in red paint or red jet pen depending upon the availability of a prominent tree or a structure facing the reference point by making a blaze of 15cm × 15cm at breast height.

1. Grid number.
2. Plot number.
3. Bearing from reference point to the plots centre.
4. Distance of plot centre from reference point in kilometre and metre.
5. Initials of crew leader.
6. Date of survey.
7. Distance and bearing from two nearby permanent trees or structures to the reference point.

Apart from the aforesaid details required to be recorded, the following noting will be made on the back side of the plot approach form.

— A free hand diagram of nearly 10cm × 10cm size showing the reference point and its surrounding prominent features. This is useful for locating the junctions of small nallahs, roads etc., which are used as reference points by crew leaders.

— A rough diagram of nearly 10cm × 10cm size showing distance and bearing from two nearby prominent trees or structures to the reference points. The names of trees should be mentioned in the diagram.

Calculation of Bearing

The bearing is an angle made by any direction/line with the north direction at a particular place. The bearing of the line joining any two points or as in the case reference points to plot center will be calculated as follows:

1. Unfold the topographical sheet or stock map and spread it on a levelled ground surface.
2. Put the silva compass on it.
3. Adjust the north-south direction of the toposheet or any printed longitude line exactly with the north-south of the needle of the compass. This process after having completed will indicate that the map is correctly oriented.

4. Ensure the reference point and the plot centre correctly on the map.
5. Without disturbing the map, place the silva compass in such a manner that its larger axis or any printed longitudinal line on it connects both the points, i.e. the reference point and the plot centre. The mirror of the compass should be adjusted towards the plot centre.
6. With a steady hand, rotate the dial of the compass in such manner that the north mark on the rim of the compass and north of the needle coincide exactly.
7. Now take out a compass and read the bearing against the index pointer of the compass.
8. All iron and magnetic articles should be kept sufficiently away from the compass in order to avoid effect of such articles on the magnetic needle and ultimately on the bearing of the plot centre.
9. A team of three persons should carry out the task of advancing towards the plot centre. One holds the silva compass and the other two carry ranging rods. The person holding a silva compass adjusts the exact bearing on the rim of the compass. Then, by holding compass in a levelled manner in one palm and stretching the hand straight in front his eyes, one settles the needle steady north-south and sights the tree or objects which exactly coincide with the threadline of the viewing slip of the compass, centre of the needle and notch at the top of the mirror. The other two persons shall keep on ranging the roads and proceeding ahead along bearing line with ranging roads and stand with vertical ranging rods in their hands at the places directed by the persons holding silva compass. Normally, small distances are such as 50 to 100 mt traversed at a time, so that the possibility of error is minimized. The persons holding silva compass frequently directs other persons holding ranging rods to stand at a sufficient distance along the bearing line. After fixing the persons holding ranging rods on two points on a bearing line, the person holding silva compass proceeds to the spots of the first person holding ranging rod and directing the person to shift from the first spot to take further position on the bearing line as viewed from silva compass and also decided by him. Likewise, this process of traversing goes on until the arrival of the plot centre.

All the distances on the map are horizontal distances. So, in the field also, there is a need to measure horizontal distance. However, in the field, the ground surface is always sloppy. Therefore, it is necessary to measure the slope and then, based on degree of slope, the horizontal distance is computed.

An instrument called hypsometer is used to determine the degree of slope between two points. A person at the first point on line views the person at the same height at the other end of the line through hypsometer and reads the angle of elevation or depression. A ready reckoner for reading horizontal distances of certain common slope distances against specific degrees of slope is also available. A corresponding horizontal distance against a definite slope distance and slope degrees may be read from the table so that a desired horizontal distance can be reached although the coverage of slope distance will be more. The difference between slope distance and horizontal distance is more in hilly areas than that in plain country. It should be borne in mind that slope correction is required to be made after every chain or rope.

After setting the compass on desired bearing and ranging and measuring desired horizontal distance, the team shall reach the plot centre. The trees along the bearing line should be blazed at breast height to facilitate supervision by senior officer on a latter day.

Lying out of Plot in the Field

The plot centre reached after covering desired distance and taking bearing from the reference point represents the centre of the plot. After reaching the plot centre, fix a pole having size of 10cm in diameter and 1.5 mt in height and write the following details on a blaze:

1. Grid number.
2. Plot number.
3. Name of the crew leader.
4. Distance and bearing from two nearby prominent trees or structure to the plot centre.

After fixing the plot centre, now the four corners of the plot are marked by measuring their distances from the centre by steel measuring tape. These four corners are then marked by poles or bamboos.

Data Collection

After demarcating the plot and getting satisfied with the precision, the crew leader shall start collecting the data. It is the personal responsibility of the crew leader to collect data properly. The crew leader shall assign the different duties relating to data collection to the subordinate staff. However, the crew leader should personally supervise, collect and record the data in the specified field formats.

The data are collected and recorded in the following field forms:

- Plot approach form.
- Plot description form.
- Plot enumeration form.
- Bamboo enumeration form.
- Sample tree form.
- Regeneration enumeration form.
- Any other form for special study.

Plot Approach Form

This form gives details such as mode of travel upto the reference point and conspiouous features observed during the journey by vehicle as well as on foot. The bearing from the reference point and the distance from the reference point to the nearest plot centre will be recorded in degrees and metres respectively. This form also indicates the time of starting from the camp and arrival at the reference point, time of arrival at the plot, time of leaving the plot and time of returning to camp. While filling this form, the crew leader should bear in mind that all information in this form is recorded in such a manner that it will help in relocating the plot during checking.

Plot Description Form

This form is filled in for every plot, which is laid on ground. An area of 2 hectares around the centre of the plot should be examined for filling up data in the form. However, there is no need to demarcate an area of 2 hectares on the ground.

Plot Enumeration Form

In this form data of trees and bamboo clumps are entered. Trees below 10cm diameter at breast height over bark and dead trees of having utility less than 70 per cent are not enumerated. This form will be filled up for every individual plot. Trees which touch the north-west and south west border line of the plot will be enumerated. However, trees that touch the borderline of the plot will be treated as "out trees" and will not be enumerated. On the same criteria as explained, counting of bamboo clumps shall commence from the western quadrant of the plot and will proceed in the clockwise direction. All bamboo clumps occurring in the plot will be serially numbered and similarly, trees will also be numbered separately and simultaneously.

Sample Tree Form

The sample tree form is filled after having completed the plot enumeration form. Here also, all trees less than a predecided (say 10 cm) diameter at breast height over bark and the dead trees are not enumerated. The data is collected from north-west quadrant of the plot. In other words, data is collected from 1/4th of the area of the plot. The data in this exercise pertains to species name, status of dominance, bark thickness, tree height, pole condition, defects and longitudinal and sectional form of bole. The quadrant can also be used for collecting information regarding plant species diversity. In that case shrubs and trees species have to be enumerated.

Bamboo Enumeration and Clump Analysis Form

In this form, data of each individual culms occurring in certain selected clumps in the plot is to be recorded. The clumps to be selected should be those which bear serial numbers 1, 9, 17, 25, 33 etc. (That is first clump and every eighth clump thereafter) of each series. Before carrying out the analysis, it is necessary to make the classification of bamboo culms as shown below:

1. Green sound culms.
2. Green damaged culms.
3. Dry sound culms.
4. Dry damaged culms.
5. Decayed culms.

Minimum Performance of Field Works

Each stock mapping group or crew should complete inventory work of one sample plot on every working day. Presuming the availability of minimum of 25 working days in a month, the monthly output should be about 25 plots per crew. Considering availability of 7 month of fair weather season in a year, the annual inventory field-work should be at least 175 plots. For an inventory work in a division, at least three crew should be engaged. This will help in archiving the targeted progress relating to resource survey in a year.

Organization of Resource Survey Report

After conducting a detailed resource survey, it is time to prepare a comprehensive report. If the resource survey is a part of the working plan exercise, then a separate chapter has to be provided in the Part I of the plan document. In addition to it, it is also required to prepare a separate detailed report about the methodology adopted and results obtained from the survey. However, sometime resource survey is conducted to assess specific forest resource with different management objective. In all such exercise, a comprehensive report is required to be prepared by the planning officer.

The resource survey report is generally organized into four parts. The first part presents background information, the second part outlines investigation and methodology and the third part covers the data processing and the fourth part consists of growing stock and yields. Besides, tables encompassing data, classified and tabulated information are annexed with the report. The chapters under different parts are as follows:

Part I :

* Need for survey.
* Name of the plan area.
* Situation and boundaries with maps.
* Climate.
* Topography.
* Geology, rock and soil.
* Land use.
* People and their socio-economic condition.
* Forests: Types, composition, classification, stratification, legal status, present management practices.

Part II :

* Investigation and methodology.
* Objective of the study.
* Reconnaissance survey.

* Inventory design.
* Pilot survey.
* Main resource survey, sampling design, stratification, sampling units, size of units, location.
* Field works.
* Format of data collection.

Part III :

* Data analysis.

Part IV :

* Growing stock and yields.
* Introduction.
* Inventory results.
* Annual yield and future management model.
* Distribution of total volume by species.
* Comparison with actual production.
* Bamboo yields.
* Cost of survey.

18

Study of Wildlife Habitat

Introduction

India ranks sixth among twelve mega-diversity countries in the world. The total number of species in the world is estimated to be around 5 to 30 million out of which, about 1.4 million species have been described. The total number of plant species in India is estimated to be about 45000. Nearly 4900 of these species are endemic to India out of which, 1500 are highly threatened (MOEF 1994). The faunal wealth of India comprises of 81000 species (5000 mollusks, 57000 arthropods, 2546 fishes, 204 amphibians, 428 reptiles, 1228 birds, and 372 mammals) out of which, 62 per cent species are endemic to India (MOEF 1994). Besides, a host of micro-organisms are yet to be described.

The Commission of National Parks and Protected Areas of IUCN classified the protected areas into the following categories. These are given below with the broad management objectives:

Category I	Strict nature Reserve/Wilderness Areas: Protected areas managed mainly for science or wilderness protection.
Category II :	National Park: Protected areas managed mainly for ecosystem conservation and recreation.
Category III :	Natural Monument: Protected areas managed mainly for conservation of specific natural features.
Category IV :	Habitat/Species Management Areas: Protected areas mainly for conservation through management intervention.
Category V:	Protected Landscape/Seascape conservation and recreation.

Category VI : Managed Resource Protected Area: Protected areas managed mainly for the sustainable use of natural ecosystem.

In India, Protected Areas (PAs) belonging to categories II, IV, V and VI are being managed. Category I areas correspond to the sanctum sanctorum of the PAs and Category III areas sometime occur incidentally in various PAs. National Parks in India fall in the category II, sanctuaries come under category IV, and Biosphere Reserve comes under category VI. The main objective in managing National Parks should be ecosystem conservation and recreation. The sanctuaries fall under habitat/species management areas, which are conserved through management intervention. Category V is areas maintained for the protection of catchment areas of rivers, sea coasts and for recreation.

Now, we shall delve into few essential concepts regarding wildlife conservation. These are wildlife habitat, succession, cover manipulation and general guidelines about wildlife conservation practices.

Successional Stages and Wildlife Habitat

The basic requirements of wild animals are food, water and shelter. Adequate availability of these factors in a habitat helps animals to attain good health, growth and population increases. These basic factors are made available to the animals by the environment.

The environment influences on the site factors such as soil, temperature, elevation, radiation, slope, aspect and rainfall. As a result these site factors influences the vegetation of the areas. The plant community is a result of evolution through a serious of complex stages that culminates into a final formation. This process of formation is called succession and various stages involved in the process are called successional stages. Different stages promote different forms of wildlife. The primary stage in the successional process provides large quantities of grass and food to herbivores, and a large number of species. Similarly, with the attainment of tree growth with varying density, a different habitat is formed to provide cover, and sheltering places to wildlife. These specific habitats are called niches.

These habitat niches are valuable to wildlife for feeding, reproduction and security. Different species need different habitat niches. Forest stands provide a habitat niche to the wild animals. More importantly, trees stands which have sufficient uniformity of composition, size, density, quality, age, pattern periodicity and spatial arrangement provide specificity to niche structure. Sometime, silviculture treatment changes the stand conditions. Any change in stand condition, composition and structure, will favour some wildlife species while adversely affecting others. Such changes can affect the number and type of wildlife species and also their use of habitats.

Edge Effect

The food, cover and other basic requirements for feeding, breeding, hiding, rearing, escaping and ambushing etc. of different species are different. Therefore, diversity in plant habitat conditions leads to promotion of varying inhabitant animal communities. The juxtaposition of plant communities, successional stages or stand conditions within communities produces edge. The area where two communities or successional stages overlaps and produces a distinct combination of plants or structure in called the 'Ecotone'. Edges and ecotones are rich habitat for wildlife because they have attributes of the edge as well as those of the adjoining communities or successional stages. The influence of this phenomenon of plant life on animal life is called edge effect. Edges contain varying composition of plant life and constitute different successional stages. Therefore, they cater to the needs of a larger number of species of animals. Increasing the amount of edge increases habitat richness, which is a measure of existence of rich wildlife within an area. Increase in interspersion enhances the degree of edge, which resultantly increases the variety of plant and animal communities. This provides an increasing number of habitats supporting more animal species.

Effects of Timber Harvesting

Timber harvesting adversely affects the natural stages of succession of plant communities. However, generally, the silvicultural systems are suitably evolved to impart minimum damage to the ecological process. By executing different silvicultural operations, with the help of manipulation of forest cover and composition, it is always considered essential to have a very good interspersion of opened up grass patches, mature tree stand cover, water bodies of small but optimal sizes placed near each other over the entire area.

In an area where an endangered species is found, while focusing on the specific management needs of that species, habitat requirements of all the species must be identified and taken into consideration. The objective of improving species richness in a forest area can be achieved by modifying silvicultural practices. This change of silvicultural practices lead to emergence of broad spectrum of habitat conditions in a forest. Nevertheless, this requires identification of special and unique habitat like riparian zones, edges, ecotone areas, cliffs, caves, talus and sites having snags, dead and down woody material. Few of such special and unique habitats are illustrated below.

Special Habitats

It is quite common to come across special habitats in the forests. They constitute potential area for promotion of wildlife conservation. Such areas do have special ecological characteristics. Few of such areas are illustrated below.

Riparian Zones

A river or stream along with the area on its banks constitutes the riparian zones. These zones are generally covered with vegetation that requires moist conditions. They are relatively more productive in terms of plant and animal biomass and constitute a critical source of diversity within the forest ecosystem. Wildlife also uses the riparian zone disproportionately more than any other type of habitat. Riparian zones frequently have a high number of edges and strata in comparatively small areas. These produce habitat for a greater number of species reflecting the diversity of plant species and complex community structure. Some of the ecological attributes of riparian zones are given below:

1. Riparian zones offer critical habitat components like food, water and cover.
2. Greater availability of moisture along with deep soil leads to increased diversity of plant species and structural diversity in the community.
3. It produces contrasts of the general forest vegetation in most of the tropical areas. It is basically attributed to the late shedding of and early sprouting of leaves in dry deciduous forest areas.
4. The micro climate of riparian zone is different from that of surrounding forest because of increased humidity, higher rate

of transpiration increased availability of shade and enhanced wind movement. Such microclimate is compatible to promotion of wildlife.

5. It provides diverse nesting and feeding opportunities, especially to bird population.
6. Riparian zones along with intermittent and permanent stream or rivers provide travel lanes to wildlife.

Riparian zones are very sensitive and vulnerable to changes. They are generally considered fragile and such belts are kept free from any harvesting activities. Looking into their significance within the broader perspectives of wildlife and ecosystem management, the following practices are found salutary.

- Avoidance of felling in riparian zones.
- Avoidance of road construction in this zone.
- Establishment of camping grounds should be discouraged.
- Minimization of recreational use of such zones.

Swamps and Wetlands

These are special macro habitats rarely found in the forests. However, whenever, such sites exist, they play an important role in the conversation of wildlife. Such sites are generally found near springs and ponds. Such sites have enormous potentiality for supporting insects, invertebrates and birds and other animals. Such sites need utmost protection.

Magic Sites

Such micro sites are found in riparian zones. They appear as an independent unit within macro habitat of riparian zone. The characteristics of such sites are adequate moistures and deep soil. Vegetation in such magic sites is completely different from the general composition and structure of the surrounding forests. Such sites are extremely significant from the point of varied ecological activity and resultant rich bio-diversity. However, over the year, such sites have been damaged with the increase in anthropogenic activities around such areas. Therefore, it is now imperative to identify and to protect such areas by prescribing suitable management prescriptions.

Unique Habitats

Such habitats are rare and uncommon. Geographic features like cliffs, talus and caves create conditions which makes the habitat unique. Such habitats provide security to some species out of which few cannot survive in habitat other than these. Destruction of such unique habitats eliminates the species that are dependent on them. Few of such unique habitats are as follows.

Cliffs

Cliffs of igneous rocks constitute one of the unique habitats. They provide physical protection to a variety of reptiles, birds and other animals. They constitute a relatively small but stable environment. Such habitat needs protection from all types of disturbances.

Talus

Talus constitutes one of the unique habitats. It is formed when broken rocks at the base of cliffs or other steep slopes are accumulated. The micro habitat in the form of talus promotes vertebrates animals like frogs, lizards, snakes, birds and small mammals. The larger the talus is, the more stable and suitable it is for becoming wildlife habitat. Talus supports an edge of herbaceous vegetation that provides the food supply for species living within it.

Caves

Caves constitute one of the unique habitats. They provide shelter from extreme weather conditions. They also provide compatible condition such as varying degree of darkness, solitude and protection from predators. Many animals like birds, bats, and insects that love darkness occupy the caves for various activities. Preservation of unique habitats plays an important role in wildlife management. Unique habitats accompanied with the continuing plant communities provide enormous potentiality for existence of a wide spectrum of wildlife. Therefore, plant communities adjacent to such unique habitats should not be disturbed.

Snags

A Snag is a standing tree from which the leaves and most of the branches have fallen off. By definition, a snag is a dead or partly dead tree above 10 cm in diameter at breast height and above 1.8 mt tall. Snags constitute an important part of the life support system for many species of plants, invertebrates, birds and mammals. Many animals depend on snags for nesting or shelter. A majority of insectivorous birds plays a pivotal role in control of forest insect pests. Birds along with insectivorous small mammals, microbes and predatory insects help control population at endemic level and exert significant control, during early stages of outbreak. Therefore, it is said that absence of suitable nest site leads to emergence of limiting factor for cavity nesting birds. When the tree dies, the snag undergoes a series of successional changes from death of the tree to final decomposition stage. Each stage in the decay process has particular value to certain group of wildlife species. Similarly, there should be prescriptions with regard to marking of tress in order to properly identify hollow trees, which harbor nests of birds, squirrels, bees or monitor lizards and to retain those trees in the forest. Therefore, potentiality of forest areas in terms of wildlife promotion should be studied comprehensively and prescriptions should be laid down with regard to retention of snags and hollow trees in the area at the time of felling, thinning and other silvicultural operations. This apparently stands contrary to the concepts of silvicultural hygiene. However, in the present context of sustainable management of bio diversity, it is indispensable to suggest prescription for promotion of wildlife since it facilitates over all improvement of the ecosystem.

Dead or Down Woody Material

Dead and down woody materials are considered by foresters as a source of serious fire hazard. Apart from this, these materials which support small vertebrates are viewed as source of pests to the forests. However, these materials in various stages of decay play an important role in wildlife habitat management. The space between the loose bark and wood of a fallen log harbors large number of invertebrates and small vertebrates. Protected micro areas beneath the fallen log provide nesting cover to small vertebrates like hares, porcupine and squirrels. It provides food and perching places for some birds. Apart from this, mineral recycling and nitrogen fixing also takes place in these micro spaces. Because of these activities, they serve as nurse logs for various herbs, shrubs and tree seedlings.

The primary function of fallen logs is to provide cover to the wildlife population and promote compatible site conditions for breeding and their multiplication. However, while suggesting prescriptions for wildlife habitat management, the following aspects should be well considered:

1. Logs should be well distributed over the ground. Too many logs or fallen trees may hinder the movement of ungulates.
2. The larger the diameter and length of the logs, the more useful it is.
3. Logs fallen along the contours of a slope are generally used by small vertebrates than logs placed across contours. Logs found along the contours, helps in checking soil erosion. The up slopeside is filled with humus which supports invertebrates living in tunnels. The down slope provides protective cover for larger vertebrates.
4. About 5 uncharred large sized logs or less important species per hectare should be retained. Logs having dimension of 90cm girth and more than 6 metres in length should be preferred due to their better habitat cover value.

Cover Manipulation

Forage, water and cover are the primary habitat factors that limit any wildlife species. For example, deers basically need relatively dense tree cover and large quantity of drinking water. Whereas, antelope prefers comparatively open areas and thrives with very little water. Therefore, the ecological requirements of such species like chetals and bark deers are more exacting and critical. So, their management strategy needs a focus on covers manipulation.

There are different types of cover present in the forests. They are:

1. Vegetative cover.
2. Thermal cover.

Vegetative cover

A cover is called vegetative when it is capable of hiding 90 per cent of a standing adult deer or animal from the view of a human at a distance equal to or less than 60 metres. Therefore, for optimum use of the cover area the radius of cover in circular patches may be kept at about 200 metres. In other words, the width of the cover strip may be kept about 400 metres to optimize the use of the cover area. Considered in this perspective, the annual coupes in different working circles may be designed to make it sufficiently effective in providing cover to the wildlife population.

Thermal cover

Covers used by animals to assist in maintaining the body temperature are called thermal covers. Animals generally take rest in such zones, which are thermally neutral. In such micro-sites the body temperature of animal and the air temperature of the site are close to each other.

It has been found that closed canopy stands are warmer at mid-night and cooler at mid-day than open canopy stands. Tree trunks and low ground vegetation reduce air movement, which protects the animal from the chilly air associated with low temperature. Similarly, at mid-day in summer months dense canopy stand creates cooler conditions. Canopy closure to the extent of 70 per cent or more provides fully effective thermal cover to the animals. Such a stretch of cover should not either have small area or too large areas. In both the situations, the effectiveness of the cover does not remain at optimum level. A patch having 12 to 15 hectare of area can be considered too small and a patch stretching around 24 to 30 hectares can be taken as a big stretch of thermal cover. However, the dimension of the stretch may vary with the change in types of forests. The size of such cover in a forest like evergreen forests and moist deciduous forests would be less than the dry deciduous forests.

Forage and Fawning Areas

Areas with dense tree cover produces very less grasses and forage. Contrary to this, man-made open areas encourages growth of more forage. Ideal grass and forage areas are basically different from optimum cover areas. Therefore, while manipulating the cover, it is important to understand the underlying concept behind maintaining forage and fawning areas.

An ideal or optimum forage area is that which produces plenty of grass and forage. Generally, an area having a radius of about 200 metres is taken as the optimum area. An area bigger than this does not seem to optimize the use of the excess central part. In other words width of forage areas should normally be kept about 400 metres. However, if antelopes use an area, then in such cases open grasslands with shady groves are considered best.

Animals, for calving and fawning generally prefer those areas, which provides plenty of succulent forage, water at close proximity and good hiding cover to protect them and the fawns from predation. Such traditional calving and fawning areas should be identified and fully protected.

Factors Inimical to Wildlife Conservation

There are many factors in the practice of forestry, which have been found inimical to wildlife conservation. Few of such factors are illustrated below.

Presence of Road

Presence of roads open to vehicular traffic prevents optimal use of that area by animals as it creates unwholesome disturbances. The type of road, its location, degree of its use markedly influences the damaging effect.

Harvesting operations

Timber harvesting operations are potent source of disturbance to animals especially to more shy species. The harvesting operations should be limited to least possible areas and also be confined to shortest possible period of time. Felling and transportation of timber and other forest produce should be completed in three and half month's period, i.e., from 1st November to 15th February as far as possible. A non-activity zone should be maintained in the forest adjoining the coupe. Besides, coupes can be laid out taking a single drainage area in a watershed at a time. This provides protection and cover to the wildlife, in spite of harvesting operation in the coupe.

Fire

Fire is a major source of threat to the wild animal. So, it is necessary to take all measures to prevent and control fire in the forests. Fire lines are essential for preventing spread of fire in a forest. Control burning in fire lines should be carried out before 15th February when the material is still moist and fire hazard is less.

Augmenting Food Availability

Taking suitable measures in the field of forest management can augment food availability to wild animals. To begin with, all such shrubs and trees should be reserved or retained which provide fruits, flower and fodder to the wild animals. Many ficus species and other fruit and flower bearing species provide fruits and flower to animals. Species like Pipal, Mahua, Kusum, Mango, Jamun, Ber, *Cordia myxa*, etc. are some of the fruit bearing trees. Trees of *Vauhinia vahli*, Dhaoda (*Annogeisus latifoelia*), Anjan, Saja (*Terminalia tomentosa*),

Cordia myxa, bamboo provide green and dry leaves which form very good food for animals like ungulates, primates etc. Minor forest produces like mahua, sal seed, honey and other wild fruits constitute the food materials for ungulates, primates, bears, birds etc. Harvesting of such minor forest produce should always be discontinues so that it promotes continuous maintenance of food, water and shelter in a forest.

Wildlife Map and Reconnaisance Report

The staff of the working plan unit traverses through all the compartments to prepare stock maps and to collect information necessary for the description of the compartments. It is at this time that the stock mapper can collect the information about wildlife and prepare wildlife habitat map and wildlife reconnaissance report of the compartment.

Apart from the silvicultural stock maps, the wildlife habitat maps should also be prepared containing all relevant information. Three habitat maps have been presented in Figs. 18.1, 18.2 and 18.3. One habitat map indicates habitat elements and physical infrastructure and the second shows the vegetation and land use pattern in the compartment and the third depicts events and management activities. These informations are generally provided in the index of the map. However, this index is only an illustration and if the Working Plan Officer feels that important information about some more plant and animal species can be conveniently indicated in the map, he may use some more signs or symbols to describe the features.

Similarly, from the point of wildlife conservation, a reconnaissance report of compartment is generally prepared. This reconnaissance report can be kept in the compartment history files. Apart from field survey, some of the information of the reconnaissance report can be collected from the offices of various departments and panchayats and by inquiry from the key local people and staff. The report should, *inter alia*, consists of a statement accounting primary burden of grazing and nistar (*bona fide* forest needs) on the forests.

Guidelines for Wildlife Management Practices

The following are few of important guidelines for evolving management strategies and practices:

1. For creation of edges and edge effects, silviculture systems aiming at even aged crops may be preferred.

WLM Compartment History

Name of the Protected Area :

Division :

Range : Scale :

Block :

Compartment No. 135 Year, last described : 1990

Area : 315 hact. Year, current description : 1998

Map A

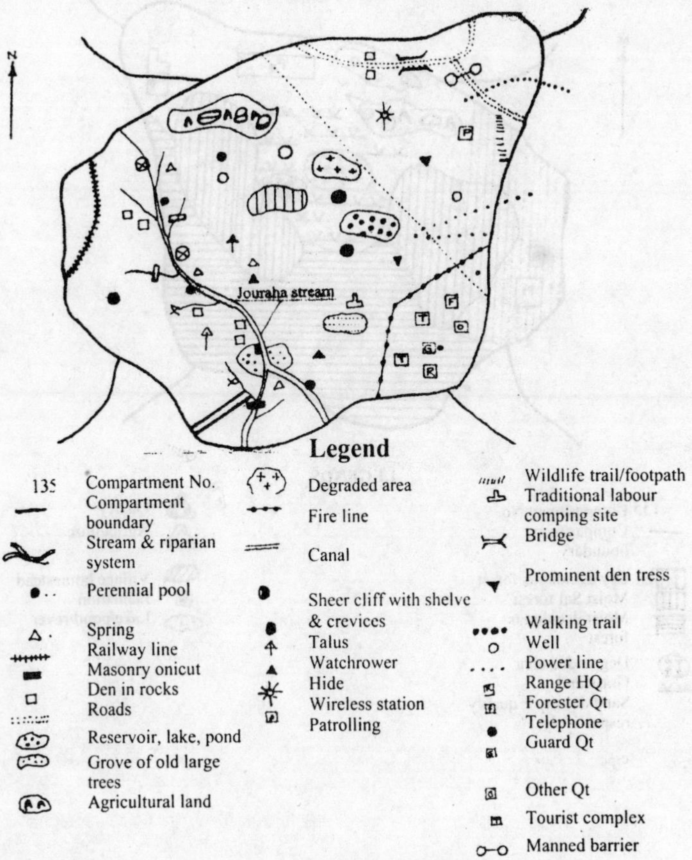

Fig. 18.1 : Map A : Habitat elements and physical infrastructure (topographic map).

WLM Compartment History

Name of the Protected Area :

Division :

Range : Scale :

Block :

Compartment No. 135 Year, last described : 1990

Area : 315 hact. Year, current description : 1998

Map B

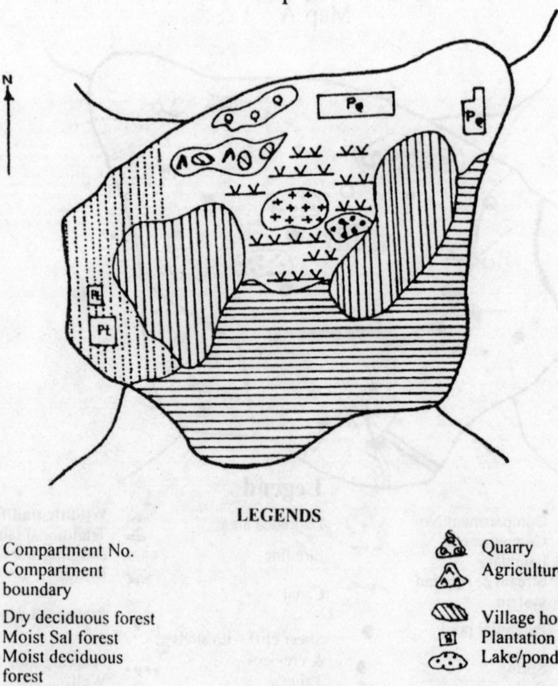

LEGENDS

- 135 Compartment No.
- Compartment boundary
- Dry deciduous forest
- Moist Sal forest
- Moist deciduous forest
- Degranded area
- Grassland
- Sand density + quality respectively
- Quarry
- Agriculture
- Village homestead
- Plantation
- Lake/pond/rever

Fig. 18.2 : Map B : Vegetation & Landuse (Topographic map).

WLM Compartment History

Name of the Protected Area :

Division :

Range :

Beat : Scale :

Block :

Compartment No. 135 Year, last described : 1990

Area : 315 hact. Year, current description : 1998

Map C

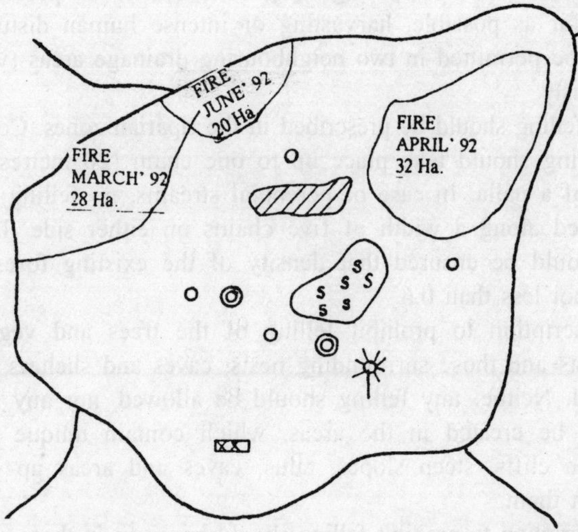

Legend

135 Compartment No.

Compartment boundary

Lantana Control

Gully plugging trenching & planting

Traditional places of tiger pugmark location

Tiger sighting

New onicut construction (partial/complete)

New wireless station/ tower

Fig. 18.3 : Map C : Events and Management activities (topographic map).

2. To optimize land use by wildlife, each annual coupe may be aimed to have a size of about 40 hectares. Coupe size may be kept larger where there is a need of minimizing the number of working areas. In the coupes, the narrower side should be kept about 400 metres long, so that at various stages of succession it may provide effective forage areas, hiding and thermal covers etc.

3. It should be ensured that in the event of felling of trees, permanent gaps in the forest are not formed. So, wherever necessary, conservative marking may be resorted to, so that after the felling operation, in any case the density does not go below 0.5. In patches with less than 0.4 density, felling of trees should not be prescribed.

4. As far as possible, harvesting or intense human disturbance should not be permitted in two neighbouring drainage areas (valleys) simultaneously.

5. No felling should be prescribed in the riparian zones. Complete ban on felling should take place up to one chain (20 metres) from the banks of a nalla. In case of perennial streams, no felling should be prescribed along a width of five chains on either side. In such strips it should be ensured that density of the existing forest strip should be not less than 0.6.

6. Prescription to prohibit felling of the trees and vegetation bearing nests and those surrounding nests, caves and shelters should be provided. Neithe. any felling should be allowed, nor any sort of disturbance be created in the areas, which contain unique habitat features like cliffs, steep slopes, talus, caves and areas up to 100 metres from them.

8. Prescription to prohibit felling should be given in those specific areas used by wildlife for mating, nesting, calving and fawning especially during the seasons of their intensive use.

9. In deciduous forests, evergreen trees, shrubs, or those, which retain green foliage during summers, should not be allowed to be felled. Similarly, in evergreen forests, deciduous trees and shrubs should be retained.

10. Trees and shrubs bearing forage for wildlife should not be allowed to be felled if they are unlikely to coppice. It should also be seen that trees of mature size are adequately left in the forests for the monkeys, which drop a lot of leaves from tall trees while feeding upon. These leaves are available to deers.

11. Another important prescription should be that trees and shrubs bearing edible fruits used by wildlife should not be allowed to be felled.

12. Bamboo is a very important fodder species and should be conserved. The areas in which the bamboo have been degraded due to maltreatment or over exploitation, prescription should be evolved to rehabilitate it.

13. In order to enhance the food availability for the wildlife population, sowing and planting of fruit and fodder species like peepal, gular, anjan, bamboo, mahua, ber, achar, acacia, cassias, dhaora, kusum etc. should be taken up wherever necessary.

14. Again if found necessary, in permanent large open areas in national parks and sanctuaries, agricultural crops can be sown over small plots not exceeding one hectare and may be left to be grazed by the wildlife.

15. Snag requirements of the area should be meticulously studied and to begin with, about ten hallow, crooked and branchy trees or snags per hectare can be retained to provide nesting places for birds.

16. About 10 thick and long logs per hectare, of economically less important species can be retained along the contours to provide cover for small mammals and reptiles etc.

17. Marking, felling and transportation of wood in a coupe should be completed within the shortest possible period. If possible, it should be completed between 1st November to 15th February.

18. To minimize human disturbance resulting on account of harvesting of all minor forest produce like sal seed, tendupatta, harra, gum etc., prescription should be given to stop such harvesting at least in game sanctuaries as their period of collection coincides with the intensive breeding activity of some of the animals.

19. Harvesting of other minor forest produces like mahua flower, bahera, ber, honey, leaves of many species must be discontinued, as they constitute food for many wildlife species.

20. There should be a total prohibition on grazing of domestic livestock in the core area. Cattle camps should not be permitted inside the Sanctuary areas. As far as possible, grazing by cattle should not be permitted in the sanctuary area. If it is not possible, a narrow strip of area on the periphery and out of the game sanctuary can be earmarked on rotational basis for free grass cutting by the neighbouring villagers.

Cattle of villages situated inside the sanctuary graze the forest. So, in order to minimize the deleterious effects of uncontrolled grazing it is necessary to draw up a grazing plan and permit those cattle on a regulated and controlled basis. Likewise, cattle in immediate neighbourhood of the sanctuary requires equal attention, if they have no other grazing area available to them. While determining the grazing potentiality of the forests, the carrying capacity may be taken as 1.5 hectare per cattle unit whereas buffalo is taken as two cattle units.

21. No labour camps should be permitted near the water holes. If unavoidable, they may be permitted not within 300 metres of the perennial water holes of the important wildlife areas with scarce water resources. Also prescribe against setting up of camps on movement routes, or dens or resting areas.

22. It should be emphasized in the management plan that deers, antlers, porcupine quills, peacock feathers and other animal parts collected by the staff or any other persons should be handed over to the forest or wildlife authority for disposal.

23. Game sanctuary or any other protected areas are potentially rich in wildlife resources. Such areas should be treated as class I areas and intense fire protection measures should be prescribed.

24. In many of the areas due to heavy overgrazing by the domestic live stock, the palatable species of grasses, herbs and shrubs have nearly been eliminated and the areas have been densely covered by unpalatable weed species like *cassia tora*, jungli tulsi, etc. Although on stopping the grazing by domestic livestock the occurrence of palatable species again starts improving, but the process is generally very slow and takes a long period of 5 to 10 years. To hasten it, steps should be prescribed to eradicate such weeds in such infested areas.

25. Standing instructions should be issued to all forest staff for recording important observations about wildlife in their diaries and compartment histories. Special proformas may be prescribed in which all the information worth noting should be enlisted.

19
Forest Maps and Mapping

Introduction

Sustainable management of forest resources necessitates formulation of a sound management plan. Planning for evolving management practices requires appropriate maps. Without maps no field-work can be accomplished. Apart from this, implementation of management plan necessitates extensive use of maps. Therefore, maps are the *sine qua non* of sustainable forest management.

Maps may be defined differently as:

- A reproduction of some portion of the earth's surface.
- A representation of whole or part of earth's surface in miniature.
- A conventional delineation of the earth's surface or a portion thereof, on a flat sheet.
- A reduced facsimile of the features on earth's surface.

Characteristics of Map

The map has the following characters:

I. All objects on the map are in the same relative position as on the ground.
II. All angles between the lines drawn on the map are equal to the angles between corresponding lines drawn on the ground.
III. The map covers much larger area and is drawn on a much smaller scale wherein detailed features cannot be represented.

Map Reading

Map reading is an exercise to get a clear and an accurate visualisation of the ground realities. It is possible by interpreting the scale, the symbols and conventional signs used in the map. Map reading requires the knowledge of the following:

Scale of the Map

Scale is the fixed relation, which every distance on the map bears to the corresponding horizontal distance on the ground. The scale of the map depends on the objective and purpose of the task. Based on the scale of a map, ground distance of any two places can be measured.

Conventional Signs

A conventional sign is the name given to the symbol used, to indicate on the map of a feature or object which occurs on the ground. Understanding of these signs is fundamental to map reading. Conventional signs are used when there is no room on paper to show the true outline or to add a written description. Smaller the scale of a map, the less is the room for description and therefore, greater the number of objects which need to be sown by conventional signs. Conventional signs are chosen in such a way that they preferably suggest the objects represented. Signs should be very simple to facilitate its drawing on the map. Such signs are not drawn to scale.

Orientation of Map

Orientation of map comprises laying out the maps in correspondence with the country it represents, such as north of the map corresponds with actual geographical north. Orientation of map is necessary for the following purposes:

1. For studying map features.
2. For finding observer's position on the map.
3. For ascertaining direction of places marked on map, such as villages, boundary pillars coup and compartment boundaries.
4. For adding fresh information.

Utility of Forest Maps

Forest maps are used for different purposes. Some of the purposes are enumerated below:

1. To undertake a forest survey.
2. To establish a pattern of ground sampling of forest conditions/attributes.
3. To enable field staff to locate the areas selected for sampling and stock mapping.

4. To use as base maps in which the different forest stands, classified with respect to type, age, density, etc.
5. To lay out the annual coup areas.
6. To identify plantation areas.
7. To lay out forest roads.

Different Types of Forest Maps

Different types of forest maps are used in forestry. The objective of the forestry task decides the type of forest maps to be used. As already said, maps are indispensable to forest management. In the maps, all forest attributes or properties are explained, analysed, and portrayed in terms of special and temporal relations (Table 19.1).

Table 19.1 : Forest Maps and Their Attributes

Sl.No.	Name of the map	Attributes to be shown.	Scale	Remarks/requirement in the course of working plan
1.	Stock map	Type of forests, Age, Site quality, Percentage of dominating crop, Quality, Density	1:15000 or 1:25000	One master set (un-cut) and three cut sets.
2.	Management map (one map for each range)	Boundaries of Compartment and forest block, working circle, number and boundaries of coups.	1:50000	One master set (un-cut) and three cut sets.
3.	Soil map	Compartment, block, soil type, depth, pH, soil texture, etc.	1:25000	Three sets prepared.
4 .	Geology map	Range, block boundaries, main roads, geology and rock conditions, towns, etc.	1:50000	Three sets prepared based on maps procured from GSI.
5.	Indicative map	Compartment, block, beat, range, road, rail, village, river, wireless station, forest check barrier, forest rest houses, depot, etc.	1:125000	Developed on the SOI maps.

There are different types of maps in vogue in forestry. They are as follows:

1. Topographic maps.
2. Management maps.
3. Stock maps.
4. Control maps.
5. Soil maps.
6. Geology maps.
7. Volume maps.
8. Indicative maps.
9. Vegetation maps.
10. Thematic maps.
11. Other maps like fire maps, regeneration maps, M.F.P. resource map, watershed map, etc.

Topographical Maps

Topographical maps are those which show such features like river, roads, village, and contours of the terrain. Such topographic maps are produced by the Survey of India, the pioneer and apex surveying and mapping institute in India on a common scale.

The country has been divided into 4*×4* sheets on 1:1000000 scale (numbered 45, 46, 47, etc). These one million scale sheets are further sub-divided into 16 equal sheets (numbered from A to P) on a scale of 1:250000. These degree sheets are further sub-divided into 16 equal sheets (Numbered from 1 to 16) on a scale of 1:50000 with a 15* extent. Again, for some areas, such 1:50000 scale map is divided into four 1:25000 sheets of 7-1/2*. They are numbered as NE, NW, SE and SW (Figs. 19.1 and 19.2). Topographic maps are available at Survey of India office at Dehradun and at sales branches of the institute in various cities.

Map constructions or preparation is a highly skilled work. Execution of large mapping preparation programme in forestry needs an organization with trained personnel like draftsmen and tracers. The scale or "representative fraction" of a map compares the apparent distance between two selected points on the map and the true distance between those points on the ground. Here, both the distances are expressed in the same unit of measurement. Sometime, confusion arises about the references to "small" or "large" scale maps. As we knew, scale is defined as the ratio of the distance on the map to the

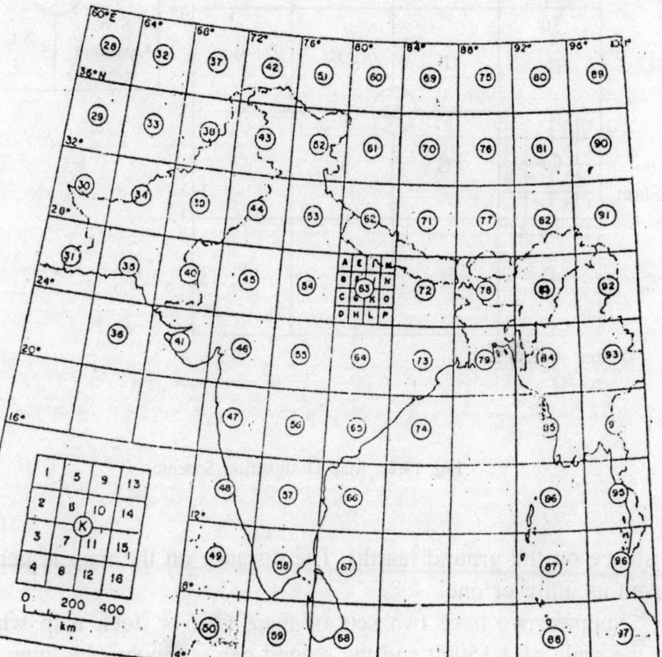

Officer in Charge. Map Record and Issue
Office Map Publication Directorate
Survey of India
Hathibarkala Estate, New Cantt. Road
Dehradun 248 001

Fig. 19.1 : Survey of India : Index Map.

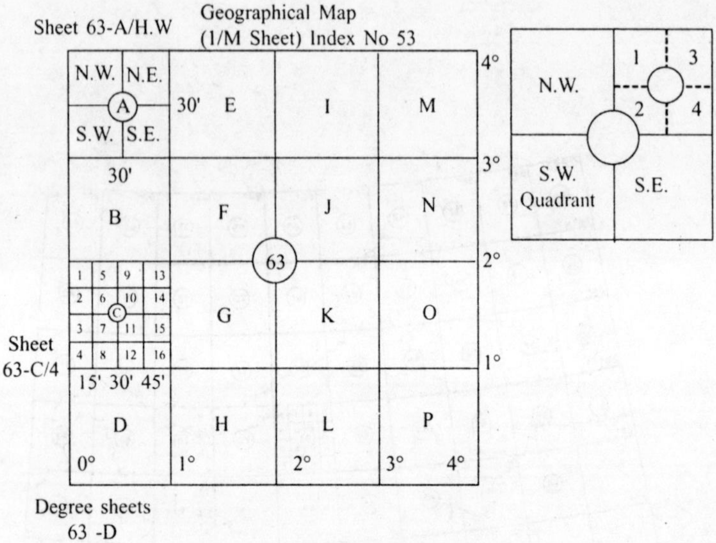

Fig. 19.2 : Map Designation Scheme.

distance on the ground (earth). The distance on the map is generally taken as unity or one.

Suppose we have two sets of map. One is stock map which is in the scale of 1:15000 and the second one is topographic map which is in the scale of 1:50000; since 1/15000 is a larger fraction than 1/50000, it follows that 1:15000 stock map is a larger scale map than the topographical sheet drawn at 1:50000.

Topographical sheets of Survey of India are generally available in the scale of 1:50000. They carry all essential features of the ground excepting the forest boundaries and conditions. They constitute the base maps. These base maps are required to be enlarged to the scale of the forest stock map. Then, the trace maps of each compartment have to be prepared. These trace maps are taken to the field by the forest rangers for stock mapping. Details regarding forest attributes like density, site quality, type of forests etc. are entered into it and these base maps are finalized. After completing the process of mapping the forest details, those are transferred to a final set of enlarged base maps. Description of attributes of forests is portrayed by using combination of different colours and symbols in accordance with prescribed legend. Three to five copies of toposheets are prepared in such fashion in view of the cost and time. Therefore, other maps like management maps, and indicative maps of the division are reproduced to generate number of copies required frequently during the course of forest administration and management. These are generally devoid of colours and are purely black and white reproduction.

Management Map

These maps are generally in the scale of 1:50000. Such maps are prepared by working plan office as a part of the working plan exercise. A good management map is always prepared on the topographical maps produced by the Survey of India. This, in addition to topographical features, shows features like working circles, felling series and annual coup boundaries. However, other features like proposed plantation sites, nursery and other relevant activities can also be shown on the map. Each working circle is given a distinct colour wash to make it conspicuous. A trace of management map is presented in Fig. 19.3.

Stock Maps

These maps are generally prepared in the scale of 1:15000. The working plan office prepares stock maps after having completed the stock mapping exercise. These maps indicate about forest types, age-class distribution, quality class, density of the crop, blank and encroachment area, old planation sites. For preparing this map, the topographical maps of Survey of India having scale of 1:50000 are taken as base maps and these base maps are enlarged to 1:15000 scale maps.

Control Maps

The working plan office while finalizing the compartment history file prepares control maps. The trace map of stock maps is called control map. Such trace map is attached to the compartment history life. The purpose of preparing such trace maps of stock maps is to show progress of various forest operations/exercises and to check and control over them in deference to working plan prescriptions. Such control maps help in assessing the following aspects:

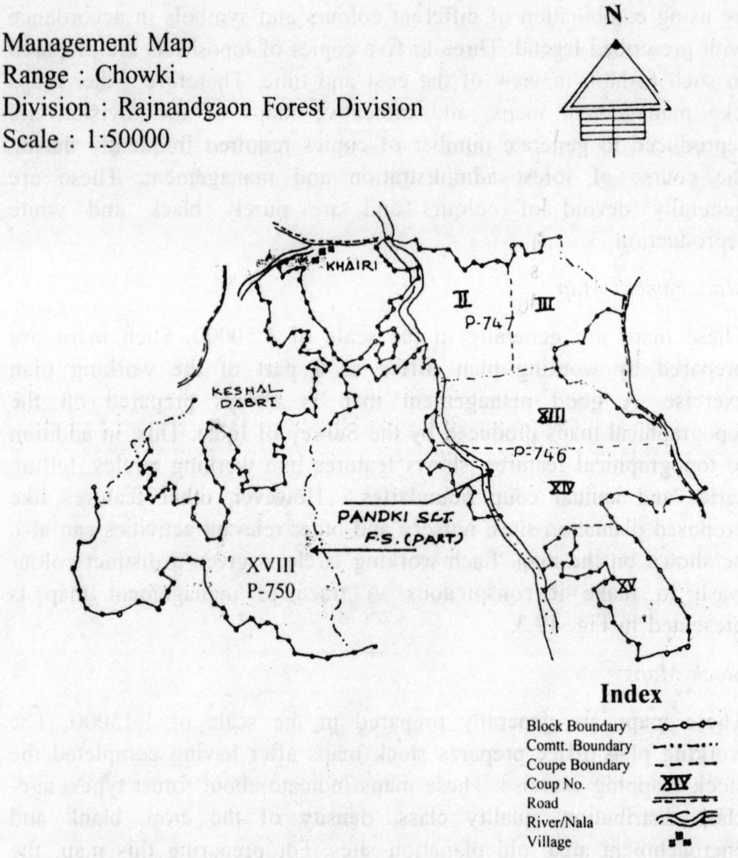

Fig. 19.3 : Management Map.

1. Recording of forest fires.
2. Artificial and natural regeneration.
3. Harvesting of all wind fallen trees.
4. Progress of soil conservation operations etc.

Various forestry operations are recorded by different colour wash and legends. Control maps are necessary for monitoring and evaluating forest works. When such maps are enclosed with control forms, they provide a full picture of past operation.

Soil Map

Soil map shows different kinds of soils and their extent on the landscape. Each delineated map units in the soil map provides information on physiography, soil characteristics and classification reflecting their potentialities and problems.

The National Bureau of Soil Survey and Landuse Planning (NBSS&LUP), Nagpur in cooperation with soil survey staff of the State Government departments prepared soil resource map of different states on 1:250000 scale, using a 3-tier approach, that is (1) image processing, (2) field surveys and laboratory analysis, and (3) cartography and printing. The soil map was printed on 1:500000 scale, without generalizing. The basic mapped units are at the level of soil family associations with dominant phases. These maps are used to optimize forest productivity by making suitable changes in management practices.

Geology Map

Geology maps shows distribution of rock features and geological conditions in the division. The scale of such maps is 1:250000.

Volume Maps

Volume Maps are maps showing mean volume of growing stock spatially. These maps are used for timber planning and management. The scale used is 1:50000.

Indicative Maps

Indicative Maps are those topographic maps, which shows the rough positions of different forest blocks and compartments on the map. The scale of the map is 1:250000. The objective of the map is to study the location of different compartments and working circle areas on a broader perspective.

Vegetative Maps

The vegetation map shows the distribution of vegetation in the forest and non-forest area of a state or country. In other words, the map shows the inventories of plant resources indicating their distribution in the landscape. The purpose is to draw attention to the extent of vegetation on a miniscule map. Few of important benefits are as follows:

1. Through proper interpretation of data mapped, it is possible to make suggestion regarding the optional land-use.
2. Used to compare the land use pattern of any two regions. Both the maps follow the standard rules of drawing it by the method of representation of physiognomy with the help of definite patterns and the climatic analogies by colours.
3. The map indicates the distribution of vegetation patterns.

Colours play an important role in the system of vegetation mapping. Generally, white indicates the practice of agriculture and area shown in colour represents natural vegetation. Different colours indicates the ecological conditions (e.g. red colour is used to depict the hot arid zone; blue indicates a high rainfall zones and green of tropical hills). The intensity of colour indicates the physiognomy. The horizontal lines on the map indicate human action on the landscape.

Thematic Maps

Thematic maps are the maps showing specific theme attributes like type of forest, density etc. These are prepared on a 1:50000 scale. Such maps are prepared by Forest Survey of India on the basis of Survey of India toposheet, aerial photographs and satellite imageries. Such maps are used to delineate different forest types (management) in a forest division. Apart from this, such maps are beneficial for forest planning and management.

The Forest Survey of India prepares thematic maps of the forest types on 1:50000 scale by interpreting aerial photographs of appropriate season. In these maps, commercially important species are delineated separately when their composition in the total mixture is more than 50 per cent. The different land use and forest types delineated in these maps are as under:

1. *Sub-tropical and temperate forests* : Alpine pastures and scrub/fir/spruce/deodar/blue pine/chir pine/khasi pines/mixed conifers/conifers mixed with broad-leaved species/oaks/mixed broad-leaved species/regrowth/plantations/bamboos/scrub.

2. *Tropical forests* : Teak/sal/mixed miscellaneous/mangrove forests/regrowth/plantations/scrub.

3. *Non-forest areas* : Agriculture/shifting cultivation/habitation/orchards/tea gardens/exposed rocks /water bodies/snow and glaciers.

The maximum area of delineation in these maps is 0.5 cm × 0.5

cm (6.25 hectares on the ground). The maximum width of strip of trees in lines is 0.3 cm. No crown density is generally reckoned for mangrove forests, scrub bamboos, plantations and regrowth. Besides, the forests are delineated into three crown-density classes *viz.*

(i) Crown closure more than 60 per cent.
(ii) Crown closure from more than 20 per cent to 60 per cent.
(iii) Crown closure from 5 per cent to 20 per cent.

Use of Thematic Maps

The thematic maps are prepared primarily for the purpose of inventorying of forest resources. These maps can also be used for preparing stock maps. For preparing stock maps, the working plan officer carry out survey in order to determine and delineate the forest types, density, site quality, and age classes. He takes nearly a year to prepare stock maps of the division. In this context, the thematic maps of Forest Survey of India are of great help to working plan exercise. Use of such maps can help to reduce the time required for preparing stock maps. Such thematic maps are prepared on the basis of 2 to 10 years old aerial photograph. In case of majority of forests, the period of a few years does not bring much change. Thus, major part of the map remains valid even after a lapse of few years. In case of the remaining areas where canopy has undergone a significant changes during the period, it is possible to identify and delineate the same on the thematic map with the help of mapper data of landsat on 1:50000 scale. After comparing thematic maps with the satellite images/data the working plan officer would be in a position to determine the area that have undergone changes and to assess the need for ground survey for revision of the stock maps. In this way, the working plan staff would be required to do the ground survey only of a small portion of the forest, say 10 to 15 per cent, instead of the whole area. The exercise of determining site quality and age of the crop can be done during inventory or by making quick visits to the compartments in a short time for each delineated patch. Apart from this, the user agency can enlarge the thematic maps to suit their own mapping scale without loss of details.

A map catalogue of the thematic maps is available with Forest Survey of India, Dehradun. The map catalouge shows the location of map sheets and the district in which they fall.

Other Maps

Beside the category of maps shown above, there are many types of maps, which are in vogue in forest planning and management. Fire maps, regeneration maps, watershed maps, M.F.P., resource maps, etc. are some of such maps which are in vogue in forestry. The scale of such maps depends on the objective of the task.

Care of Maps

There should be a suitable system for storing these valuable maps. Now-a-days iron filing cases are used to keep these maps. Maps are stored flat, rather than rolled. The filing system should be insect proof and they should be so arranged that any specific map could be retrieved with no time.

PART - IV

DATA MANAGEMENT, REPORT ORGANIZATION AND TECHNICAL WRITING

20

Collection, Classification and Presentation of Data

Introduction

Data are basic to statistics. They are collected from the field through statistical surveys. Statistical surveys and analysis are part of the forestry practices. Besides, experimental data are collected through special studies related to different individual phenomenon in the forests. In a forest survey, data are collected from the entire forest in which the individual effect of various factors affecting the phenomenon or event under study can not be controlled or isolated. For example, when we collect information about the density and crop volume of a forest, it is difficult to isolate the factors affecting the phenomenon. We simply keep on collecting information with regard to the density and volume without taking into account the other attributes like factors of locality, anthropogenic pressures and management input, which composedly affect it. Contrary to it, in an experiment we collect data, in which the individual effects of various factors influencing the phenomenon under study can be easily isolated and controlled. In forestry practice, different experimental plots are laid in forests to study different phenomenon like growth, survival percentage and diversity. Individual tree experiments are also conducted.

STATISTICAL SURVEY

Collection of quantitative data is done with the help of statistical survey. Statistical survey is an elaborate field exercise. It needs careful and meticulous planning and organization.

Organization of statistical survey consists of two phases. They are:

I. Planning a survey.
II. Executing a survey.

PLANNING A SURVEY

Planning is an important component of statistical survey. It consists of the following aspects:

(i) Object and scope.
(ii) Sources of information.
(iii) Types of inquiry.
(iv) Statistical units and its definition.
(v) Degree of accuracy.

Object and Scope

The object of the statistical survey is generally determined in the beginning. This helps in organizing, collecting and analyzing the information gathered. Besides, it helps in improving the accuracy of the collected data. In addition to it, while determining the objective, the scope of the statistical investigation should be well analysed. The extent to which the data should be collected and can be useful for the said purpose of the survey should be thoroughly examined before embarking upon the work of data collection. Sometime, insufficient data lead to erroneous inferences. Likewise, enormity of data leads to a situation where it becomes unmanageable to deal with the information and consequently difficult to draw relevant conclusions. Considered in this backdrop, it is necessary to determine the objective and the scope of the statistical survey at the beginning of the exercise distinctively.

Sources of Information

There are two major sources of information. They are:

— Primary sources.
— Secondary sources.

Primary data are those informations, which are collected for the first time. For collection of primary data, we need to build a appropriate field organization and detailed planning. Now, the said organization gathers the data from the field. Preparing inventory of resources requires data from primary source. Secondary data in case of working plan process are those, which have been collected by others and are available in the forms of district statistical record, district Gazatteers, technical journals, old working plans and research publication. Population of the district, state and division and growth data of different tree species are example of some of the secondary data. While using secondary data for preparing a report, one is required to be cautious, otherwise results of the survey or field-work will be inaccurate and incongruous. All such secondary data should be critically analyzed before its application. It has been seen that site-specific growth data with regard to miscellaneous crop is not sufficiently available in forest department and research centres. While making the diametric growth projection in relation to rotation of the crop, it is required to know the growth rate of the particular species, in a particular division. Generally, in the absence of site-specific data, statistics of growth for same species for different areas are used. In this backdrop, it is necessary to see that the forest area, which contains some growth data more or less, tallies with the forest area for which the data is being applied. If this precaution is not followed, application of secondary data will lead to inaccurate conclusion with respect to growth of the species at different age.

Types of Inquiry

Objects of the study determine the type and scope of inquiry. The object of the resource survey in a forest division is generally to find out the growing stock, diametric distribution of crop, age distribution of crop, regeneration status of the forests, bio-diversity status of the managed forests, composition of the species in the growing stock. However, the objects may be more widened according to the need of the management. In order to achieve the above said objectives, it is necessary to evolve a certain type of inquiry. However, if the objective is enlarged, then the type and scope of inquiry or investigation will also change accordingly. Therefore, we find that a decision about the type of inquiry suitable for the particular forest study depends largely on the object and scope of the investigation.

A field study with respect to preparation or revision of working plan involves the following kinds of inquiry:

1. Sample and census.
2. Direct or indirect.

In a sample type of inquiry, selected representative units are studied. In a census type of inquiry, all the units are taken into account. Generally, in a tiger population study, the census type is resorted to. This census type of inquiry is a costly affair and also, needs a lot of time. With respect to inquiry of tree growth in a small plantation, this census method can be adopted. But, it is not possible to apply this method to a large tract of forests. In such case, the sample type of inquiry is generally adopted.

Statistical investigation may be direct or indirect. Sampling enumeration of tree resources with respect to number of trees, site qualities, volume of trees, diameter distribution in a unit area involves direct quantitative measurement. Such type of investigation or survey is called direct inquiry. However, certain study involves quantitative data and based on it, certain attributes are evolved. Poverty level and the state of development in forest villages require quantitative data of the villages. With the help of such data, the attributes of poverty and development are found indirectly. This type of study is an indirect inquiry.

Statistical Units

The collection of statistical data needs measurement. Therefore, it is necessary that the units in which the data are to be collected should be distinctly defined. The unit of measurement should be uniform throughout the working plan study. The characteristics of a sound statistical unit are:

1. It should be specific and unambiguous.
2. It should be stable.
3. It should be uniform and homogeneous

If statistical units are broadly categorized, they are of two types. They are:

1. Units of collection.
2. Units of analysis.

Units of collection are those units in which data relating to a particular attribute are either enumerated or estimated. For example, the number of trees in a sample plot is enumerated and their height and growth data are recorded. Based on this record, the mean volume of trees in the sample plot is estimated. Units of measurement may be either simple or composite. Units like number, cubic metre, square metre, quintal, metre or centimetre are all simple to define. Therefore, such units are called as simple units in forestry. A composite unit is one, which is formed with an addition of a qualifying word. This makes the definition difficult. In forestry, the units like notional tonne, row-kilometre, and man-days are such composite units.

Units of analysis are those units with which statistical information are analyzed, and interpreted. Such units are percentage, ratio, per hectare, growth rates like MAI/CAI and co-efficient. Such units are useful for making a critical comparison of forests and its growing stocks, in different sites and in different forest types.

Degree of Accuracy

Accuracy is the corner stone of any scientific study. Therefore, it is necessary to ensure a reasonable degree of accuracy in the study. Before commencement of the field-work, one is required to decide a level of accuracy and so, he tries to maintain it. However, it is a known fact that absolute accuracy is impossible to be achieved. In such statistical exercise, it is only necessary to achieve a reasonable standard of accuracy. This provides sufficient reliable data to draw a dependable picture of the forests. The degree of precision required in sampling while undertaking resource survey has been explained in separate chapter in detail.

EXECUTING A SURVEY

Collection of data requires field exercises and survey like resource survey, socio-economic survey, stem and stump analysis, joint forest management studies, etc. After making detailed planning of such survey and exercises, it is now necessary to execute these plans. Execution of survey plans and exercises involves the following steps:

1. Setting up a divisional organization.
2. Selecting field staff.
3. Imparting necessary training to field level investigators.
4. Supervision of field-work.
5. Collection and analysis of data.
6. Preparation of the report.

Setting up of a small divisional office provides the administrative infrastructure for the field works. The office should consist of ministerial as well as technical staffs. Selection of field staff is pivotal for the success of the mission. It is necessary to see that only such staffs who have basic attribute and potentiality are put in the job. Imparting relevant training to such selected staff is the next step. Since field surveys are discontinuous work, therefore, it is necessary to provide the basic training frame work to the staff. Such staff should be well conversant with the objective and methodology of the study, use of instruments, appliances, and data collection procedures. It is also imperative to supervise the work closely. This removes the possibility of unreasonable errors in such statistical investigation. This also provides direction to the field work.

Analysis of data starts after the process of collection is complete. Data are fed into the computers for tabulation and classification. Then such classified data are properly analyzed and interpreted in the backdrop of the avowed objective of the study. Finally, the technical report like working plan in prepared after having gone through a succession of stages with respect to documentation.

DATA COLLECTION

After organizing the survey, the next important task is data collection. This in the most important step because, the entire result depends upon the quality of this exercise. As we have already said it is necessary to undertake data collection with a meticulous planning. Planning as already explained encompasses the aspects of scope, objective, methodology and other details.

There are various methods of collecting primary and secondary data. Preparation or revision of working plan needs both primary and secondary data. However, the method of inquiry should be correctly chosen. There are two major factors, which influence the selection of methods. They are:

1. Availability of time.
2. Availability of funds.

Method of Collecting Primary Data

The propriety of a particular method primarily depends upon the object and purpose of the inquiry. There are two methods of inquiry being practiced in forestry. They are:

(i) *Direct Personal Inquiry*

The investigator has to collect the required information personally from the site itself. This exercise may involve a group of investigators also. This method demands time as well as money. This method is used for inventory preparation, stem and stump analysis and many such forest studies.

(ii) *Schedules and Questionnaires*

It is also an important method of data collection. This generally used for socio-economic study of the people living in and around forests. In this method, detail schedules and questionnaires are prepared and information is collected from the fields in such questionnaires after eliciting answers from respondents. This method is employed in socio-economic survey and case studies whose results act as an input to evolving working plan prescriptions.

Here, the investigators shall go to the villagers with the questionnaires and record their replies. The investigator plays an important role in collecting such data. He explains the aims and objects of the investigation to the villagers/people and emphasises the necessity and usefulness of correct answers. He acts as facilitator in such inquiries. This is a reliable method of inquiry, so far socio-economic studies and evaluation exercises are concerned. However, what is important in this method is the fact that the questions should be simple and few in number and the investigator should be very courteous, polite and properly trained. Generally, range officers and their subordinate staffs do such investigations. It is necessary to select suitable people out of such staff for that task.

Drafting a Questionnaire

Success of this method depends to a large extent on the proper drafting of questionnaire. There is no set guidelines for preparing such questionnaire. It needs basic skill and lot of field experience in the field of forestry and rural development. However, the questionnaire should satisfy the following aspects:

1. It should be simple and clear.
2. The question should be arranged in logical sequence.
3. It should require answers, which are definite, brief and easy to treat statistically.

4. It should avoid questions whose answers require calculation.
5. It should avoid any personal questions, which hurt the sentiments of the people.

For preparing inventory of tree resources, and also socio-economic survey, it is required to adopt sample investigation method. In case of inventory, a sample in a forest is a representative of that forest. If the sample and its sizes are properly decided, then whatever holds good for the sample should hold good for the forests. In other words, a sample gives representative data for the whole of forests.

Selection of the sample is the most important point in the whole sampling exercise. There are, broadly, two methods, by which samples are selected. They are as follows:

1. Deliberate or purposive sampling.
2. Random sampling.

Deliberate or Purposive sampling

In this method, the investigator selects from the forests, few units, which according to his estimates are best representatives of the population. His selection is based on his observation and selected perception about the representativeness of the sampled units. These units are then studied and necessary data collected. This method is generally employed in site quality determination in teak plantation, yield and growth measurement in plantation, monitoring thinning by putting check-plots in teak plantations. This technique is beset with many drawbacks. It runs the risk of biases of the investigator and thereby, influences the selection. When the investigator is prejudiced, then it is quite possible that he may select such sampling unit which would give conclusions that suit his requirements. Apart from this, such type of method do not provide any idea about the degree of accuracy of the inquiry.

Random Sampling

In this method, selection of sample is done in such a manner that the chance of selection of each unit of the forests is the same. Selection is completely based on chance. In other words, sample taken by random sampling method is fairly representative of the whole area of forests. This method is based on the mathematical theory of probability. The theory states that "if from a very large number of items, a moderately large number of items is selected at random, than the smaller group so selected is likely to have the characteristics of the larger group from which the smaller group is selected".

Random sampling has much merit. They are as follows:

1. It is possible to have an idea about the errors of estimation.
2. It is possible to assign limits within which the true value of a measure of forests must invariably lie. More on this topic has been discussed in separate chapter on inventory.
3. Any biases or prejudices do not affect selection of sample.

Random sampling method is widely used for making forest inventories. However, many variations have been affected in this random method to suit to particular forest conditions. Such methods with necessary variations have been categorized as follows:

1. Simple random sampling.
2. Stratified random sampling.
3. Multi-stage sampling.
4. Multi-phase sampling.

The above types of random sampling methods are discussed in detail in the separate chapter of "forest inventory". However, the most common type is the stratified random sampling method. This method is employed in working plan preparation while making resource survey or inventory preparation.

Generally, forests encompass a number of features, all of which require to be represented in the sample. In this situation, the best way is to divide the forests into few strata based on special features like density, growing stock or mean volume per hectare or intensity of management and then to select at random number of item or units from each of such strata to form a sample and study it. This is known as stratified random sampling. The principal feature of a sample taken randomly is that the items so chosen have an equal chance of being selected. The methodology of designing stratified random sampling is discussed in the chapter "Forest inventory". Generally, this method is resorted to, when the forest is heterogeneous in nature. Such heterogeneous units should be represented in the sample in the same proportion in which they occur in the forest. This provides picture of representativeness and reliability of data. This study shall be applicable to the whole forests of the division for which the inventory is being prepared.

Accuracy and Errors

In such statistical methods, counting, estimations and probabilities are involved. Measurements may not always be accurate. Therefore, it is necessary to lay down the standard of accuracy desired. However, the degree of accuracy desired is a relative term depending upon the objective and the nature of inquiry.

While dealing with accuracy and approximation, it is obvious to refer to errors. The word "error" has special significance in statistics. It is different from the world "mistake" which implies a wrong interpretation, conclusion or calculation. The world "error" in statistics is used to explain the difference between the true or accurate value and estimated or approximated value of an item.

Measurement of Errors

In statistical investigations, errors are imperative. It is, therefore, necessary to measure their magnitude and to find out the limit or range within which our survey results can vary. From this standpoint, there are two types of errors. They are:

1. Absolute error.
2. Relative error.

Absolute error is the difference between the actual or true value and the estimated value of an item. It may be positive or negative. Relative error is the ratio which the absolute error bears to the estimate of the item. It is sometime expressed as a percentage error to make the relative error more significant apparently. The above classification of error is based on the way of expressing the error. However, from the point of investigator or enumerator, the error can be categorized into two types. They are:

1. Biased error.
2. Unbiased error.

Biased error arises on account of a bias or prejudice on the part of enumerator or on account of defect in measuring instruments. Unbiased errors are those which arise without any conscious efforts on the part of the enumerator or without any defect in the measuring instrument. Such errors occur because of genuine mistake in estimating the girth or height of the tree. It also arises on account of not exercising care on the part of the enumerator. Measuring the growth of tree over a flute is one of such mistakes.

The effect of biased error is disastrous, because it is of cumulative in nature. Ultimately, it may lead to under-estimation or over-estimation. The effect of unbiased error is compensating in nature. It is not very harmful, because the error has the effect of compensating itself and in the ultimate analysis, it results into a minor error. Therefore, while undertaking sampling, it is necessary to guard against biased errors.

Collection of Secondary Data

As we have already discussed, the secondary data are those which have already been collected and analyzed by someone else. Secondary data may be either published or unpublished. Published data are found in Government records, old working plan documents, technical journals etc. Unpublished data are found with research workers, field workers of other working plan unit or other department.

It is necessary to scrutinize and verify secondary data and to edit them to find inconsistencies, errors and omissions. The secondary data should have the following attributes:

1. The data should be reliable.
2. The data should be suitable for our purpose.
3. The data should be adequate.

Therefore, it is always necessary to use secondary data after due examination and verification.

CLASSIFICATION OF DATA

Data generated during the field work exercise do not give an idea about the problems of forest under study. They cannot be taken up directly for analysis and interpretation. For example, the figures of enumeration in a sample plot as collected in the field do not indicate the stand density in a forest stratum or the growing-stock in the forests. Unless all such enumeration data are classified and tabulated, it is not possible to draw any inferences from them. For the purpose of analysis and interpretation, it is necessary to divide the data into homogeneous groups. This process is called classification. In other words, it is a process of arranging data in-groups or classes according to resemblances and similarities. This brings homogeneity in the heterogeneous elements of the data. In this process, units having a common characteristic are grouped or divided into a number of classes.

The main objective of classification is to simplify the complex data and render it easy to convey a meaning. Classification also helps in comparing two groups of data. Therefore, classification is an important process and is a precursor to tabulation.

In a working plan exercise, generally data are expressed in numerical quantities. Hence, they are arranged according to class interval. For example, classification of trees are available in a sample plot according to girth classes like 10 to 20 cm, 21 to 40 cm, 41 to 60 cm, 61 to 80 cm, 81 to 100 cm and 101 to 120 cm and above 120 cm. However, the trees can also be classified according to diametric class-wise and species-wise. The diametric or girth class range is called class interval. The difference between the upper limit and the lower limit of a class interval is known as the magnitude of the class-interval. When this difference is same throughout the various classes in the class intervals, the magnitude is known as uniform. If the difference changes in various class intervals, then it is known as un-uniform magnitude.

In forest inventory, both the inclusive and exclusive method of framing class-interval is adopted. In inclusive method, the class-interval does not overlap; whereas in an exclusive method the class-interval overlaps. The class-interval of diameter of trees will appear as 10 to 15 cm, 16 to 20 cm etc. in exclusive method.

Before embarking on classification of data, one should lay stress on the following points.

- Classification must be exhaustive, i.e. it should be possible to include all data.
- Classification should be unambiguous.
- The number of classes should be neither too large nor too small.
- The magnitude of class-interval should be uniform throughout the classification

In forestry data, there is another process complementary to

classification. That is seriation. This is closely associated with classification. According to L.R. Conner "if two variable quantities can be arranged side by side so that the measurable differences in the one correspond to the measurable differences in the other the result is said to form a statistical series". Seriation can be carried out on the basis of time, space, or some condition.

Tabulation of Data

Tabulation of data is defined as systematic arrangement of data in columns and rows. It aims at preparing data for the purpose of analysis and statistical inferences. In other words, the basic objective of this exercise is to simplify the process of presentation of data for the purpose of analysis•and statistical inferences. In other words, tabulation helps making complex and varied data comprehensive. It is also a process of condensing classified data in such a manner that it can be easily analysed and understood and so that any comparison among data can be made to draw any inferences. Tabulation is an art. Continuous observation, learning and practice make one to develop this art. Good tabulation makes an impact on the reader and conveys the message.

In general, a statistical table is divided into five parts. They are:

(i) *Table number*: Each table should be numbered for easy identification and easy reference.

(ii) *Title of the table*: Each table should have a distinct and short title at the top of the table in the centre. A title indicates the nature of data. The title should be in prominent letters.

(iii) *Caption*: This refers to the headings of the vertical columns. A caption may have one heading or a number of sub-headings.

(iv) *Stub*: This refers to the headings of the horizontal rows. They are positioned on the left side of the rows.

(v) *Body*: It is that part of the table, which contains the data. This is the most important part of the table. The data in the body are arranged according to captions and stubs.

Basic Principles of Tabulation

There are no hard and fast rules for ideal presentation of tabulation.

Preparing good table needs understanding of the subject, objective and practical experience. They help in designing the structure of good table. The basic principles of tabulation with respect to forest data are as follows:

- Table should be short and precise.
- When data are enormous, these can be presented in a number of tables. Items in a table should be arranged in a logical order.
- Caption, heading, sub-headings should be self-explanatory.
- Units of measurement under each heading or sub-heading should be distinctly indicated.
- In case of a comparative analysis, columns containing data should be arranged side by side.
- Source of data should be indicated in the footnote of the table.

Types of Tabulation

There are two types of tabulation. One is simple and the other is complex. Both types of tabulation process are illustrated below:

1. *Simple Tabulation*

It is a process, which involves arrangement of one set of related data and seeks to address one or more groups of an independent investigation. In a simple table, two factors or characteristics are placed in relation to each other. The following table is a product of simple tabulation. It shows the distribution of stems of different species in one hectare of forest. This is also called a one-way table (Table 20.1).

Table 20.1 : Stem distribution in a miscellaneous forest

Name of the tree species	No. of trees per hectare
Teak	18
Bija	12
Saja	32
Tendu	13
Haldu	3
Salai	8
Karra	78
Miscellaneous	82

2. *Complex tabulation*

It is a process of tabulation, which involves arrangement of data pertaining to a number of coordinate factors or features. When there are two co-ordinate factors, the table is called two-way tables or double table. When there are three co-ordinate factors involved in the table, it is called three-way table or the process is called treble tabulation. When it is a case of more than three co-ordinate groups, the table is called multiple tables or the process of tabulation is called multiple tabulation. In a multiple table, one can go for posting details of stem distribution in different categories of diametric classes.

PRESENTATION OF DATA

Tabulated data extends over a large number of columns. It makes comparison of features difficult. So, there is a necessity of application of visual methods for representation. Every intelligent mind readily grasps the comparison and magnitude of data from the visual aids at once. Besides, such visual methods leave a lasting impression on the mind of readers. Graphs and diagrams are the two visual methods of representation of data. With the help of diagrams and graphs, complex data are presented in a simple form. They also render the comparison easy. It is for this reason, such visual methods are becoming popular day by day.

As discussed above, there are two main categories of visual methods. They are:

1. Diagrammatic presentation of data.
2. Graphical presentation of data.

Diagrammatic Presentation of Data

Diagrammatic presentation of data means providing a pictorial representation of the homogenous quantitative data to make them readily comprehensible. They provide requisite emphasis on different characteristics of data. There are certain basic principles for drawing the diagrams. One should focus on title, size, proportion of length and breadth, scale and index of the diagram. Diagrams should be simple and attractive. Apart from this, what is most important is selection of proper and appropriate diagram. Not all types of diagrams are suitable for all types of data. Therefore, one should be very careful in selecting the diagram to convey an impression about any phenomenon in question.

Types of Diagrams

There are many types of diagrams used to express an explicit meaning and message. They are:

1. *Dimensional Diagrams*

1. *One-dimensional diagram*: It is a type of diagram where only one-dimension changes in size according to changes in the variables. Examples of such one-dimensional diagrams are lines or bars (either simple or sub-divided bars.) are illustrated below:

— *Simple bar diagrams*: Simple bar diagram indicates that the height of the bar varies according to the data. All the bars are raised on a common base and there should be some gap between two bars to make the bars distinguishable. Bar diagrams are suitable for presentation of quantitative data of special series. The bars may either be shaded or coloured or be left blank. Generally, values are mentioned in the top of the bar (Fig. 20.1).

— *Sub-divided bar diagrams*: They are of two types. One is sub-divided bar-diagram on percentage basis. The other one is a sub-divided bar on absolute basis. This gives an idea of totality as well as its individual components in a same diagram.

— *Sub-divided bars on percentage basis*: Comparison of data is always made on equal footings. Items are expressed as percentage of the aggregate. The heights of the bars are equal because they represent 100 in each case. From these bars, sub-divisions are made according to the percentage they bear to the total. This makes comparisons of various items or features very simple and clears. Various factors constituting the aggregate can be compared in their proper perspective (Fig. 20.2).

— *Sub-divided bars on absolute basis*: This also gives a total picture as well as idea about each component on a relative basis. However, each component does not show the relationship in which they stand in comparison with the total aggregate (Fig. 20.3).

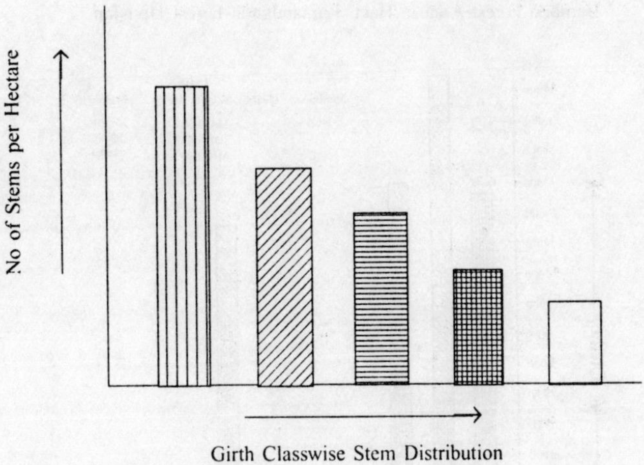

Fig. 20.1 : Simple Bar Diagram.

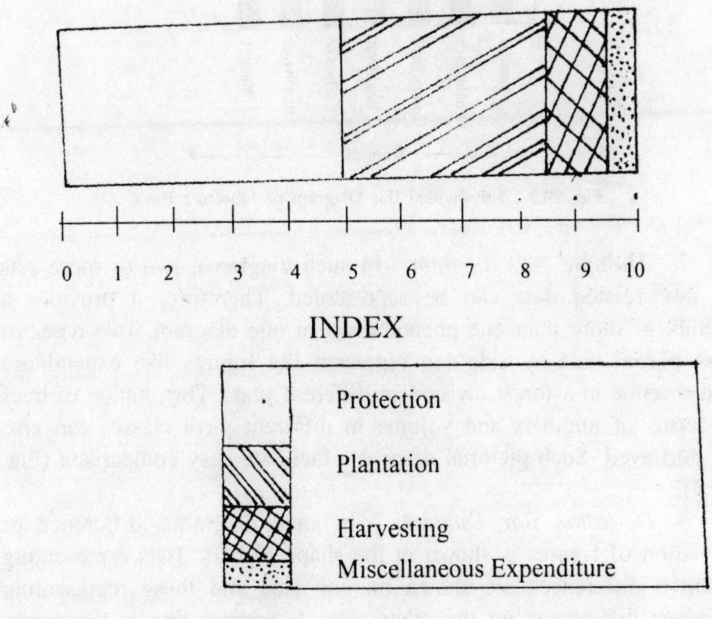

Fig 20.2 : Expenditure on Different Parts of Forestry in a Division.

Rangewise and Qualitywise Distribution of
Bamboo Forest Area in Hact. Rajnandgaon Forest Division

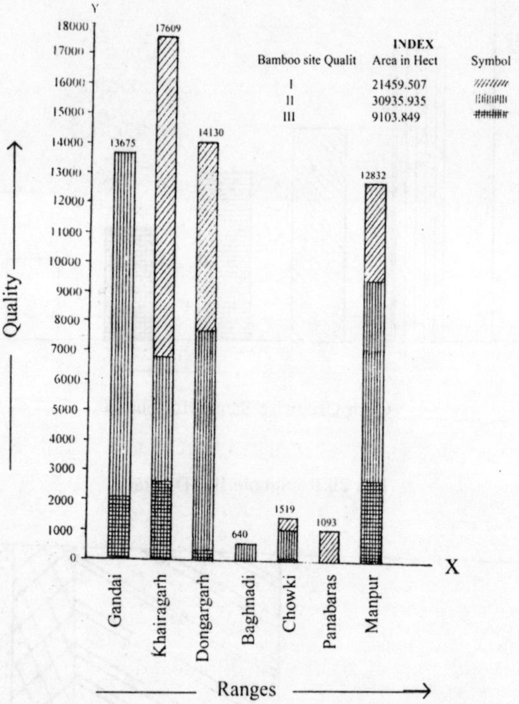

Fig. 20.3 : Sub-divided Bar Diagram on Obsolute Basis

2. *Multiple bars diagram:* In such diagrams, two or more sets of inter-related data can be represented. Therefore, it provides a picture of more than one phenomenon in one diagram. Two types of bars placed side by side can represent the figures like expenditure and revenue in a forest division in different years. Distribution of trees in terms of numbers and volume in different girth classes can also be portrayed. Such pictorial diagrams facilitate easy comparison (Fig. 20.4).

3. *Deviation Bar Diagram*: In such diagrams, difference or deviation of figures is shown in the shape of bars. Bars representing positive differences are shown on one side and those representing negative differences on the other side. A vertical line in the centre represents the point of equilibrium and on the right-hand side positive

Girth Classwise No. of Trees and Volume Pre Hact. in CMT
Working Circle Sci. Rajnandgaon Forest Division

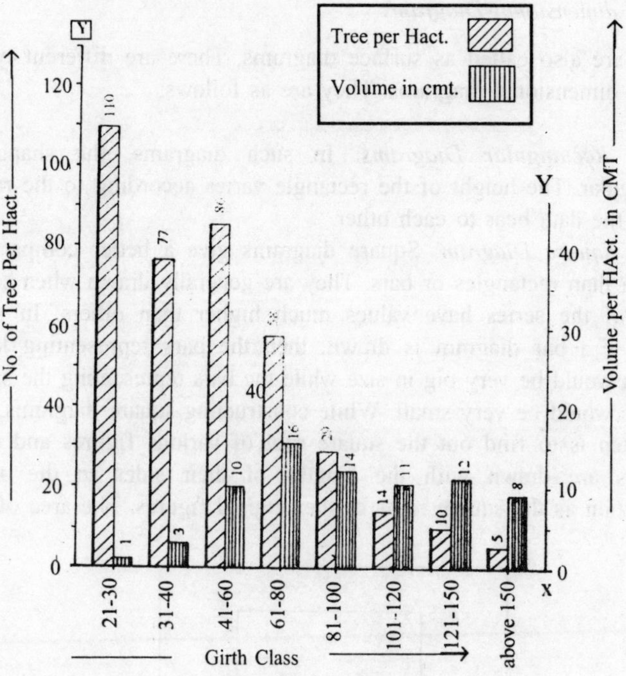

Fig. 20.4 : Multiple Bar Diagram.

Change in Forert Cover

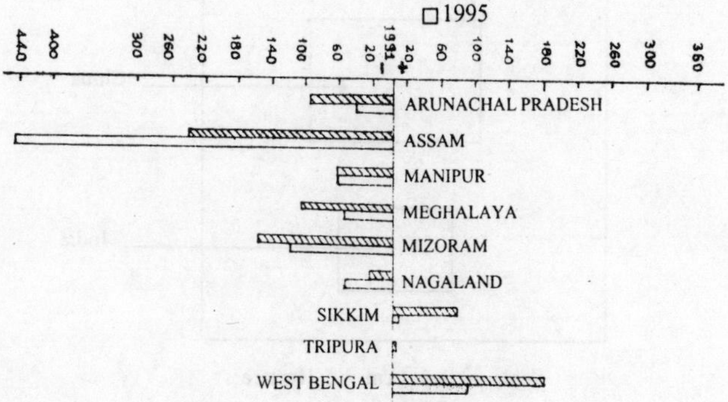

Trand of Forest Cover Change in 1993 & 1995 Assesments

Fig. 20.5 : Foprest cover change (+ or -) in Sq. km.

differences and on the left-hand side negative differences are shown (Fig. 20.5).

2. *Two-dimensional Diagrams*

These are also called as surface diagrams. There are different types of two-dimensional diagrams. They are as follows:

1. *Rectangular Diagrams*: In such diagrams, the shape is rectangular. The height of the rectangle varies according to the ratio, which the data bear to each other.

2. *Square Diagram*: Square diagrams give a better comparison of data than rectangles or bars. They are generally drawn when some items of the series have values much higher than others. In such cases, if a bar diagram is drawn, then the bars representing large figures would be very big in size while the bars representing the small figures would be very small. While constructing square diagrams, the first step is to find out the square root of various figures and then squares are drawn with the lengths of their sides in the same proportion as the square roots of the original figures. The area of the

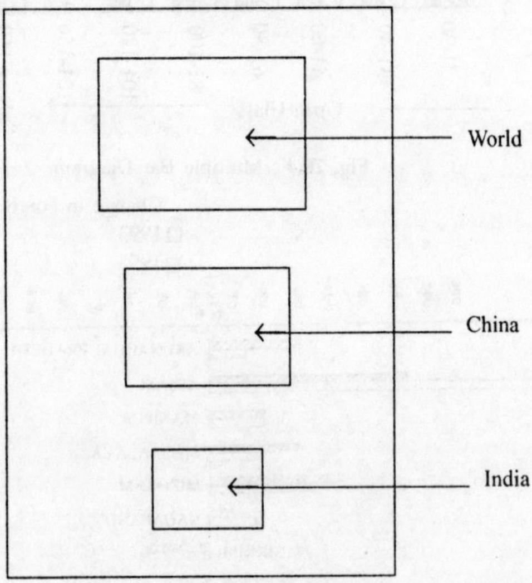

Scale — 1 sq. cm = 5 Hectare

Two Dimensional (Square) Diagram

Fig. 20.6 : Forest Area per Capita

squares would be in the same proportion as the ratio of original figures (Fig. 20.6).

3. *Circle or Pie Diagrams*: Circles are very important and popular two-dimensional diagrams. The area of the circle is directly proportional to the square of its radius. Therefore, if the radius of a circle were four times the radius of another circle, its area would be sixteen times the area of the other circle. In other words, in case of circle their radii are kept in the ratio of the squares of their radii. Data can be represented in the forms of circle in all those situations where square diagrams are used. (Fig. 20.7).

Quality wise Bomboo Forest of Rajnandgaon DN. in Hact.

Index

Quality	Area in Hact	Symbol
I	21459.507	
II	30935.935	
III	9103.849	

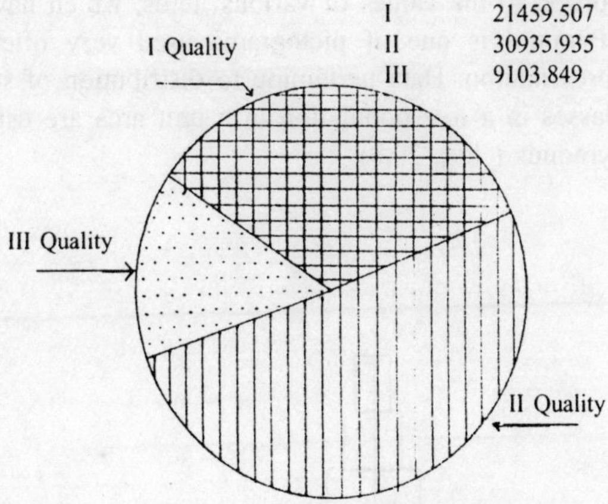

Fig. 20.7 : Pie Diagram

In case of circles, the aggregates can also be represented by a big circle and the various components by sectors is cut in side it. Such diagrams are known as Angular Diagrams. The areas of various sectors or parts are in proportion to the angle, which they make at the centre of the circle. The total angle of 360 degrees can be divided into a number of smaller angles whose degrees would be in proportion to the values of the components.

When two aggregates and their components are to be compared, two circles have to be drawn with areas proportionate to their ratio of the two aggregates. Each circle would then be divided into sectors on the basis of values of its components.

3. *Three-Dimensional Diagrams*

Three-dimensional diagrams are used to project the difference in ratios between two sets of data, more prominently. In such diagrams length, breadth, and height are considered. Therefore, these diagrams are called volume diagrams. Cylinders, spheres, cubes, etc. are some of the three dimensional diagrams. Population in different states, forest areas in different states or districts can be shown by using such diagrams. A comparison of growing stock of different forest types in a state can also be represented by using such three-dimensional diagrams.

1. *Pictograms*: Pictograms are diagrams where the value of items is represented by picture. The size of different number of pictures drawn is in proportion to the values of various items, which have to be represented. Pyramid is one of pictograms used very often in technical data representation. Data pertaining to distribution of stems of various age classes in a tree population in a unit area are usually represented by pyramids (Fig. 20.8).

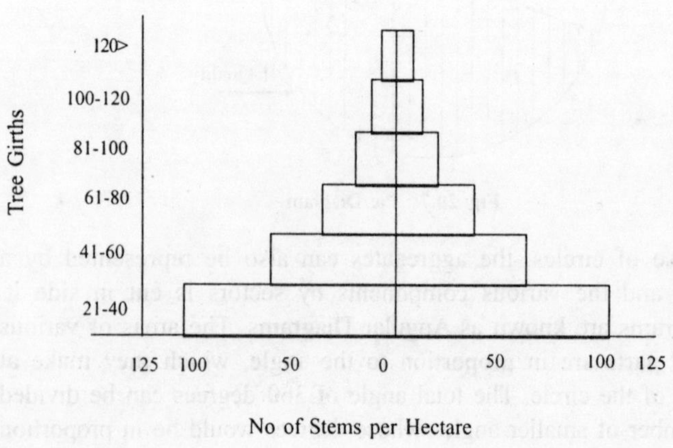

Fig. 20.8 : Pictogram (Stem Distribution per Hectare in the Shape of Pyramids).

2. *Cartograms*: It is a type of diagram where the distribution of items is represented in the form of dots over a map. The items like distribution of rainfall, population density, etc. can be shown on a piece of map by putting down dots representing rainfall or number of people. In a forest division, the grazing pressure or fire incidence can be represented on an indicative map or vegetational map by using cartograms. The population density in a district or a forest division with range boundaries can also be shown on a map by adopting cartogram methods. Likewise, distribution of resources of non-timber products can be shown on a divisional map (Fig. 20.9).

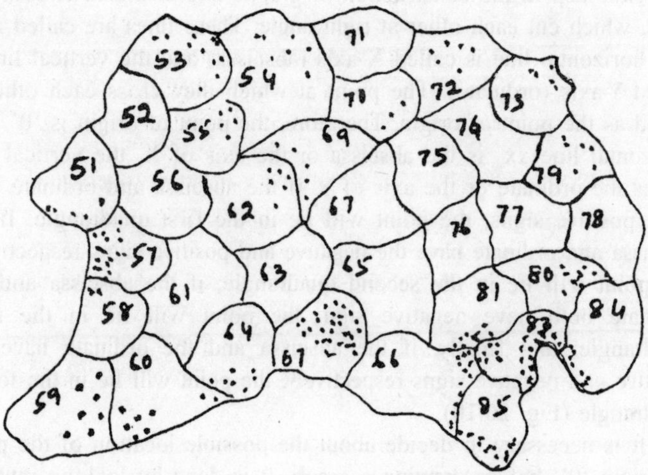

Fig. 20.9 : Cartogram (Fire Incidences in a Range).

From the above discussion, it is clear that there are enormous possibilities for effective representation of data. However, in order to convey a desired impression it is necessary to select proper diagram. Otherwise, the diagram will not make the desired impact on the mind of readers. Beside this, while constructing various diagrams and pictures, due attention should be paid to the aspects of simplicity, neatness and display.

Graphical Presentation of Data

Data in the form of diagrams are easily grasped. It leaves a lasting impression on the mind of readers. But diagrams have their own limitation. They only give an approximate and rough idea about the phenomenon. In this backdrop, graphs and charts have better applicability in scientific fields. Graphs are generally used to study the relationship between two variables. What is more significant is the fact that graphs are accurate and precise. Graphs are generally useful for studying time series and frequency distribution.

Construction of Graphs

The first step in the construction of graphs is drawing of two simple lines, which cut each other at right angle. These lines are called axis. The horizontal line is called X-axis (abscissa) and the vertical line is called Y-axis (ordinate). The point at which they cross each other is called as the point of origin. Therefore, the point of origin is '0'. The horizontal line xx′ is the abscissa or the axis of X, the vertical line yy′ is the ordinate or the axis of Y. If the abscissa and ordinate both have positive signs, the point will lie in the first quadrangle. If the abscissa and ordinate have the negative and positive signs respectively, the point will lie in the second quadrangle; if the abscissa and the ordinate both have negative sign, the point will lie in the third quadrangle; and finally, if the abscissa and the ordinate have the positive and negative signs respectively, the point will lie in the fourth quadrangle (Fig. 20.10).

It is necessary to decide about the possible location of the point of origin '0', before drawing a graph. It is done by looking into the negative or positive nature of data or numerical. If the quantities or data to be plotted are all negative, it is required to place the point of origin '0' along the left-hand margin of the graph paper.

It is an usual practice to put the values of independent variable on the axis of X, i.e. if a time series is required to be represented, the time will be measured on the abscissa and the other variable on the ordinate. In case of a frequency distribution, the measurements or data will be shown on the abscissa (X-axis).

Choice of scale plays an important role in construction of graphs. The scale should be chosen in such a way that it should accommodate the whole data. However, no rules can be laid down regarding the correct selection of scale. The vertical and horizontal scales are

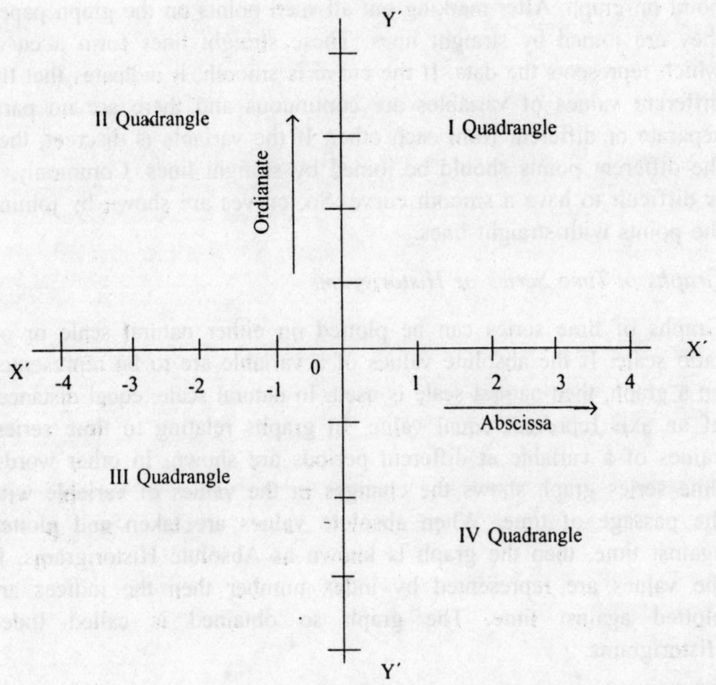

Fig. 20.10 : Graph Construction.

divided at point's equivalent to each other. Generally, they are equivalent to some multiples of 5 to 10. It is always necessary to put figures or units at short intervals. This makes the scale neat and clear.

After deciding and putting a scale, the points are plotted on the graph paper. As we have already learnt, a point on the graph paper is a function of two variable measurements. Therefore, plot a point on the X-axis and then calculate the corresponding measurement on the Y-axis and then draw a perpendicular line from the said point on the X-axis and another perpendicular line from the said point on Y-axis. The point at which these two perpendiculars meet will be the point on graph. After marking out all such points on the graph paper, they are joined by straight lines. These straight lines form a curve which represents the data. If the curve is smooth, it indicates that the different values of variables are continuous and there are no parts separate or different from each other. If the variable is discreet, then the different points should be joined by straight lines. Commonly, it is difficult to have a smooth curve. So, curves are shown by joining the points with straight lines.

Graphs of Time Series or Historigrams

Graphs of time series can be plotted on either natural scale or on ratio scale. If the absolute values of a variable are to be represented on a graph, then natural scale is used. In natural scale, equal distances of an axis represent equal value. In graphs relating to time series, values of a variable at different periods are shown. In other words, time series graph shows the changes in the values of variable with the passage of time. When absolute values are taken and plotted against time, then the graph is known as Absolute Historigrams. If the values are represented by index number then the indices are plotted against time. The graph so obtained is called Index Historigrams.

Absolute Histograms - With One Variable

The graph indicates change of one variable over a span of time. The quantity of timber produced in a forest division in the year 1990 to 1997 is given in Table 20.2.

Table 20.2 : Production of Timer from 1990 to 1997 in a Forest Division (in 000 CMT)

S. No.	Name of the Year	Quantity of timber in cubic mt.
1.	1990	31500
2.	1991	28200
3.	1992	21100
4.	1993	16200
5.	1994	15100
6.	1995	15700
7	1996	15000
8	1997	14800

The graph is plotted in Fig. 20.11. From the said graph, the production of timber in each year can be known. The graph also indicates the changes in the production from year 1990 to year 1997. There is another graph showing number of trees in a unit area, presented in Fig. 20.12. Therefore, absolute historigrams tell us about the changes in real values or quantity. The index historigrams are also used to indicate the relative changes. Such changes generally expressed in terms of index or percentage. In the index historigrams, the production in the base year will be taken as 100 and in relation to it, the production figure in subsequent years in the form of indices are plotted which results into a graph. This graph is called index historigrams (Fig. 20.13).

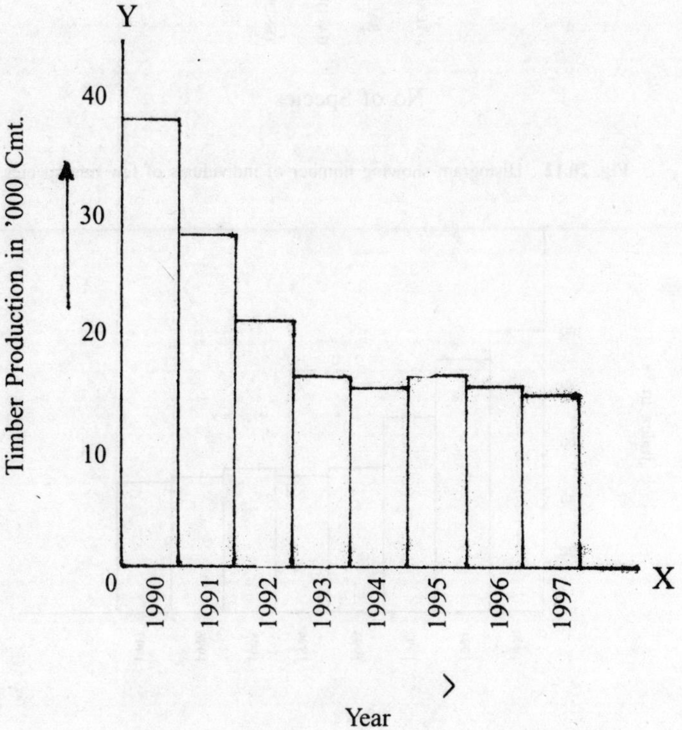

Fig. 20.11 : Absolute Histogram.

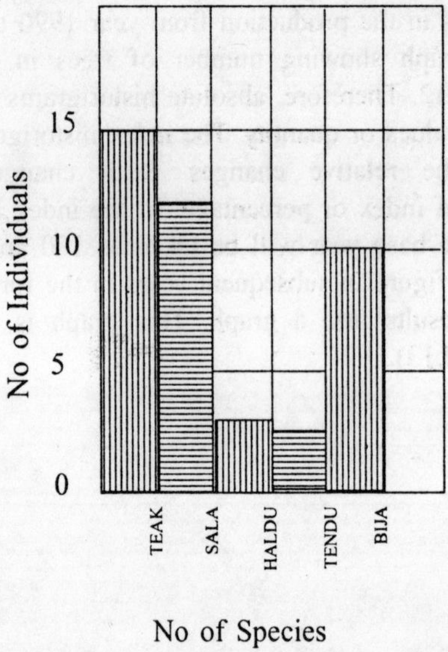

Fig. 20.12 : Histogram showing number of individuals of few tree species.

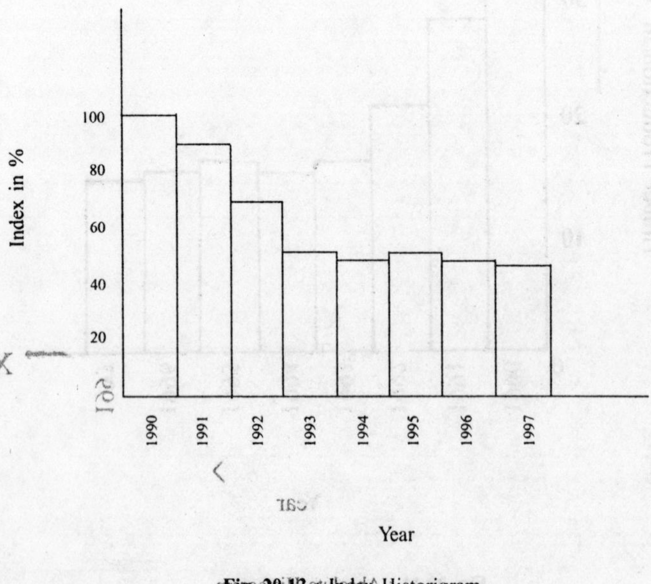

Fig. 20.13 : Index Historigram.

The graph discussed above was a historigram with one variable. However two or more variables can be shown on a graph. These variables are expressed in the same unit. In such case, both the horizontal and vertical scales are common for all the variables. The procedure of construction of such graphs is same as that of the previous one. Here two curves will emerge. The advantage in having two or more curves simultaneously on a single graph is that we can show the difference of two related variables or the sum of two related variables by plotting another curve along the original curve in the same sheet of graph. Such graphs are used generally to indicate the magnitude of difference between two variables over a time series. For example, the income and expenditure of a forest division can be shown by plotting two different curves. A third curve adjoining the original curve can be drawn to show the difference between income and expenditure.

When the units in which two different variables are expressed differently, in such case two or more scale of different units have to be shown in the same graph separately. Here one variable can be shown on the vertical scale of the left-hand side and another variable on the vertical scale of the right-hand side of the graph sheet. In such cases, chosen scales are adjusted to provide a good comparative picture of the different aspects of a same phenomenon.

Such graphs have wide relevance to presentation of different aspects of forestry. Variation in production of forest produces in different years, revenue-expenditure and their differences can be graphically represented by resorting to such graphs.

Range Zone Graph

Zone graph shows the range of variation. Range zone graphs are used in such situations where the variation between maximum and minimum values of a given set of variables are required to be projected prominently and distinctly. However, a false base line is taken in case where the difference between maxima and minima is not very much (Fig. 20.14). On the contrary, where the difference is large, there is no need to take a false base line. In Fig. 20.15, the minimum and maximum values with respect to sale of teak wood per cubic metre in different years have been plotted. The difference between the maximum and minimum price values with respect to each year has been made prominent by drawing thin bars between these values. The size of such bars indicates the range of variation in teak

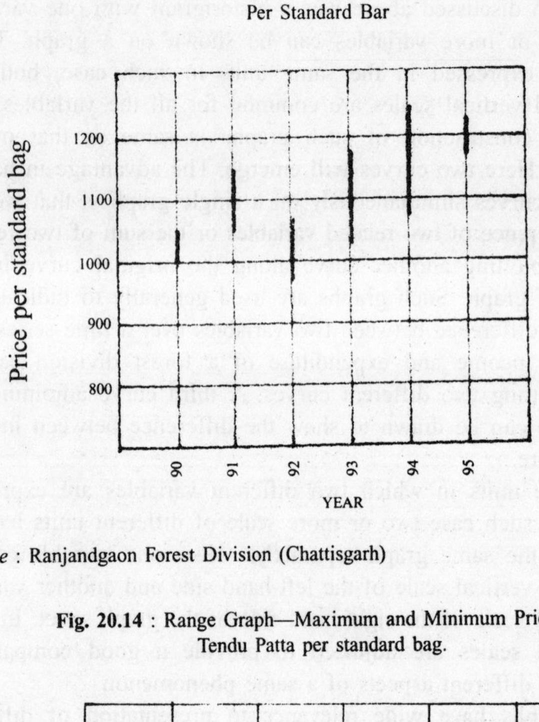

Source : Ranjandgaon Forest Division (Chattisgarh)

Fig. 20.14 : Range Graph—Maximum and Minimum Prices of Tendu Patta per standard bag.

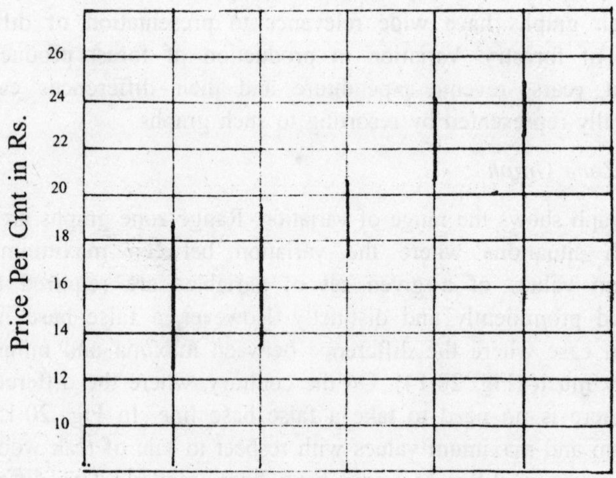

Source : Chichari Depot of Rajnandgaon forest division (Chattisgarh) Year : 1992-96

Fig. 20.15 : Maximum-Minimum Value Graph. Highest and Lowest Prices of Teak Wood per cmt.

wood prices in different years. The data regarding price variation over a period of five years is presented in Table 20.3.

The data relating to maximum and minimum can also be represented by drawing two historigrams. One historigrams shall represent the maximum values and the other the minimum values. The space between these two curves represents the range of variation. Sometime the area between these two historigrams are shaded to underline the prominence of variation.

Table : 20.3 : Maximum and Minimum Price of Teak Wood (Grade III) per cu.mt.

S. No.	Year	Maximum prices of timber in Rs.	Minimum prices of timber in Rs.
1	1992	18500	12400
2	1993	19400	13700
3	1994	20500	18700
4	1995	24100	20300
5	1996	24600	22600

Source: Chichari sale depot of Rajnandgaon forest division. (Chhattisgarh)

In forest planning and management, such graphs are used to show the rainfall, humidity and other aspects of meteorological attributes of an area. The price range of different forest produces can also be represented distinctively by using such range graphs.

Balance Graphs

Balance graphs are used in such situation where it is required to show two sets of variables and balances on a point of time. In such case, a historigram can also be resorted to, but things are better represented in the form of balance graphs. As for example, the revenue, expenditure and balance position of a forest division, either in the form of deficit or surplus can be represented in such balance graphs (Fig. 20.16). The data related to the graph is given in Table 20.4.

Band Graphs

Band graph is a type of line graph used to indicate the total quantities over a successive time period, but the total quantities are broken up into various component parts of the total quantities. Various component parts are plotted over one another. There will be as many bands as the number of component parts. This type of graph is useful

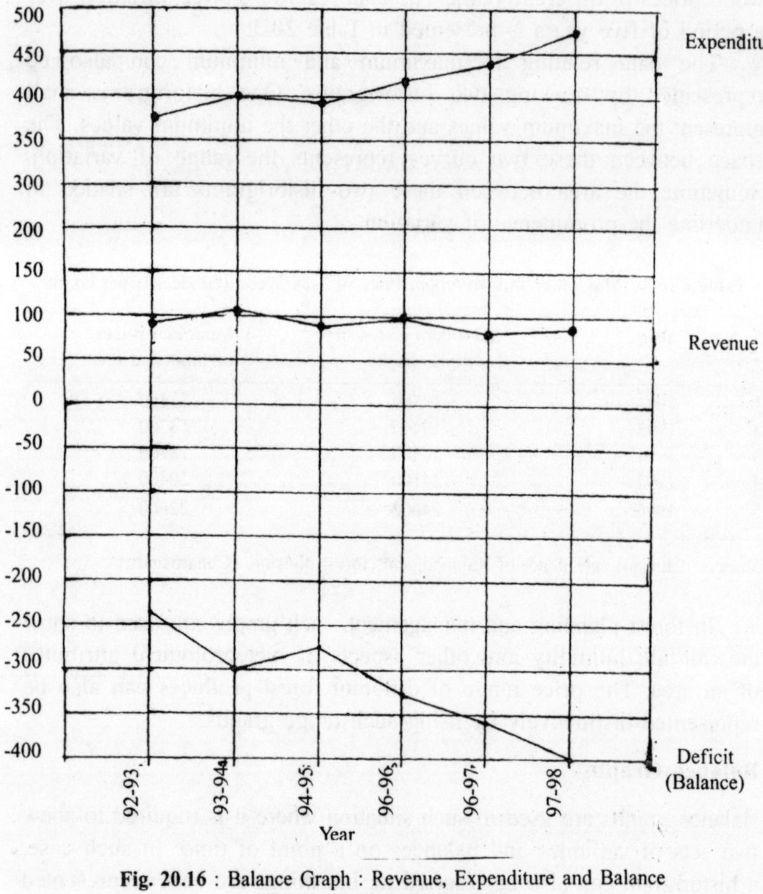

Fig. 20.16 : Balance Graph : Revenue, Expenditure and Balance Position of a Forest Division.

Table 20.4 : Revenue and expenditure of a Forest Division

S.No.	Year	Revenue in Lakh (Rs.)	Expenditure in Lakh (Rs.)	Surplus or Dificit in Lakh (Rs.)
1.	1992-93	92	337	-245
2.	1993-94	103	402	-299
3.	1994-95	92	371	-279
4.	1995-96	100	420	-320
5.	1996-97	80	436	-356
6.	1997-98	85	476	-391

for studying total cost on plantation divided into various component parts like pre-planting, planting, post planting or other operations. This graph can also be used in studying in total cost of harvesting divided into different components like felling, logging, transporting and sale. As already discussed, such graph indicates the trend as well as of its components. In such graphs percentages of the actual figures can also be plotted to make such a band graph.

Zee Chart or Z Graph

Z graphs are a type of graphs which assumes the shape of the alphabet Z. Generally, such graphs show the entire picture over a long period. It has very limited general use in presentation of forestry data.

Graphs of Frequency Distribution

Frequency distributions are represented by means of graphs. Such graphs are called as frequency graphs. There are two types of frequency distribution graphs. They are as follows:

- Discrete series.
- Continuous series.

Discrete series is one in which an item cannot assume any value in a particular class interval. The value of each item is fixed and definite. A continuous series is one in which item can assume any value in a particular class interval.

1. *Discrete Series Graphs*

Such graphs are indicated by line or bar frequency diagrams. To draw such graphs, first the size of items is taken on the vertical scale. The data related to number of encroachment sites in a forest division are graphically represented in Fig. 20.17 and Fig. 20.18. The lines showing the frequencies may be made more prominent by making the lines thicker and darker. The data on which the graph has been drawn is given in Table 20.5.

Table 20.5 : Number of Encroachment sites in Forest Compartments

Number of encroachment sites	No. of forest compartments
1	270
2	165
3	110
4	75
5	45

2. *Continuous Series Graph*

Continuous frequency curves are used to depict such data which gives continuous series and in which the frequency can assume value of the variable but it always remains within a particular class limit. In such case the class intervals are of uniform size. The data given in Table 20.6 pertains to stem distribution of trees in a unit area.

Table 20.6 : Stem distribution in a Unit Area

Girth class (in cm)	No. of trees or individual
21-30	76
31-40	57
41-50	45
51-60	35
61-70	25
71-80	20
81-90	15
91-100	7
101-110	5
111-120	4
121>	3

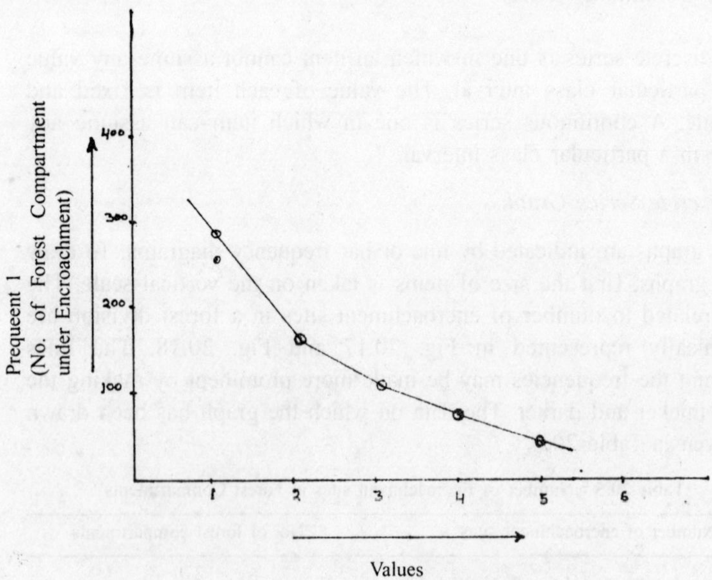

Fig. 20.17 : Values of Variables and their Frequency.

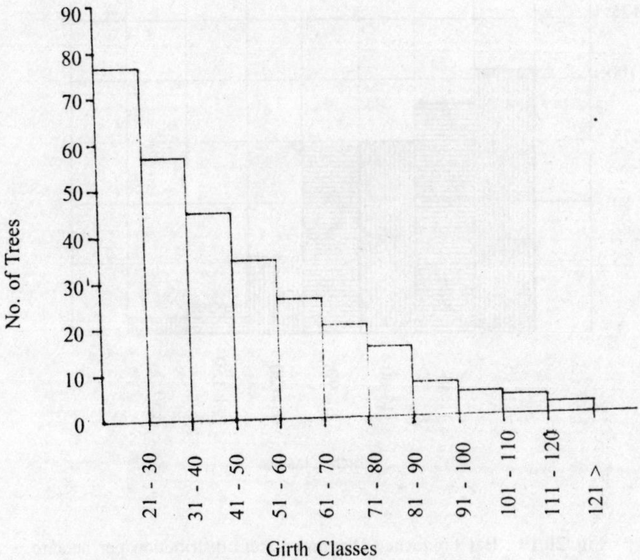

Fig. 20.18 : Bar Frequency Diagram : Stem Distribution Per Hactare in Teak Forest

The above data are either plotted in the form of frequency curve or frequency polygon or bar frequency diagram (Fig. 20.19 and Fig. 20.20). In drawing a frequency polygon, various bars are joined together and no space is left between two bars. This provides an impression of continuity of distribution of stems. Frequency curve can also be drawn on the basis of frequency polygon. This is possible when the mid-points at the top of each bar are joined together by straight lines. However, frequency curve can also be drawn on the basis of data separately. Generally, frequency curves are made smooth and regular by removing the angularities. However, it is necessary to take precaution against distortion of original shape of the curve in the name of smoothness. In a case of extremely irregular forest, the frequency curve indicating stem distribution cannot be smothered beyond a limit, otherwise, the nature of distribution of trees in different girth classes will not be properly represented. In such situations, only micro angularities can be removed to bring the curve into a presentable form.

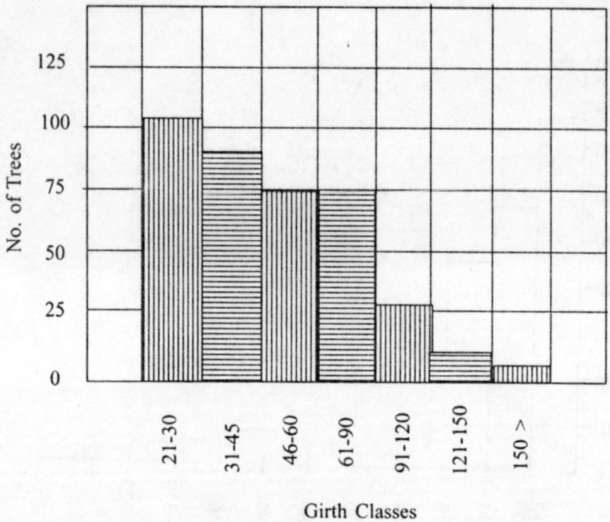

Fig. 20.19 : Bar Frequency Diagram : Stem distribution per hectare in selection-cum-improvement WC.

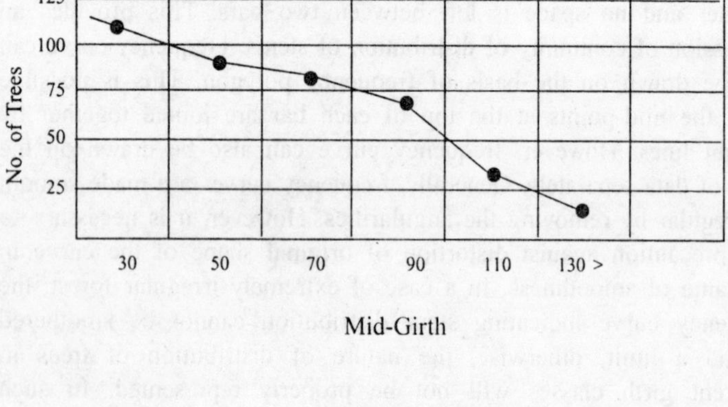

Fig. 20.20 : Frequency Polygon : Stem distribution per hectare in selection-cum-improvement WC.

DATA INTERPRETATION

After collecting and analyzing data, the working plan officer has to accomplish the task of drawing inferences followed by report writing. This has to be done very carefully, otherwise misleading conclusions may be drawn and the whole purpose of plan preparation will go haywire. It is only through interpretation that the planner can reveal relationship and process that underlie his inferences. Now, the question is what does interpretation means. Interpretation refers to task of drawing inferences from the collected facts after an analysis. It is central to any planning process. It is the most difficult task in the whole planning process. It requires great expertise and experiences on the part of planning officer. Generally, an officer who learns through continuous practice and experience over years puts his knowledge and skill in the critical process of interpretation. The technique of interpretation involves the following steps.

1. The planning officer should provide reasonable logic and explanation to the relation he finds with respect to any change in the forest resource status like change in growing stock, change in density (area-wise), forest composition, change in bio-diversity indices, management status, etc.

2. Extraneous information collected during exercise should be considered and examined along with the collected data during the course of field study.

3. It is always prudent to consult senior officers/academicians who have sufficient insight into the changing phenomena in forests. This helps in improving logical development of arguments that come in the course of analysis of a problem.

21

Presentational Organization of Working Plan (Management Plan) Reports

Introduction

Reports are the major products of any kind of study. The revision work of management plan remains incomplete without preparation and submission of reports to review committees for discussion and necessary action. Therefore, report writing is an integral part of the revision process of any plan. The result of the revision works is presented in the plan report. The reports are organized on the basis of its objective and types of field works. The organization of forest management or working plan reports differs from state to state. However, there is lot of uniformity in the organization of different chapters in the plan documents all over India. The writer has presented the pattern of organization followed in the forest department of Madhya Pradesh.

In the working plan revision process, the following reports are generally prepared at different stages of its preparation:

1. Basic Information Report (BIR).
2. First Preliminary (Working Plan) Report.
3. Second Preliminary (Working Plan) Report.
4. Final Working Plan Report.

Basic Information Report

Before commencing the working plan revision, the forest division is required to gather basic information. The divisional forest officer of the concerned territorial division prepares the Basic Information Report (BIR). This report is presented to the working plan officer for initiating the preliminary revision work of the forest division.

1. *Organization of Basic Information Report* (BIR)

The report contains information as given below:

1. *Name and situation of the division*: Name of the forest division and geographical position and area of the division should be given.

2. *Administrative units*: The information regarding administrative units like division, range and sub-range is provided. These informations are: -

(i) Name of division and its head quarter.

(ii) Notification number and date.

(iii) Head quarter.

A detailed list containing names of sub-divisions and their head quarters with notification number and date, names of the ranges and their head quarters with notification number and date, name of range-assistant circles, their head quarters with name and area of forest beats is prepared and provided in the report.

3. *Present status of Management :* The working plan, which is under-operation, needs to be mentioned here. The name of the working plan officer that prepared the plan should be indicated. Besides, the duration of the plan, extension given with details of government order and the area included in the plan/plans should be mentioned. If the government or any other organization has put any restriction with respect to management practices prescribed in the plan, then a detailed note should be given.

4. *Position of forest area :* The legal position of forest area like reserved forest, protected forest, and un-classified forests should be provided separately.

5. *Forest Area Registers* : The position of the forest area register should be mentioned. It is required to indicate whether the area register has been properly updated with area details along with subsequent notification or de-notification of different forest blocks. It should be verified to ensure the correct entry of the notification numbers and date. Before handing over the forest area register to the working plan officer, the divisional forest officer (Territorial) should ensure, that the said register is upto date and complete in all respects.

6. *Details of Forest Compartments and Blocks:* The information regarding the forest area is prepared with respect to each forest block.

Therefore, the information is prepared as shown below:

(1) Legal status of RF/PF, (2). Name of Forest block, (3) Name of compartments included in the block, (4) Notified area of the block in hectares, (5) Section of IFA-1927 under which notification issued and (6) Notification number and date.

7. *Position of forest area denotified or transferred to other department* : The following information should be collected.

1. Encroachment area settled and notified by the government.
2. Details of area transferred to other departments and notification by the government.
3. Area temporarily transferred for non-forestry operation.

Details of such areas with respect to compartment, block, date of transfer, notification number should be provided.

8. *Position of area received from other departments* : Details of area, and their notification number, date should be given in this part of the report. If compensatory afforestation has been done on any such area, then, details of plantation, year, species planted, should be provided.

9. *State of forest boundaries and pillars* : Under this head, the information regarding forest boundary should be given as fallows:

1. Length of forest boundary Natural/Artificial and Internal and External.
2. State of forest boundary.
3. Total number of boundary pillars and their condition (Pacca, Kacha, Damaged/Absent).
4. Total amount of budget received in the last five years and the position of financial and physical achievement.

10. *Position of maps* : The position of divisional maps should be clearly stated. Details of maps with respect to different types of forest maps like stock map, management map, and topographic sheets of Survey of India (Scale 1:50000), available in the division should be found out and an inventory should be prepared.

11. *Implementation of existing working plan and evaluation of its effect on the forests* : It deals with the implementation of prescriptions under different working circles and evaluation of their impact on the forest. Therefore, first the position of implementation

should be given in the form of number of felling series and number of coupes worked under different working circles. Second, the treatments, which could not be carried out in a working circle, should be indicated under every working circle. A list containing coupes written off and coupes not undertaken for treatment should be prepared.

12. *Plantation and Rehabilitation of degraded forests:* A detailed list of plantations and rehabilitation works undertaken in the last plan period should be prepared. The following details with respect to each plantation site should be provided:

1. Name of plantation sites.
2. Year of plantation.
3. Name of the scheme/project.
4. Area of plantation sites.
5. Name of the species and number of plants planted.
6. Name or the range, compartment, working circle, felling series and coup number.
7. Status of present plantation (Growth and survival percentage).
8. Availability of up-to-date plantation journals.

13. *Development work* : A detailed account of developmental works undertaken in the forest division during the last plan period should be prepared. Such developmental works are done by different agencies. Information regarding such works is available in the District Rural Development Agencies (DRDA). Therefore, details of developmental activities/works like road, buildings, agriculture, irrigation, soil conservation, drinking water supply, education, health facilities etc. should be collected.

14. *Wildlife conservation* : This part requires information regarding National Park, sanctuaries, or protected areas. Apart from this, name of the wild animals found in the division and their number (census) should be given. An account of extent of damage made by wild animals to crop and livestock in the vicinity and compensation paid for it should be prepared. An inventory of rare and endangered wild animals and plants including tree species should be prepared along with the description of the habitat.

15. *Forest and revenue village* : Forest and revenue villages are found in the forest division area. Name of such forest villages and revenue villages, which are in proximity to the forest (normally with

five km. distance), should be listed. Socio-economic profiles of these villages should be prepared.

16. *Forest produce* : An account of forest produces of the division in the last plan period should be prepared. The items generally included are timber, fuel-wood, poles, bamboos, other woody produces and all other minor forest produces. The production should be expressed in terms of prevalent measuring units. Quantity of production and the revenue collected should be indicated.

17. *Forest concession and privileges* : A detailed account of practices relating to concessions and privileges accorded by the state government should be prepared. It should encompass name of all nistar and consumer depot and prices of forest produces in the last plan period.

18. *Nursery* : The location of different nurseries available in the division along with other details like capacity, suitability, performance in the last five years should be given.

19. *Forest offences and biotic pressures* : This constitutes important part of the report. Detail account of forest offences like illicit felling, illicit transportation, poaching etc. should be prepared. The position of forest offences and their status of investigation should be stated clearly rangewise. The state of fire protection should also be included in the report. An analysis with respect to fire protection in the last five years should be made. For this, the actual requirement of fire protection work, and budget availability should be taken up for detailed analysis. The grazing pressure should also be determined by analysing the records and the field conditions. The practice of grazing passes and its applicability at present should be analysed.

20. *Miscellaneous Information* : Leaving aside the items mentioned above, all other activities having relevance to forest management should be mentioned in the report. The subject of forest road, forest rest houses, location of sample plots, preservation plot, seed production area should be discussed.

21. *Control Register* : Status of following control register should be described:

1. Compartment history.
2. Area Register.
3. Coup Control Register.
4. Forest divisional book.
5. Others.

First Preliminary (Working Plan) Report

Introduction

First Preliminary report is the first report submitted by the working plan officer to the review committee set up for revision of the working plan. This report is generally submitted before initiating field work. Normally, a working plan officer takes one to three months to prepare and submit this report. This report is primarily based on the Basic Information Report (BIR) of the concerned forest division and observation of ground conditions and preliminary assessment of the working plan officer with respect to effects of last plan on forest and the existing and emergent problems of forest management in the said division.

Objective of the Reports

The objective of this report are as fellows:

1. To determine and delineate forest areas for inclusion in the working plan.
2. To undertake monitoring and evaluation of the effects of treatments meted out to forests during the period of last plan.
3. To decide about the field studies required to hammer out strategy for management practices and their methodology.
4. To determine the institutional frame work and infrastructure facilities necessary for carrying out the working plan exercise.

Basis of Report

The report, as said earlier, is based on the following aspects:

1. Basic Information Report (BIR).
2. Reconnaissance survey of forest and observation of ground conditions by the working plan officer and assessment made by him and his team.

After having been posted, the working plan officer generally go through the last working plan thoroughly and examine the progress provided by the BIR furnished by the division critically. After studying the prescriptions of the last plan and its detailed implemention results, the working plan officer undertakes a reconnaissance survey of the forests. During the course of his visit to the forests, he crystallizes his observation and makes notes in the end of the day at the camping site. On the basis of data and observation, he makes a critical assessment of the forest resources and management practices.

Organization of First Preliminary Report of Working Plan

1. *Introduction :*

   1. Details regarding name of working plan officer, date of posting of working plan officer, date of joining.
   2. Details about the last plan like name of the working plan officer, period of plan, extension if given.

2. *Working plan area*

   1. Geographical position. Longitude and latitude with map (1: 250000 scale).
   2. Forest Area:
      (i) Legal statute. (RF/PF/Unclassed forests).
      (ii) Area covered under last plan: Number of blocks and area in hectares.
      (iii) Area reduced or increased during the last plan (Number of blocks and area).
      (iv) Present position.

3. *Forest area based on management types.*

4. *State of forest block boundaries and pillars :*

   (i) Length of boundaries (Artificial/Naturai).
   (ii) No of pillars.
   (iii) No of pillars constructed and repaired during the last plan period.

5. *Encroachment*:

   1. Area shown under encroachment.
   2. Encroached area settled legally during the plan period; notification number.
   3. Status of present unsettled encroachment.

6. *Classification of forests (based on Champion and Seth)* and evaluation of past management :

Details of treatment given to different working circles under the existing/last plan. Information is provided in two formats.

Format No. 1 : Details of work done (Progress monitoring):

The following items constitute the column headings:

1. Name of the working circle.
2. Number of felling series and areas.
3. Number of coupes and areas.
4. Total area of the working circle.
5. No. of felling series.
6. No. of coupes worked.
7. Area treated.
8. General comment.

Format No. 2 : Evaluation :

The following items constitute the column headings:

1. Name of working circle.
2. Details of treatment prescribed.
3. Weather implemented or not.
4. If no, give reason.
5. If yes, enumerate effects of treatment on the crop.

7. *Result of Joint Forest Management:*

Details of area allotted under Joint Forest Management (JFM), works done, and their critical evaluation.

8. *Concession and privileges:*

Their effects on forests and responses of people.

9. *Forest villages*:

Number of villages, areas, works undertaken during the plan period, population trend, involvement and participation level in JFM.

10. *State of important records:*

The condition of the following records is examined:

1. Basic Information report received from territorial DFO.
2. Area register.
3. Coup control form.
4. Plantation and Nursery journal.

11. *Re-numbering of compartments, if required.*

12. *Maps:*

The position of availability of different maps in the division and the requirement be indicated. If fresh maps have to be prepared or maps have to be enlarged, then the details regarding the works mentioning about the time and money required be shown.

13. *Requirement of basic infrastructure for undertaking revision mark:*

The requirement with respect to staff, building, vehicle, store articles, survey appliances should be estimated and shown with the position of present availability. Apart from this, other constraints and problems, which are expected to come up during the course of field works, should be discussed.

14. *Proposed field work and their methodology:*

The field works like, pilot survey, inventorying, stock mapping, stem and stump analysis, bio-diversity study, soil resource study, JFM case studies and other field studies which are required to generate information should be mentioned.

15. *State of bio-diversity and wildlife management:*

Information relating to bio-diversity, wildlife conservation in different habitats of the forest division should be gathered and analyzed. Based on this, an assessment should be made about the status of bio-diversity and wildlife conservation and their prospects.

16. Forest extension programme implemented in the plan period should be evaluated and an assessment should be proposed.

17. *Timetables for different field and office works:*

A detailed timetable should be proposed with respect to completion of different field and office activities. Apart from this, the probable date of submission for second preliminary should be mentioned.

Second Preliminary (Working Plan) Report

This is a report prepared by the working plan officer after analysing and interpreting the data collected from primary and secondary sources. Generally, it takes twelve to fifteen months from the date of submission of first preliminary report. At this stage, the working plan officer is in a position to visualise the problems and constraints of management. Based on his assessment and evaluation of last plan, he presents a strategy for sustainable forest management of the forest division.

The second preliminary report, which contains analysis, evaluation and future strategy, is submitted and presented to a review committee. The review committee consists of senior officers of the state forest department as well as Ministry of Forest and Environment, Government of India. The committee discusses different aspects of proposed management practices in the context of problems and constraints. The committee after a long deliberation draws up the broad strategy necessary for evolving management plan prescriptions.

Organization of Second Preliminary (Working Plan) Report

The contents of the plan document are organised in two parts.

Part I

1. *Introduction*

   1.1 Introduction

   1.2 Action taken report with respect to points raised in the review meeting of first preliminary meeting.

2. *Working Plan Area:*

Details of plan area should be given as in the format below:

(A) Classification of area based on legal status.

Legal Status	No. of forest blocks	No. of compartments	Area in hectares	
			Notified	Plannimetered
RF				
PF				
Unclassified Forest				

(B) Classification of area based on difference between notified area and plannimetered area.

This information should be given in detail.

(C) Classification of area on the basis of terrain.

1. Up to 10 per cent slope
2. 10 to 30 per cent slope
3. 30 to 40 per cent slope
4. 40 per cent and above slope.

3. *Details of Forest*

1. Classification of forests based on Champion and Seth.
2. Types of forests based on management, *e.g.* Teak, Sal, Miscellaneous forests.
3. Species composition in different forest types.
4. Bio-diversity.
5. Wildlife.
6. Non-wood forest products.

4. *Forests Resource Survey*

1. Results of Pilot Survey.
2. Methodology adopted for resource survey.
3. Analysis of results of forest resource survey.
4. Applicability of the analysis in formulating the proposed management.
5. Resource status of different working circles of the last plan. This status should be with respect to number of trees and volume per hectare in different girth classes and other parameters.

5. *Status of Natural Regeneration*

Information about regeneration of coppice as well as seed origin seedlings of different important tree species should be provided. Such information should be arranged as per old and newly proposed management or working circlewise. Regeneration status of other NWFP species should also be provided.

6. *Statistics of Growth and Yield*

The following information should be provided:

1. Growth data of different species in different sample plots maintained by Forest Research Institute, data collected from stem and stump analysis, secondary growth data of species found in the division.

2. Growth data of plantation undertaken in the division.
3. Production data of timber and non-timber forest produce of the last plan period.

7. *Socio-economic Survey Results and Requirement of Forest Produce*

The following details will be provided:

1. Socio-economic profile of the forest divisions/district.
2. Methodology of survey adopted.
3. Assessment of anthropogenic pressure on the forests.
4. Requirement of forest produces.

If the division's production falls short of the requirement, then the alternative option of biomass production should be discussed.

8. *Results of Past Management* :

Results of the management practices undertaken during the last plan should be analyed and evaluated critically. This evaluation should be working circlewise. This exercise should encompass the following items:

1. Change in crop composition.
2. Change in site quality.
3. Change in crop and crown density.
4. Change in growing stock.
5. Change in crop condition and crop hygiene.
6. Effect on regeneration status.
7. Other aspects like soil and water conservation, regeneration work, plantation, rehabilitation of degraded forests, etc.

PART II

1. General objectives of management and strategy adopted.
2. Proposed Management Practices and Working Circles:

The management or working circles should contain the following information:

1. Name of proposed working circle.
2. Area of the working circle (Planimetered area).
3. Number of felling series.
4. Weather this working circle was existing in the last plan or not.

3. Details of individual working circle:

The following details should be provided with respect to individual management or working circle:

1. Specific object of management of working circles and their basis.
2. Forest types, composition, density, site quality, age class, and terrain of the area of the circle.
3. Name of proposed felling/treatment cycles and their basis of proposal.
4. Determination of exploitable girth.
5. Details of yield calculation formula and reason for its application should be given.
6. Proposed treatment method.
7. Forest produce and its computation.
8. Subsidiary silvicultural operation proposed.
9. Prescriptions regarding fire and general protection and control of grazing.

4. Proposed treatment for bio-diversity and wildlife.

5. Forest extension, availability of area, extension services, plantation models, social acceptability.

6. Joint Forest Management. Assessment of JFM related works. Determination of suitable area, and proposed strategy for further enlargement of the programme.

7. Appraisal of plantation results and prospects of all successful plantations. Identification of area, species selected and technique to be adopted in future.

8. Industries based on forest produces.

9. Eco-tourism.

10. Proposed seed production area, seed orchard, etc.

Final Working Plan Report

On the basis of decision of review committee, the management regime for the forests is finalized. Now, detail structures of the treatments along with logic should be elucidated. All these discussions and prescriptions take the shape of final report. The final report is organized in four parts. They are as follows:

First Part: Introduction: Basis of proposal.
Second Part: Future management: discussed and prescribed.
Third Part: Appendices relating to information of the division.
Fourth Part: Forest Resource Survey Report along with inventory information in the form of appendices.
Fifth Part: Forest Soil Resource Inventory.

Organization of Working Plan Report (Final).

Part I

Content
Preface/Introduction
Proposal
Location Map (1:2500000 Scale) of the final decision.

List of local, General Name and Botanical names of plant species and animal species.

Glossary of local and technical names used in working plan arranged alphabetically.

Meaning of technical words used in the plan.

Basis of Proposal : Facts on which Proposals are Based

CHAPTER 1 : Tract Dealt with :

1. Name and Situation.
2. Description of land extent of forest.
3. State of forest boundary.
4. Legal status of forests.
5. Rights and Concession.

CHAPTER 2 : Factors of locality to which the Forest is subjected:

1. Topography.
2. Geology, Rock and Soil.
3. Climate (with analysis).
4. Availability of water.
5. Fire.
6. Grazing.
7. Injury from man.

8. Natural factors like forest fire, wind, drought, flood, etc.
9. Erosion.
10. Injury from animals.
11. Injury from insects, pests and other diseases.
12. Injury from climbers, epiphytes etc.

CHAPTER 3 : Forest Resources:

1. General description of forest resources of the division.
2. Types of forests (According to Champion and Seth classification).
3. Types of forests (management types).
4. Composition of forests.
5. Valuable species.
6. Forest resources, survey and result.
7. Regeneration status of valuable species.
8. Harvesting of Minor Forest Produce (M.F.P.) and its results on forests.
9. Stock mapping and procedure adopted.
10. Comparison of forest area of the plan area with the estimate made by Forest Survey of India (FSI) in State of Forest Reports.

CHAPTER 4 : Statistics of Growth and Yield:

1. Results of stock-mapping and Resource Survey and comparison with the previous stock mapping and resource survey.
2. Growth data of important species and their graphical representation.
3. Stem and stump analysis, Procedure adopted (in brief) and results.

CHAPTER 5 : Bio-diversity and Wildlife:

1. Plant bio-diversity.
2. Animal bio-diversity.
3. Endangered and rare species: Identification and description of their habitat.
4. Natural habitat of wildlife.
5. Factors injurious to wildlife and their habitat.
6. Results of Bio-diversity Survey.

CHAPTER 6 : Demand and Supply of Forest Produces:

1. General demographic position of the division/district.
2. Socio-economic survey: Procedure and basic information.
3. Analysis of socio-economic information/data, primary and secondary.
4. Estimation of demand/supply/consumption of forest produces per family in rural and urban areas.
5. Estimation of present annual requirement of forest produces.
6. Estimation of raw materials required for forest based industries.
7. Estimation of requirement of forest produces for supply to other organization/department.
8. Position of demand and supply position of forest produces required for households (nistar).
9. Forest production trends and main markets and trade of forest produce.
10. Analysis of prices of forest produces— Then and now.

CHAPTER 7 : Forest Extension and Social Forestry:

1. Evaluation of forest extension and social forestry work in the last plan and its results.
2. Evaluation of plantation done in private sector and estimation of prospect for forest plantation in private sector.

CHAPTER 8 : Past System of Management: Terminal Analysis and Evaluation:

1. Short history of past system of management.
2. Description of working circles and treatment carried out in the current working plan.
3. Result of past management: Analysis and evaluation with respect to working circle and forests as a whole.

PART II

Future Management : Discussed and Prescribed

CHAPTER 1 : Basis of Proposal:

1. Policy environment and general objects of management.
2. Factors supposed to influence the objective.

3. Working circles: Basis of allotment of forest areas to different management or working circles.

CHAPTER 2 : Working Plans for Different Working Circles:

1. Objectives of working circle.
2. General constitution of working circle.
3. Method of treatment.
4. Treatment cycle.
5. Division of working circle areas into treatment series and coupes (and sequence of treatment operation).
6. Determination of exploitable girth and its basis.
7. Method of applying the treatment.
8. Subsidiary cultural operation.
9. Fire protection and grazing control.
10. Yield projection and its regulation.
11. Probable expenditure on different proposed treatments under the working circle.

CHAPTER 3 : Bio-diversity and Wildlife Management:

1. Objective.
2. Present state of management.
3. Proposed Bio-diversity management strategy.
4. Evaluation and monitoring. Methods: adopted and results.

CHAPTER 4 : Joint Forest Management:

1. Introduction and objective.
2. Government resolution.
3. Present status of participation.
4. Evaluation and monitoring.
5. Proposed strategy.

CHAPTER 5 : Forest extension and social forestry:

1. Introduction/objective.
2. Present state of social forestry.
3. Formulation of scheme.
4. Evaluation and monitoring.
5. Strategy.

CHAPTER 6 : Management of Non-wood Forest Products (NWFP):

1. Introduction/objective.
2. Present status of management.
3. Evaluation and monitoring.
4. Proposed strategy.

CHAPTER 7 : Miscellaneous Rules:

1. Maintenance of forest boundaries and pillars :
   It requires a discussion about the state of forest boundaries and pillars in the plan area. It is a convention to draw a proposal for a five-year demarcation and reparing of forest boundary and pillars.
2. Nistar (Supplying of bona fide needs of villager) :
   (i) Present policy of Nistar, (ii) Present status, (iii) Effect of Nistar on the forest, (iv) Proposed Nistar.
3. Control grazing :
   (1) Statutory provision with regard to grazing, (2) Present status of grazing in the forest and carrying capacity of the forests, (3) Effect of grazing on the forests, (4) Proposed strategy for control and regulation of grazing.
4. Fire protection :
   1. Evaluation of effects of fire on forest and regeneration.
   2. Fire protection plan.
   3. Delineation of fire lines and its maintenance.
   4. Integrated fire protection strategy.

   It will be a good work if fire prone areas, identified during the course of stock mapping, are marked on the map to prepare a fire map and measures are proposed to prevent and control fire.

5. Subsidiary Cultural Operation :
   Thinning and other subsidiary cultural operations including climber cutting are analysed. It is imperative to evaluate the extent to which the prescription of the plans have been given effect to and what suggestions are offered to give effect to the prescriptions in respect of this important work.
6. Nurseries :
   1. Status of existing nurseries.
   2. Selection of suitable species.

3. Seed collection and treatment.
4. Nursery technique.
5. Future strategy.

7. Plantation :

1. Status of old plantations and summary evaluation of their growth.

2. Plantation technique:
   - (i) Site selection.
   - (ii) Project report.
   - (iii) Identification of areas.
   - (iv) Clearing of area.
   - (v) Site preparation.
   - (vi) Planting and cultural operation.
   - (vii) Protection.

3. Preparation of various Plantation models and their details.
4. Pasture and Silvi-pasture development.
5. Assisted Natural Regeneration (ANR).
6. Irrigated and high-tech plantation: Prospect and possibilities.

8. Forest villages:
   1. Present status of forest villages.
   2. Basic socio-economic profile.
   3. Evaluation of developmental works carried in the last plan.
   4. Resource development of forest villages: Strategy.

9. Building:
   1. Availability of buildings in the plan area.
   2. Requirement of buildings.

10. Forest Road:
    1. Present status of forest road.
    2. Requirements of additional forest road.

11. Wild-life tourism or eco-tourism :
    1. Scope, prospects and strategy.

12. Sample plots, preservation plot, and seed production area:
    1. Present status.
    2. Proposed work.

13. Forest based industries:
    1. Present status.
    2. Prospects of new forest based industries.
    3. Proposed schemes.

14. Harvesting of forest produce:
    1. Major Forest Produce. Marking, felling, logging and extraction, transportation and disposal.
    2. Minor Forest Produce: Collection, transportation, and marketing.
15. Irregular exploitation:
    1. Present status.
    I. Fuel-wood and timber.
    II. Gregarious flowering of bamboos.
    III. Exploitation of forests/trees for research and education purposes.

16. De-reservation of forest land and their transfer.
17. Illicit felling:
    1. Present status. Including summary of forest offence at least for five years.
    2. Strategy to control the menace.
18. Encroachment:
    1. Present status. Forest area encroached after the commencement of last plan.
    2. Strategy to control.
19. Soil and water conservation:
    1. Present status of erosion in the different forest pockets of the plan area.
    2. Strategy for soil and water conservation.
    3. Preparation of models.
20. Water supplies, Public health services, reorganization of administrative units, maintenance of maps, preserved trees.

CHAPTER 8. Control and Records :

1. System of control.
2. Area register.
3. Control forms.
4. Compartment history.
5. Divisional Note Book.
6. Range Order Book.
7. Beat books and beat maps.
8. Plantation and nursery journals.
9. Record of rainfall and temperature.
10. Deviation statement-format.

CHAPTER 9. Establishment and labour :

1. Establishment.
2. Training.
3. Labour.
4. Labour welfare.

CHAPTER 10 : Financial forecast of the plan and expenditure incurred on the working plan preparation:

1. Financial Forecast.
2. Revenue estimate.
3. Expenditure estimate.
4. Cost incurred on preparation of plan.

CHAPTER 11 : Period of the plan:

The plan period should be mentioned in this section.

22

Elements of Technical Report Writing

Introduction

This chapter deals with basic elements of technical report writing. It aims at introducing foresters to different facets of professional and technical writing. It provides basic framework which helps them develop the necessary knowledge and skills to produce good, clear and effective writing. Emphasis has been given on such basic concepts, principles, skills which have strong bearing on technical report writing.

Writing is an important aspect of every professional forester. At every stage of his carrier, he is required to prepare technical reports. Such reports are prepared to explain the results of past management, analyse and solve problems related to administrative and technical works of the forest department. The basic objective of developing writing skill is to communicate effectively and confidently. This is one of the basic requirements for developing professional competency among forest officers.

Writing is an exercise in communication. Communication is the act of transmitting an idea from one person to another. In other words, writing is a form of communication. It has two components. One is the writer itself and the second one is reader. The writer creates information and the reader receives the information. So, while writing, the writer is required to understand the outlook and needs of the reader. The information, which the writer wants to share with the reader, is called document. The document may be a report, essay or write-up or a blueprint on any subject with a clear objective. All the information in the document should be structured in such a fashion that the reader easily understands it. The information may be arranged by using medium or language which is commonly understood by the target readers.

Technical writing is a practical writing that professional people do as part of their job. It leads to generation of document in the form of report, action plan, work plan, micro-plan, research papers, technical books, and periodicals. Some documents may be brief like character roll report or micro plans. Documents like forest working plan or management plan are lengthy. They run into several hundred pages.

Purposes and Characteristics of Technical Writing

Technical writing has four basic purposes. They are as follows:

(1) To inform.
(2) To analyse.
(3) To instruct or direct.
(4) To persuade and convince.

The forest management plan or working plan informs the forest officers, other government officer and the public about the general outlook of the forests, management status and strategy to manage the forest resources on the principles of sustainability. It provides information with regard to physical, cultural, ecological and socio-economic aspects of the forest division. The description of the forests provides the complete picture about the available resources along with its accompanying constraints and problems. The plan provides the time schedule for carrying out different silvicultural operations in different forest blocks. The working plan also analyses different existing situation with regard to management options and draws a logical conclusion. Based on such inferences, different prescriptions for management of forest resources are evolved. For every distinct prescription, cogent reason is provided. Therefore, forest management plan proposes strategy to manage forest area and resources to attain goals decided in the forest policy. However, in a management plan or working plan document, all these aspects of providing information, analysis, logical discussion and direction are so intimately intermingled, that it is not possible to visualize things in isolation. Therefore, forest management plan or working plan is an integrated technical package of information, discussion and prescription with regard to management of forests, past, present and future.

Drafting a Management Plan

Drafting of management plan or working plan involves technical writing. It consists of four primary characteristics. They are:

- It addresses a specific audience.
- It uses plain and objective language.
- It lays emphasis on presentation.
- It optionally uses visual aids.

Forest management plans or working plans are prepared by a senior practicing forest officer. Over the years, management plans or working plans have been accepted as an wholesome technical document containing management analysis and prescriptions for sustainable development of forest resources. Off late, many public officers, administrative officers, politicians, researches, students, judicial officers and the environmentalists have started to refer the document for understanding the forest resources and its management options in a forest division. With the increase in public awakening towards forest conservation, prescriptions provided in the management plans are looked and accepted as comprehensive guidelines for better forest management. Over the years, on account of enunciation of new forest policy, promulgation of Forest Conservation Act (1980) and increased activism in judiciary, there is manifold rise in the number of audience.

Working plan document is prepared in plain and objective language. It uses technical language in a way, which a scientific reader can understand. It lays emphasis on specific relevant facts. It revolves round the core subject of forest management. Narration should not be used to convey emotional or unreasonable interpretations of any phenomenon or events.

Presentational organization of the plan document should be such that it enables the readers to assimilate the information at a glance. The plan document should be structured in such a fashion that the rationale and strategy for forest management are understood easily. Towards this end, the steps generally followed are naming the topic, listing its sub-divisions, using obvious repetition, employing visual aids, using marginal indentation and bold headlines. It has been proved beyond doubt that visual aids improve the presentational organization of the plan document. Such distinct aids like graphs, tables, drawings and maps helps the reader to assimilate the information and understand the logic easily and quickly.

Process of Technical Writing

The act of writing is a process. It involves continuing activities of different nature. Besides writing, the process comprises such other activities like analysing data, researching technical facts, selecting visual aids. Therefore, the exercise of writing encompasses certain activities, which are generally grouped in three main phases. They are:

(1) Prewriting planning.
(2) Writing, drafting and revising.
(3) Post-writing and finishing.

Planning is chalking out steps required for the write-up and collecting all relevant information about the working-plan. Drafting involves in selecting and arranging all the elements in the working plan. Finishing is all about editing the plan document into a final shape. Grasping the gist is the key to successful drafting of the plan-document. The gist can be grasped by understanding basic facts, outline, strategy, effective outline and proper tone of the document. Once, such aspects are comprehended properly, the plan document can be developed with ease.

Determining the basic framework of facts for the working-plan is the key to planning part of the writing activity. It encompasses the following primary steps:

- Reviewing the technical information related to the crop of the area.
- Collection facts through field survey.
- Selection and organization of the information that satisfies the technical requirement.

All the steps are equally important. Nevertheless, more important is the fact that the information should be so neatly organized and presented in the document that the audience should get the intended meaning of each aspect of the management plan or working-plan. Effective outlining is also an important activity in writing. However, management plans or working plans have more or less fixed outlines. This aspect of organization has been dealt separately. The question of using a proper tone in the text is also very significant. Tone means writer's attitude towards the subject and the audience. Other subsidiary aspects of planning before beginning of writing work are illustrated below :

- Selecting and choosing the format.

- Choosing the visual aids.
- Selecting a production schedules with respect to time and manpower.

Drafting and Revising

In the course of writing, drafts are produced. This happens because of immature and un-systematic thinking. In spite of extensive planning, drafting and redrafting normally takes place. This process leads to the outcome of a good draft. Drafting has two primary aspects:

1. Clarifying.
2. Discovering.

Because of these two aspects, drafts are continuously refined and improved in order to improve the contents and to bring more clarity. The basic guidelines for good drafting are as follows:

1. State the objective explicitly.
2. Use set outlines. If new chapter is added in the plan document, then make the topic of each section and paragraph clear. Put the topic sentences at the beginning of the paragraphs.
3. Use the same general and accepted terminology through out the draft.
4. Avoid explaining the basics of forestry.
5. Write clear and shorter sentences.
6. Use active voice and employ parallelism.
7. Make your writing and discussion logical and interesting. The write-up can be made interesting by using visual aids, which help the reader to visualize the situation in an imaginative way. Comparison, brief description, narration, visual aids including graph, map and diagram helps the reader to create a picture in his mind.
8. The use of brackets and parentheses should be kept to the minimum. Avoid redundancy.
9. Facetious remarks, satire and slang should not be used. They should only be used where great emphasis is desired.
10. Keep everything in a logical sequence.
11. Tedious details on petty matters in the document should be

avoided, while undertaking a discussion with reasonable length. Status of the forest resources and results of past management are generally discussed in the Part I of the management plan. Part II of the plan document contains operational prescriptions only. Nevertheless, some facts with respect to miscellaneous regulation and management of forests can also be included in the Part II.

12. Brevity and simplicity is the hallmark of a good plan document. Actual prescription should be illustrated and elucidated in a lucid and simple language.
13. Chapter heading should be given in block letters, sectional heading in Italics and marginal headings in plain script.
14. All bulky tabular statements should appear as appendices and not in the body of the plan. All appendices should be serially numbered by roman numerals.

Finishing or Completing the Plan Document

This is the end stage of the writing exercise. This stage consists of two types of activities:

(i) Editing and
(ii) Producing the document.

Editing means developing a consistent, logical and accurate text. This sets the first activity. Here, one checks spelling, punctuation, basic grammar, formatting and to some extent accuracy of the facts described in the draft. Likewise, producing a document sets the next activity. It consists of the physical completion of the plan-document. Physical completion includes typing or printing the final document. When the document is voluminous, it is obvious that it shall take enormous time. This factor should be well considered while planning the document writing.

Guidelines for Good Technical Writing

Technical report writing is a meticulous job. It needs concentration, patience, command over the thinking process and language. Apart from this, it necessitates professional proficiency. It requires a good writing skill too. However, one can strive to draft a technical report successfully, if he applies the basic principles of writing and employs his skill. The job of technical writing involves certain elementary aspects. Such aspects have been dealt with in the previous chapter succinctly. However, certain aspects again need elaboration. Hence, such aspects are again presented in detail below:

- Technical writing style.
- Organizing the plan-document or report.
- Researching the technical information.
- Using visual aids judiciously.
- Taking precaution while drafting management plan.

1. *Technical writing style*

Management plan or working plan is a technical report or document. Technical reports should be clear and smooth in reading. It should be written in such a style that it allows the target readers to understand the technical theme and instructions clearly. Two important aspects for writing the working plan are as follows:

(i) *Write clear sentences* : The sentences should have adequate clarity. The working plan officer should strive for systematic and logical thinking. It is because of the fact that logical and systematic thinking leads to clarity. This holds key to effective writing. To begin with, place the main idea in the beginning of the paragraph. Using parallelism is also another method to bring clarity in the text. It means similar structure of sentences should be used to describe similar situations. Besides, it is also necessary to avoid repetition of facts.

(ii) *Revising for clarity*: Revision of draft leads to more clarity. While revising the draft, one looks for language and those parts, which leads to confusion and does not convey the message, which the writer intends to do. Secondly, the paragraphs should not be burdened with excessive wordiness. Pruning of excessive and superfluous words should be undertaken to avoid redundancy and repetition.

2. *Organizing the plan documents or reports*

If the plan document is systematically organized, then it helps the reader to comprehend the theme easily. Systematic organization of the plan document is done in the same fashion as sentences and paragraphs are structured. The following points are relevant to the task of organizing the document:

(i) *Use context setting introduction* : Introduction provides an overall framework of the subject one intends to say. It can be approached in many ways. It can be done by defining terms and by telling why this subject is being taken. It is also done by explaining the objective behind the subject or topic undertaken.

(ii) *Place important material first* : Putting the important material in the beginning of a paragraph or chapter emphasizes its significance and relevance in the context of the overall subject. Generally, important information, and statements are put at the beginning.

(iii) *Using Structural Parallelism* : Structural parallelism means that similar section or chapters of the document follow the same organizational pattern. The following two chapters of working plan report have same structure.

CHAPTER 2 : Selection-cum-Improvement Working Circle :

— Objective and Basis.
— Forest composition.
  Quality classes, Density, Age classes.
— Status of forest resources.
— Method of calculating treatment cycle.
— Determination of exploitable girth.
— Proposed treatment.
— Calculation of expected yield.

CHAPTER 3 : Improvement Filling Working Circle :

— Objective and basis.
— Forest composition.
— Quality classes, Density, Age classes.
— Status of forest resources.
— Method of calculating treatment cycle.
— Determination of exploitable girth.
— Proposed treatment.
— Calculation of expected yield.

The above points, if followed, helps the writer to draft the working plan document effectively. Besides, there is a need to adopt an appropriate tone. To achieve an appropriate tone the writer needs to use active and passive voice, nouns and pronouns, judiciously and effectively. More importantly, it is imperative to organize the different subjects, topics, discussion and chapters effectively as it constitutes the hallmark of a good technical report.

3. *Researching the technical information*

Researching generates enormous information. It is an important aspect of job of the working plan officer. When plan revision starts, it is imperative on the part of planning officer to search information and facts relevant to the forest division. Out of all such information, a pattern with respect to every phenomenon in the forest crop or sector is discovered and its significance and impact on the crop is studied. Take for example the changing population of plant communities and their diversity. This is a phenomenon. One has to go deep into the analysis to identify dimension of the phenomenon. Then he has to start studying the impact of this phenomenon on the forest crop.

The main objective of researching is to collect, analyse and interpret the information. This aspect is not taken care of during the field work exercise. Field work with regard to working plan exercise generates information about inventory, socio-economic pressure on the forest, growth data of few species, growing stock, density and quality classes. However, growth data and yield statistics have to be collected from the outside sources. The population data have to be collected from the district statistical offices. Geological and meteorological data are collected from their respective departments. In other words, preparation of working plan needs enormous data and their analysis. Therefore, it is important to chalk out a strategy to collect relevant data and information. The main aspects of the strategy will be to:

— Identify the information and facts one needs. This is possible by going through old working plans, various forest records and studying the present requirement.
— Consult the right sources.

It is necessary to consult the right people who have accessibility to such information. Generally, officers who are in the research field or the officers who are working in the forest division or officers of other department operating in the area of the same division are the most suitable resource persons. Besides, one should consult the technical library or forest research institute, if available in the vicinity for generating relevant information.

(i) *Collect and store the information in an orderly and systematic manner*:

It is necessary to store information in such a way that it can be retrieved at anytime. This requires developing file system with assigning different name to different topic. For this a data base can be prepared on the computer.

(ii) *Find key documents and keys terms:*

Identifying the key documents is one of the most common and fundamental works for the revision or preparation of management or working plan. Generally, old working plan documents constitute the key documents. Such old and distinguished working plan document contains not only logical discussion about different features and changes in the forests but also strategy adopted to improve forest crop. Reading through such discussion helps to comprehend the topic well.

Key words are those words or terms, which have specific significance and relevance in relation to present time. Such terms helps the writer in elucidating certain phenomenon effectively. In the field of forestry, sustainable growth, sustainable yield and sustainable development are some of the key words. Apart from this, conservation, bio-diversity, and protected area are some of other key terms used in elucidating forestry. Therefore, writer should search for such key words, understand their meaning with proper perspective, and use them in proper place.

Using Visual Aids

Using visual aids is an important step towards effective presentation of data. They are the essential part of the working plan document. With the advent of computer application in forestry, it is very increasingly easy to create visual aids and present it with the narration and discussion. Presently bar graphs, line graphs pie charts and miniscule maps are commonly used to create the desired picture and impact in the mind of the reader. Visual aids are used in working plan reports for the following purposes :

- To summarize and explain data.
- To provide readers an easy understanding of data.
- To give readers more opportunity to explore data.
- To provide a different entry point into the discussion.

Summarizing is a way of presentation of data in such a manner, which is not only concise but also readily acceptable to the mind of the reader. The graph in Fig. 22.1 indicates the girth classwise distribution of trees and their volume in two successive inventories carried out in Rajnandgaon forest division in two sccessive inventories. The reader easily grasps the pattern of distribution of trees and volumes in different girth classes. This apart, the reader also comprehends the change in the tree distribution and volumes over the gap of two successive inventories. Such a picture can also be given in tabular form (Table 22.1)

Table 22.1 : Distribution of trees in different girth classes in two successive inventories

Girth Classes in cm	Distribution of trees in no		Distribution of volume in cmt.	
	Inventory - 1981	Inventory - 1997	Inventory - 1981	Inventory - 1997
30 – 45	127.914	96.870	6.824	5.806
45 – 60	40.285	42.292	6.555	4.643
60 – 90	44.786	66.458	17.034	20.048
90 – 120	17.325	22.083	14.675	15.085
120 – 150	4.613	7.708	7.237	9.611
150 >	1.846	4.584	5.91	7.602

Source : Bebarta, K.C. (1999) Working Plan for Rajnandgaon Forest division.

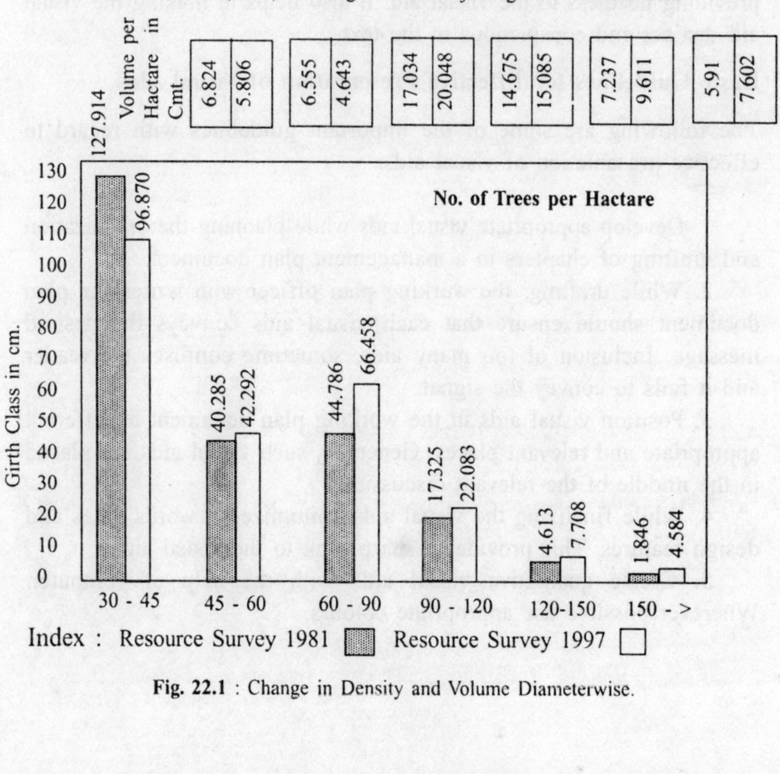

Fig. 22.1 : Change in Density and Volume Diameterwise.

Providing a miniscule map of the forest division *vis-a-vis* the state in the beginning of the text or the chapter orient the reader to the subject before any discussion is initiated. Sometime presenting a graph or table in the initial discussion of the chapter helps to provide an entry point into the discussion. However, it is necessary to decide upon the right visual aids. To create the right visual aids, it is necessary to follow the basic communication process. The main components of the process are : (1) plan, (2) draft, and (3) finish.

Planning is an important component of the creation process of visual aids. Different visual aids have different values of applicability. Graphs generally have an emotional impact on the mind. Lines that slope up towards right gives positive emotions and lines that slope down up to the right gives negative emotions. Beside this, there are other aspects like arranging different inputs in different column and the time constraints.

Drafting is a serious and careful exercise. It is done in the same manner as the main text is done. Revision of visual aids continues until they appear to be effectively presentable. Finishing means providing neatness to the visual aid. It also helps in making the visual aid distinct and conspicuous in the text.

Basic Guidelines for Effective Presentation of Visual Aids

The following are some of the important guidelines with regard to effective presentation of visual aids:

1. Develop appropriate visual aids while planning the organization and drafting of chapters in a management plan document.

2. While drafting, the working plan officer who writes the plan document should ensure that each visual aids conveys the desired message. Inclusion of too many aids, sometime confuses the reader and it fails to convey the signal.

3. Position visual aids in the working plan document at different appropriate and relevant places. Generally, such visual aids are placed in the middle of the relevant discussion.

4. While finalizing the visual aids, minimize on words, lines and design features. This provides a sharp look to the visual aids.

5. Create qualitative visual aids with the help of computer. Wherever possible use appropriate colours.

Different Types of Visual Aids

There are different types of visual aids used in technical reports. These types of visual aids are purported to draw the attention of the reader to the essence and substance of the data. Most important of them are:

- Tables.
- Line graphs.
- Bar graphs.
- Pie charts.
- Charts.
- Photographs.
- Map and drawings.

The visual aids mentioned above helps in focussing the subject very effectively. More importantly, illustrations in the form of photographs and drawings including miniscule maps constitute the most potent visual aids. A good photograph offers many advantages. It is memorable as well as easy to refer to. It also shows certain line of relationship among various components. Generally, such relationships cannot be represented with the aid of table and graphs. In a forest, the stand density, the crown coverage, the ground floor conditions and special features of trees like buttressing are some of the attributes, which can be represented in the photograph very effectively. Such photographs provide an imprint on the mind of readers. Likewise, the maps and their miniscule forms also provide a bird's eye view of the geographical representation of the attributes like forest area, forest types, density etc. Drawings also represent an item and its relationship to others. Drawings are used to show the specific parts of tree, soil profiles, civil structures like road, bridges, countour trenches etc. They enable the readers to focus on details, which the writer wants to show with reference to his discussion on the subject.

Presentation of Data

Tabulated data extends over a large number of columns. It makes comparison of features difficult. So, there is a necessity of application of visual methods for representation. Every intelligent mind readily grasps the comparison and magnitude of data from the visual aids at once. Beside this, such visual methods leaves a lasting impression on the mind of readers. Graphs and diagrams are the two visual methods of representation of data. With the help of diagrams and graphs, complex data are presented in a simple form. They also render the comparison easy. It is for this reason, such visual methods are becoming popular day by day.

As discussed above, there are two main categories of visual methods. They are:

1. Diagrams.
2. Graphs.

Diagrams: These are meant to provide a pictorial representation of the homogeneaus quantitative data to make them readily comprehensible. They provide proper required emphasis on different characteristics of data. There are certain basic principles for drawing the diagrams. One should focus on title, size, proportion of length and breadth, scale and index of the diagram. Diagrams should be simple and attractive. Apart from this, what is most important is selection of proper and appropriate diagram. Not all types of diagrams are suitable for all types of data. Therefore, one should be very careful in selecting the diagram to convey an impression about any phenomenon in question.

Types of Diagrams

There are many types of diagrams found to be used to express an explicit meaning and message. They are:

I. *Dimensional Diagrams*

One-dimensional diagram

It is a type of diagram where only one-dimension changes in size according to changes in the variables. Examples of such one-dimensional diagrams are lines or bars (either simple or sub-divided bars) are illustrated below:

Simple bar diagrams indicates that the height of the bar varies according to the data. All the bars are raised on a common base and there should be some gap between two bars to make the bars distinguishable. Bar diagrams are suitable for presentation of quantitative data of special series. The bars may either be shaded or coloured or be left blank. Generally, values are mentioned in the top of the bar.

Sub-divided bar diagram: They are of two types. One is sub-divided bar-diagram on percentage basis. The other one is a sub-divided bar on absolute basis. This gives an idea of totality as well as its individual components in a same diagram.

Sub-divided bars on percentage basis: Comparison of data is always made on equal footings. Items are expressed as percentage of the aggregate. The heights of the bars are equal because they represent 100 in each case. From these bars, sub-divisions are made according to the percentage they bear to the total. This makes comparisons of various items or features very simple and clear. Various factors constituting the aggregate can be compared in their perspective.

Sub-divided bars on absolute basis: This also gives a total picture as well as idea about each component on a relative basis. However, each component does not show the relationship in which they stand in comparison with the total aggregate.

Multiple Bars Diagram.

In such diagrams, two or more sets of inter-related data can be represented. Therefore, it provides a picture of more than one phenomenon in one diagram. Two types of bars placed side by side can represent the figures like expenditure and revenue in a forest division in different years. Such pictorial diagrams facilitate easy comparison.

Deviation Bar Diagram

In such diagrams, difference or deviation of figures is shown in the shape of bars. Bars representing positive differences are shown on one side and those representing negative differences on the other side. A vertical line in the centre represents the point of equilibrium and on the right-hand side positive differences and on the left-hand side negative differences are shown.

Two-dimensional Diagrams

These are also called as surface diagrams. There are different types of two-dimensional diagrams.

1. *Rectangular Diagrams*: In such diagrams, the shape is rectangular. The height of the rectangle varies according to the ratio, which the data bear to each other.

2. *Square Diagram*: Square diagrams give a better comparison of data than rectangles or bars. They are generally drawn when some items of the series have values much higher than others. In such cases, if a bar diagram is drawn, then the bars representing large

figures would be very big in size while the bars representing the small figures would be very small. While constructing square diagrams, the first step is to find out the square root of various figures and then squares are drawn with the lengths of their sides in the same proportion as the square roots of the original figures. The area of the squares would be in the same proportion as the ratio of original figures.

3. *Circle or Pie Diagrams*: Circles are very important and popular two-dimensional diagrams. The area of the circle is directly proportional to the square of its radius. Therefore, if the radius of a circle were four times the radius of another circle, its area would be sixteen times the area of the other circle. In other words, in case of circle their radii are kept in the ratio of the squares of their radii. Data can be represented in the forms of circle in all those situations where square diagrams are used.

In case of circles, the aggregates can also be represented by a big circle and the various components by sectors which are cut inside it. Such diagrams are known as Angular Diagrams. The areas of various sectors or parts are in proportion to the angle, which they make at the centre of the circle. The total angle of 360 degrees can be divided into a number of smaller angles whose degrees would be in proportion to the values of the components.

When two aggregates and their components are to be compared, two circles have to be drawn with areas proportionate to their ratio of the two aggregates. Each circle would then be divided into sectors on the basis of values of its components.

Three-Dimensional Diagrams

Three-dimensional diagrams are used to project the difference in ratios between two sets of data, more prominently. In such diagrams length, breadth, and height are considered. Therefore, these diagrams are called volume diagrams. Cylinders, spheres, cubes, etc. are some of the three dimensional diagrams. Population in different states, forest areas in different states or districts can be shown by using such diagrams. A comparison of growing stock of different forest types in a state can also be represented by using such three dimensional diagrams.

II. Pictograms

Pictograms are diagrams where the value of items is represented by picture. The size of different number of pictures drawn is in proportion to the values of various items, which have to be represented. Pyramid is one of pictograms used very often in technical data representation. Data pertaining to distribution of stems of various age classes in a tree population in a unit area are usually represented by pyramids.

III. *Cartograms*

It is a type of diagram where the distribution of items is represented in the form of dots over a map. The items like distribution of rainfall, population density, etc. can be shown on a piece of map by putting down dots representing rainfall or number of people. In a forest division, the grazing pressure or fire incidence can be represented on an indicative map or vegetational map by using cartograms. The population density in a district or a forest division with range boundaries can also be shown on a map by adopting cartogram methods. Likewise, distribution of resources of non-timber products can be shown on a divisional map.

From the above discussion, it is clear that there are enormous possibilities for effective representation of data. However, in order to convey a desired impression it is necessary to select proper diagram. Otherwise, the diagram will not make the desired impact on the mind of readers. Beside this, while constructing various diagrams and pictures, due attention should be paid to the aspects of simplicity, neatness and display.

Classification of Data

Data generated during the field work exercise do not give an idea about the problems of forest under study. They cannot be taken up directly for analysis and interpretation. For example, the figures of enumeration in a sample plot as collected in the field do not indicate the stand density in a forest stratum or the growing-stock in the forests. Unless all such enumeration data are classified and tabulated, it is not possible to draw any inferences from them. For the purpose of analysis and interpretation, it is necessary to divide the data into homogeneous groups. This process is called classification.

Classification is a process of arranging data in groups or classes according to resemblances and similarities. This brings homogeneity in the heterogeneous elements of the data. In this process, units having a common characteristic are grouped or divided into a number of classes.

The main objective of classification is to simplify the complex data and render it easy to convey a meaning. Classification also helps in comparing two groups of data. Therefore, classification is an important process and is a precursor to tabulation.

In a working plan exercise, generally data are expressed in numerical quantities. Hence, they are arranged according to class interval. For example, classification of trees are available in a sample plot according to girth classes like 10 to 20 cm, 21 to 40 cm, 41 to 60 cm, 61 to 80 cm, 81 to 100 cm and 101 to 120 cm and above 120 cm. However, the trees can also be classified according to diametric classwise and species-wise. Here, the diametric class range is called class interval. The difference between the upper limit and the lower limit of a class interval is known as the magnitude of the class-interval. When this difference is same throughout the various classes in the class intervals, the magnitude is known as uniform. If the difference changes in various class intervals, then it is known as un-uniform magnitude.

In forest inventory, both the inclusive and inclusive method of framing class-interval is adopted. In inclusive method, the class-interval does not overlap; whereas in an exclusive method the class interval overlaps. The class-interval of diameter of trees will appear as 10 to 15 cm, 16 to 20 cm etc. in exclusive method.

Before embarking on classification of data, one should lay stress on the following points:

- Classification must be exhaustive, i.e. it should be possible to include all data.
- Classification should be unambiguous.
- The number of classes should be neither too large nor too small.
- The magnitude of class-interval should be uniform throughout the classification

Tabulation of Data

Tabulation is a process of systematic arrangement of data in columns and rows. The basic objective of this exercise is to simplify the process of presentation of data for the purpose of analysis and statistical inferences. In other words, tabulation helps in making comprehension of complex and varied data. It is also a process of condensing classified data in such a manner that it can be easily analysed and understood and so that any comparison among data can be made to draw any inferences. Tabulation is an art. Continuous observation, learning and practice make one to develop this art. Good tabulation makes an impact on the reader and conveys the message.

In general, a statistical table is divided into five parts. They are:

- *Table number* : Each table should be numbered for easy identification and easy reference.
- *Title of the table* : Each table should have a distinct and short title at the top of the table in the centre. A title indicates the nature of data. The title should be in prominent letters.
- *Caption*: This refers to the headings of the vertical columns. A caption may have one heading or a number of sub-headings.
- *Stub*: This refers to the headings of the horizontal rows. They are positioned on the left side of the rows.
- *Body*: It is that part of the table, which contains the data. This is the most important part of the table. The data in the body are arranged according to captions and stubs.

Basic Principles of Tabulation

There are no hard and fast rules for ideal presentation of tabulation. Preparing good table needs understanding of the subject and practical experience. They help in designing the structure of good table. The basic principles of tabulation with respect to forest data are as follows:

- Table should be short and precise.
- When data are enormous, these can be presented in a number of tables. Items in a table should be arranged in a logical order.
- Caption, heading, sub-headings should be self-explanatory.
- Units of measurement under each heading or sub-heading should be distinctly indicated.
- In case of a comparative analysis, columns containing data should be arranged side by side.
- Source of data should be indicated in the footnote of the table.

Types of Tabulation

There are two types of tabulation. One is simple and the other is complex. Both types of tabulation process are illustrated below.

1. *Simple tabulation* : It is a process of tabulation which involves arrangement of one set of related data and seeks to address one or more groups of an independent investigation. In a simple table, two factors or characteristics are placed in relation to each other. The following table is a product of simple tabulation. It shows the distribution of stems among different in one hectare of forest. This is also called a one-way table.

Name of the tree species	No. of trees per hectare
Teak	18
Bija	12
Saja	32
Tendu	13
Haldu	3
Salai	8
Karra	78
Miscellaneous	82

2. *Complex tabulation* : It is a process of tabulation, which involves arrangement of data pertaining to a number of coordinate factors or features. When there are two co-ordinate factors, the table is called two-way tables or double table. When there are three co-ordinate factors involved in the table, it is called three-way table or the process is called treble tabulation. When it is a case of more than three co-ordinate groups, the table is called multiple tables or the process of tabulation is called multiple tabulation. In a multiple table, one can go for posting details of stem distribution in different categories of diametric classes.

Interpretation

After collecting and analyzing data, the working plan officer has to accomplish the task of drawing inferences followed by report writing. This has to be done very carefully, otherwise misleading conclusions may be drawn and the whole purpose of plan preparation will go haywire. It is only through interpretation that the planner can reveal relationship and process that underlie his inferences. Now the question is what does interpretation means. Interpretation refers to task of drawing inferences from the collected facts after an analysis. It is central to any planning process. It is the most difficult task in the whole planning process. It requires great expertise and experiences on the part of planning officer. Generally, an officer who learns through continuous practice and experience over the years puts his knowledge and skill in the critical process of interpretation. The technique of interpretation involves the following essential steps.

1. The planning officer should provide reasonable logic and explanation to the relation he finds with respect to any change in the forest resource status like change in growing stock, change in density (areawise), forest composition, change in bio-diversity indices, management status, etc.

2. Extraneous information collected during exercise should be considered and examined along with the collected data during the course of field study.

3. It is always prudent to consult senior officers/academicians who have sufficient insight into the changing phenomena in forests. This helps in improving logical development of arguments that come in the course of analysis of a problem.

PART - V

NEW TOOLS OF PLANNING

23

Geographical Information System and its Application

Introduction

Geographical Information System abbreviated as GIS is now extensively used in management of natural resources world-wide. Before we attempt to delve further into the subjects, it is necessary to understand the concept underlying the geographical information system. Nevertheless, before understanding the composite concept of GIS, it makes sense to comprehend the meaning of information and system separately. To begin with, information is derived from the interpretation of data, which are symbolic representation of features. The value of information depends on many things including its timeliness, the context in which it is applied, and the cost of collection, storage, manipulation, and presentation. Likewise, a system is a group of connected entities and activities, which interact for a common purpose.

The information system is a set of processes executed on raw data, to produce information, which will be useful in decision making. Geographical data describes objects from the real world in terms of (a) their position with respect to a known co-ordinate system, (b) their attributes that are unrelated to position (such as colour, cost, incidence of disease etc.) and (c) their spatial inter-relation with each other (topological relations), which describe how they are linked together or how one can travel between them (Fig. 23.1).

A Geographical Information System uses geographically referenced data as well as non-spatial data. It undertakes operations, which support spatial analysis. The basic objective behind application of GIS into any resource management system is to facilitate decision making. In other words, this technology is widely applied to manage use of land, natural resources, transportation, oceans, or any spatially distributed entities. The connecting link between the elements of the system is geography, *e.g.* location, proximity, spatial distribution. In

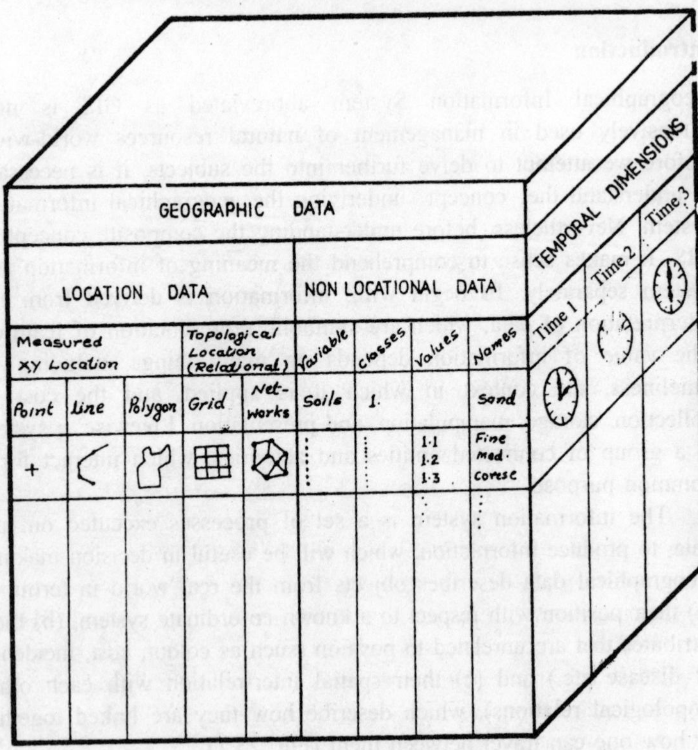

Fig. 23.1 : Three conceptual components of a geographic information system.

this context GIS can be seen as a system of hardware, software and procedures designed to support the capture, management, manipulation, analysis, modelling and display of spatially-referenced data for solving complex and management problems.

Definitions

Geographical Information System has been defined by various people. Few of such definitions are as follows :

- A system for capturing, storing, checking, manipulating, analysing and displaying data which are spatially referenced to the Earth (DOE: 1987:132).
- Any manual or computer based set of procedures used to store and manipulate geographically referenced data (Arnoff. 1989:39).
- An information technology, which stores, analyses, and displays both spatial and non-spatial data (Parker, 1988:1547).
- A special case of information systems where the database consists of observations on spatially distributed features, activities, or events, which are defined in space as points, lines or areas. A GIS manipulates data about these points, lines, and areas to retrieve data for ad hoc queries and analyses (Dueker, 1979:106).
- A database system in which most of the data are spatially indexed and upon which a set of procedures operated in order to answer queries about spatial entities in the database (Smith *et al.* 1987: 13).
- An automated set of functions that provides professionals with advanced capabilities for the storage, retrieval, manipulation, and display of geographically located data (Ozemoy, Smith and Sicherman: 1981:62).
- A powerful set of tools for collecting, storing, retrieving at will, transforming and displaying spatial data from the real world (Burrough 1986:6).
- A decision support system involving the integration of spatially referenced data in a problem-solving environment (Cowen, 1988:1554).

Classification

There are three basic approaches for classifying GIS, *viz.*

(i) the process or function oriented approach.
(ii) the application approach.
(iii) the database approach.

The process oriented approach lays emphasis on the information handling capabilities of GIS. The applications approach categorises the systems on the basis of the problems they seek to address. They are classified into transaction processing systems and decision support system. In transaction processing systems, emphasis is placed on recording and manipulating the occurrence of operations. This is applied in the field of banking and airline reservation systems. Decision support systems are used for applications such as tactical warfare, market analysis, environmental impact assessment and environmental management planning. They are normally retrieval oriented and are required to be able to operate in a flexible manner. The database approach is also extensively used, because of the of the fact that database theory and practices influences Geographical Information System.

Contributing Disciplines and Technologies

GIS is a convergence of technological fields and traditional disciplines. It is called an enabling technology because of the potential it offers for the variety of disciplines, which deals with spatial data. It encompasses the following disciplines.

1. *Grography*:—It has a long tradition in spatial analysis and hence provides techniques for it and a spatial perspective on research.
2. *Cartography*:—It is concerned with the display of spatial information. Maps are the main source of input data for GIS as well as output from GIS. Computer cartography (also called "digital cartography", "automated cartography" provides methods for digital representation and manipulation of cartographic features and methods of visualisation.
3. *Remote Sensing*:—Images from space and the air are a major source of geographical data. Remote sensing consists of techniques for data organisation and processing anywhere on the globe at low cost, and consistent update potential. Moreover, integrated data from a remote sensing system can be managed with other data layers in a GIS.

4. *Photogrammetry*:—It uses aerial photography and techniques for making accurate measurements from them. Photogrammetry is the source of most of data on topography (Ground surface elevation) used for input to GIS.
5. *Surveying*:—It provides high quality data on position of land boundaries, buildings, etc.
6. *Statistics*:—Many models built using GIS are statistical in nature and use statistical techniques for analysis. Statistics is important in underscoring issues of error and uncertainty in GIS data.
7. *Operation Research*:—Many applications of GIS require use of organisation techniques for decision making.
8. *Computer Science*:—Computer-aided Design (CAD) provides software and technique for data input, display and visualisations. Database Management Systems (DBMS) contribute methods for representing data in digital form, developing procedures for system design, and handling large volume of data.
9. *Mathematics*:—Several branches of mathematics, especially geometry and graphs are used in GIS system to design and analyse spatial data.
10. *Civil Engineering*:—GIS has many applications in transportation and urban engineering.

Components of a GIS

The following are the components of a Geographical Information System:

1. Computer hardware.
2. Data and sets of application process.
3. An organisational context.

I. *Computer Hardware*

The GIS operates in an environment defined by the hardware i.e. the machinery, including a host computer, ranging from a micro-computer to a large mainframe computer supporting many users. The general hardware components of a geographic information system are presented in Fig. 23.2.

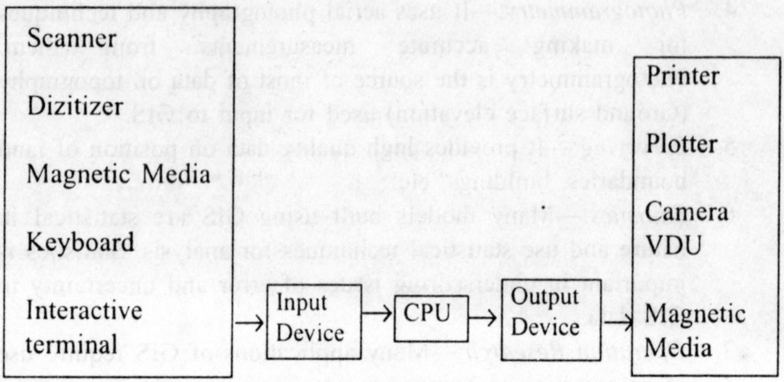

Fig. 23.2 : Computer Hardware Components.

The two important hardware components of computer are as follows:

1. Central processing unit.
2. Input and output devices.

Central Processing Unit

It is the part that executes the programmes and controls the operations of the hardware. It comprises of the Arithmetic and Logic Unit (ALU), the control unit and the primary storage unit. Arithmetic and logic unit is a part of CPU where all mathematical and logical functions are performed. The control unit interprets any instructions it receives from memory, directs the sequence of events necessary to execute the instruction, and establishes the timing of these events. The primary storage unit refers to the internal storage unit of computer, where programmes and their data are stored. The memory can either be the main memory, which stores all data and instructions before it can be processed by the computer; or secondary/auxiliary memory for large, permanent and semi-permanent files. Data input is a major bottleneck in the application of GIS technology. Apart from being labour intensive, tedious and error prone, the cost of inputs is also prohibitive.

Mode of Data Input

The GIS comprises different data input devices which are mentioned in Fig. 23.3. Data entry can be done through the following modes:

- — Keyboard entry for non-spatial attributes and occasionally locational data.
- — Manual locating devices in which the user directly manipulates a device whose location is recognized by the computer, *e.g.* digitizing.
- — Automated devices that automatically extract spatial data from maps and photographs, *e.g.* scanning.
- — Data conversion from other digital sources.
- — Voice input has been tried particularly for controlling digitization operations.

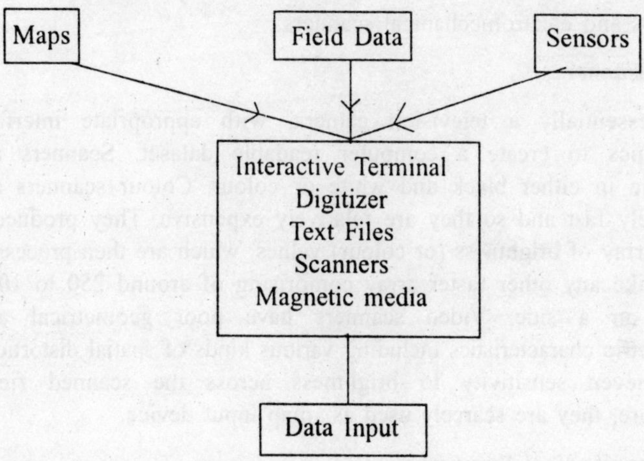

Fig. 23.3 : Data Input Devices.

Digitizer

Digitizer is the most common device used for extracting spatial information from maps and photographs. Digitizing the map contents is done in two different ways. One is the point mode and the other is stream mode. In point mode, the graphics identify the points to be captured explicitly by pressing button. In stream mode, points are captured at set time intervals, typically 10 per seconds. In the point mode the operator selects points subjectively. Stream mode, on the other hand generates large number of points out of which many may be redundant. Most of the digitizing work is done in point mode.

Problem with digitizing maps arises because most maps have not been prepared with an objective to digitize. Besides, paper maps are generally unstable. Each time the map is removed from the digitization table, the reference points must be re-entered when the map is fixed to the table again. If the maps are stretched or shrunk during the course of dizitization, then the newly digitized points will be slightly off in their location when compared to previously digitized points. Other errors, which generally occur while preparing maps, enter the database.

Scanners

Scanners are automated devices and can be broadly classified as video scanners and electromechanical scanners.

Video Scanner

It is essentially a television camera, with appropriate interface electronics to create a computer readable dataset. Scanners are available in either black and white or colour. Colour scanners are extremely fast and so they are relatively expensive. They produce a raster array of brightness (or colour) values, which are then processed much like any other raster array comprising of around 250 to 1000 pixels on a side. Video scanners have poor geometrical and radiometric characteristics including various kinds of spatial distortions and uneven sensitivity to brightness across the scanned field. Therefore, they are scarcely used as map input device.

Electromechanical Scanners

These types of scanning systems are more expensive and slow in yielding results. However, they can create better quality products. One common class of scanners involves attaching the graphic to a drum. As the drum rotates about its axis a scanner head containing a light source and photo-detector reads the reflectivity of the target graphic, and digitizing this signal, creates a single column of pixels from the graphic. The scanner head moves along the axis of the drum to create the next column of pixels, and so on through the entire scan. This controls distortion by bringing the single light source and detector to position on a regular grid of locations on the graphic. An alternative mechanism involves an array of photo-detectors that extract data from several rows of the raster simultaneously. If good quality source documents are available, scanning can be an efficient and time saving mode of data input.

Raster Scanners

The raster scanner works on the simple principle that a point on any part of a map may have one or two colours, black or white. The scanner incorporates a source of illumination and a television camera with a high-resolution lens. The scanner is mounted on rails and can be moved systematically backwards and forwards over the surface of the source document.

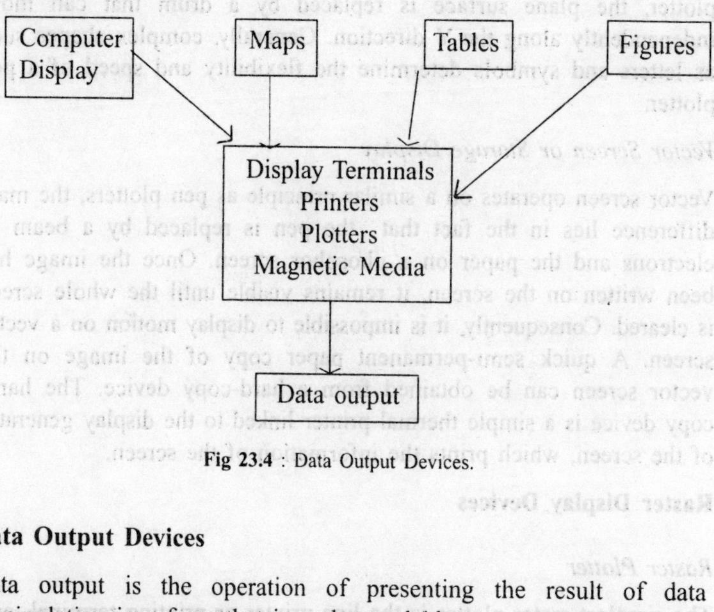

Fig 23.4 : Data Output Devices.

Data Output Devices

Data output is the operation of presenting the result of data manipulation in a form that is understandable to a user and also in a form that allows data transfer to another computer system. The organization of different data output devices is presented in Fig. 23.4. People compatible output is found in the form of maps, graphs, and tables. Likewise, computer compatible output is in the form of a magnetic tape that can be fed into another system, or it may involve some form of electronic data transmission over communication network. People compatible output devices can be classified into those that produce ephemeral displays on electronic screens and that produce permanent images (hard-copies) on paper or other material. The electronic displays are known as VDUs or graphic terminals, the hard copy devices as plotters. Both VDUs and plotters can further be classified according to whether the display is produced by a raster scanning technique or by a vector line drawing technique.

Vector Diplay Devices

Pen Plotters

Pen plotters are robot draughtsmen. They are essentially a plane surface on which paper can be laid, and a pen holder that can be moved independently in the X and Y directions. Known as the drum plotter, the plane surface is replaced by a drum that can move independently along the Y direction. Generally, complex shapes such as letters and symbols determine the flexibility and speed of a pen plotter.

Vector Screen or Storage Display

Vector screen operates on a similar principle as pen plotters, the main difference lies in the fact that the pen is replaced by a beam of electrons and the paper on a phosphor screen. Once the image has been written on the screen, it remains visible until the whole screen is cleared. Consequently, it is impossible to display motion on a vector screen. A quick semi-permanent paper copy of the image on the vector screen can be obtained from a hard-copy device. The hard-copy device is a simple thermal printer linked to the display generator of the screen, which prints the information of the screen.

Raster Display Devices

Raster Plotter

The simplest raster plotter is the line printer or printing terminal, and the earliest computer mapping systems were designed to use these tools for output. Multi-coloured printing is available on some matrix printer by using a three-colour ribbon. The simplest way is to make a coloured thematic map using a matrix printer used to print each map class separately as a fully black image and use the printout to make a series of lithographic colour masters. Much better resolution raster printing coupled with fast printing speeds can be obtained from matrix plotters that make use of the electrostatic principles of xerography. The main advantage of matrix plotters is their speed.

Optical Film Writers

Very high quality of rasterized data is achieved in colour or in black and white by using a special film recorder. The print process is very similar to that of the drum scanner except that the right source is focused and modulated to expose a point on the surface of a photographic film mounted on a high speed revolving drum.

Raster Display Screen

The simplest raster display screen use conventional television technology in the sense that the image is built up from lines scanned across the screen. The difference with conventional television comes from the fact that the strength of the scanning electron beam is not controlled by an analogue or continuously variable signal but by a bit pattern in a special computer memory called a frame buffer.

II. *Data and Sets of Application Process*

GIS application process encompasses the following aspects:

1. Assessment of data need and identification of its source.
2. Data collection and evaluation.
3. Building a geographic database.
4. Geographic analysis of data.

1. *Assessment of data need and identification of sources*

Geographic data has three major features or aspects. They are as follows:

1. Geographic position.
2. Attributes or properties.
3. Time or dynamics.

Geographical position can again be categorized into two types. They are (1) absolute position and (2) relative position. Likewise, attributes or properties can be of the following categories:

1. *Normal*:—Values that establish identity, *i.e.* attribute to which a name is given.

2. *Ordinal*:—Values that establish an inherent order.
3. *Interval*:—With natural sequence and equal intervals on a scale with an arbitrary zero point.
4. *Ratio*:—With same characteristics as interval but with a natural zero or starting point.

The data related to aspect of time component can be categorized into the followings:

- Time duration.
- Temporal resolution.
- Temporal frequency.

2. *Data Collection and Evaluation*

Data collection is a time consuming exercise. Besides, it involves enormous cost. However, the overall performance and success depends on, among other things, the collection of quality data. Therefore, it is imperative to pay adequate attention to data collection. Data are collected either from primary or secondary sources. Field work done afresh generates primary data. Data relating to published maps, stock maps, thematic maps, census report, working plan statistics etc. constitute the secondary source of data.

In a Geographical Information System (GIS) data layers are generated from such information gathered from primary and secondary sources or existing layers one on other. As for example, in a forest, erosion prone area can be identified and delineated by overlaying layers of forest cover, slope and soil erosion status.

In a Geographical Information System, data is either spatial, *i.e.* those with geographical location or attributes, *i.e.* those descriptive in nature or both. Spatial data are represented by area or polygon features, line features and point features. The kind of data that fall in those aforesaid categories is presented in Table 23.1.

Table 23.1 : Spatial data

Area or polygon features	Line features	Point features
Land covers or land use.	Streams Roads	Important land marks, Wells.
Soils Geology surface and sub-surface. Terrain features. Political and administrative boundaries.	Contours	Elevation points, Location of buildings.

The most important source of spatial or analog data is the map. Map data is in the vector form. Data is also available in digital form. The common example of such type of data is remote sensed data. Remote sensing is perhaps the most important source of digital data. Data collected from satellites sensors have been widely used as a means of updating natural resource and topographic maps. In the field of surveying, global positioning system based on navigation satellites are also capable of developing digital location data. Satellite data differs from nearly all other geographical data in their consistency, high positional accuracy, high spatial and temporal resolution and low level of human interpretation. GIS requires raster capabilities to store and analyses large volumes of these data with minimum loss of resolution.

Attributes or descriptive data are complimentary to the spatial data. Such data are obtained from the spatial data. Such data are obtained from primary as well as secondary sources. Census data are also available in computer readable form at NIC/NET centres at every district. This constitute a potent source of secondary data.

Identification of sources of data plays as an important role in data collection exercise. Many relevant data are available in different departments or organizations. Table 23.2 and Table 23.3 provide an insight about the availability of some spatial and attribute data and their sources.

Data Evaluation

After having collected the necessary data, the next step is to evaluate the data with respect to required format, scale, data, medium, classification categories. Determination of map scale is an important aspect of data evaluation. Scale determines the smaller area that can be represented on a map. In a map having a scale of 1:50,000, a line having length of 25 mt cannot be properly represented. It will come about 0.5 cm. in length over the map. Thus, many roads and streams/ rivers having width of less than 25 mt cannot be shown on 1:50,000-scale map. Therefore, scale influences the quantum of information available in the thematic and topographical maps.

In a forest stock map, it is necessary to establish 'minimum mapping unit'. Any feature having less than the mapping unit cannot be represented on the map. In such cases, features will be merged with neighbouring areas or features. Table 23.4 gives the smallest area that can be shown and read on a hard copy output.

Table 23.2 : Types of data and their method of capture

Source of information	Types of data	Method of capture
Field survey	Boundaries, slopes, altitude, latitude and longitude	Theodolite, Electronic distance measurement, GPS
Telemetry	Flows of river or streams.	Remote meters and counters, balloons.
Socio-economic survey	Household characteristics, social stratification	Data loggers retrieval from census and official survey and PRA
Remote sensing	Multispectral information	Satellites
Aerial photographs	Topographic information	Photogrammetry scanner
Maps	Topographical and thematic information	Dizitizers, scanners and image processing

Table 23.3 : Data and its sources

Data	Source of data
Forestry statistics, thematic map data, working plans information, locational position of forest boundaries.	Divisional forest office, Forest Survey of India. Forest Research Institute.
Remote sensing data.	National Remote Sensing Agency, (NRSA), Hyderabad.
Aerial photographs.	Survey of India.
Topographic maps.	Survey of India.
Geological and mineralogical information.	Geological Survey of India. State Geology and Mining Department.
Soil maps	National Bureau of Soil Survey and Land Planning, Nagpur; State Agriculture Departments.
Socio-economic, landuse, land ownerships, political and administrative boundaries, roads.	Government reports, working plans, District statistical book, census report. NIC/NET service centres.

Table 23.4 : Scale and minimum mapping unit

Scale	Resolution (m)	Detection (m)
1:10,000	05	10
1:50,000	25	50
1:1,00,000	50	100
1:2,50,000	125	250
1:10,00,000	500	1000

While assessing the data, certain aspects should be considered. They are as follows:

- Vegetation and land use data need frequent updating.
- Accuracy of maps should be evaluated before the data is used.

3. Building a Geographic Database

Building a geographical database is an important step towards developing Geographical Information System (GIS). It is a gradual process. The output of one process becomes the input for others. The different steps involved in building a database are enumerated and explained below:

1. Digitization of maps.
2. Editing and cleaning.
3. Constructing topology.
4. Adding attribute information.
5. Coordinate transformation.

Digitization of Maps

Existing maps provide data for developing thematic layers. These layers constitute the geographical database. Now, these layers are converted into a computer compatible form for integration within the Geographical Information System. This conversion is accomplished through the process of 'digitization', when spatial or geographic features on the map are transformed into digital format. Basically, digitization involves obtaining x and y coordinates for point, line and area features on a map. This process of finding out the appropriate coordinates can be visualized, if a map is put over a graph paper and the origin is overlaid in the lower left corner. Now, every feature is overlaid in the lower left corner and shall have *x* and *y* coordinate values. These coordinate values define the location and spatial extent of the point or feature. A single point in the map will be represented by a single *x*, *y* coordinate pairs like 9, 5. Representation of lines shall require a series of *x*, *y*, pairs like 0,3; 2,3; 3,2; 5,2. Here the first and last *x*, *y*, pairs indicate the starting and ending points of the line. Representation of polygonal area features requires a series of *x*, *y*, coordinates pairs but identical first and last pairs indicate a closed area (Figs. 23.5 and 23.6). Now, after having digitized the entire spatial data, the computer can draw or reproduce the map by using these *x*, *y* coordinates.

Digitization is done with the help of digitizer. The map that is digitized is pasted on the digitizer board and features are traced with the digitizer cursor. The cursor, keyboard, when used, identifies the coordinate value, transmits, and stores such values in the computer. Digitization is a time consuming and tedious job.

Editing and Cleaning

Notwithstanding systematic and meticulous planning and care, digitisation results into a number of errors. Such errors require to be eliminated before the map coverage is used further. Some of the common errors include dangling arc, duplicate line work, duplicate level points, inaccurately shaped arcs and missing data. Such discrepancies and incongruities need to be eliminated to make the digitised coverage a true representation of the original map. Such error are described briefly below.

Dangling Arcs

This error results when it is intended to connect an arc to another at an inter-section point but the connecting arc either falls short or extends beyond the inter-section point. When the connecting arc falls short of the inter-section point, it is called an undershoot. If it extends beyond, it is called an overshoot. Both the errors produces either open polygons or extra unwanted areas.

Duplicate line work :—This type of error results when an area is digitised twice.

Duplicate label points:—This type of error results when a level point is digitised twice.

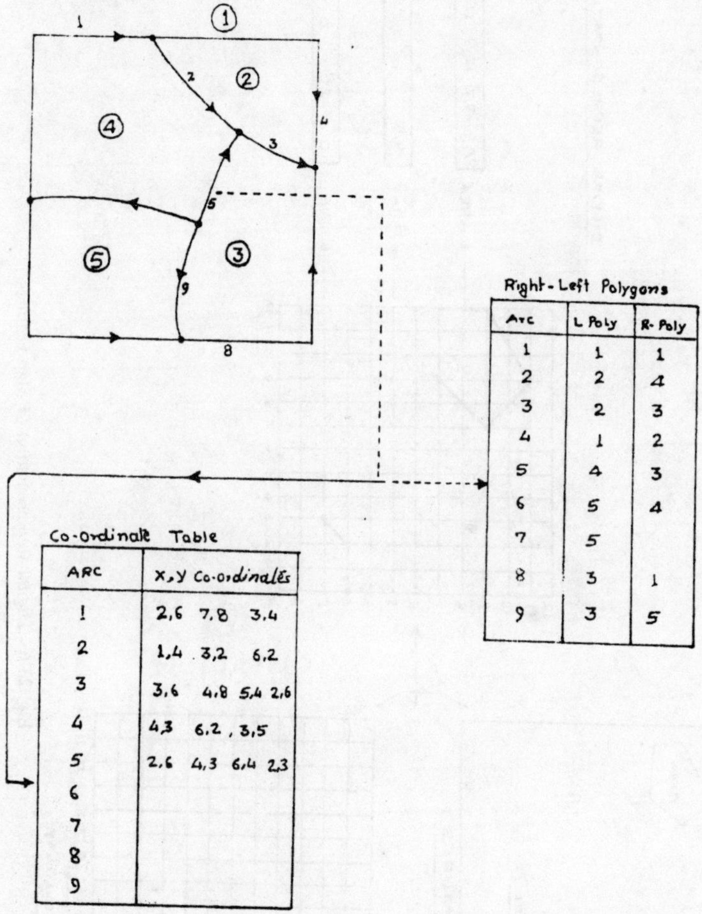

Right-Left Polygons

Arc	L Poly	R-Poly
1	1	1
2	2	4
3	2	3
4	1	2
5	4	3
6	5	4
7	5	
8	3	1
9	3	5

Co-ordinate Table

ARC	X, Y Co-ordinates
1	2,6 7,8 3,4
2	1,4 3,2 6,2
3	3,6 4,8 5,4 2,6
4	4,3 6,2 , 3,5
5	2,6 4,3 6,4 2,3
6	
7	
8	
9	

Fig. 23.5 : Polygon Typology in ARC INFO.

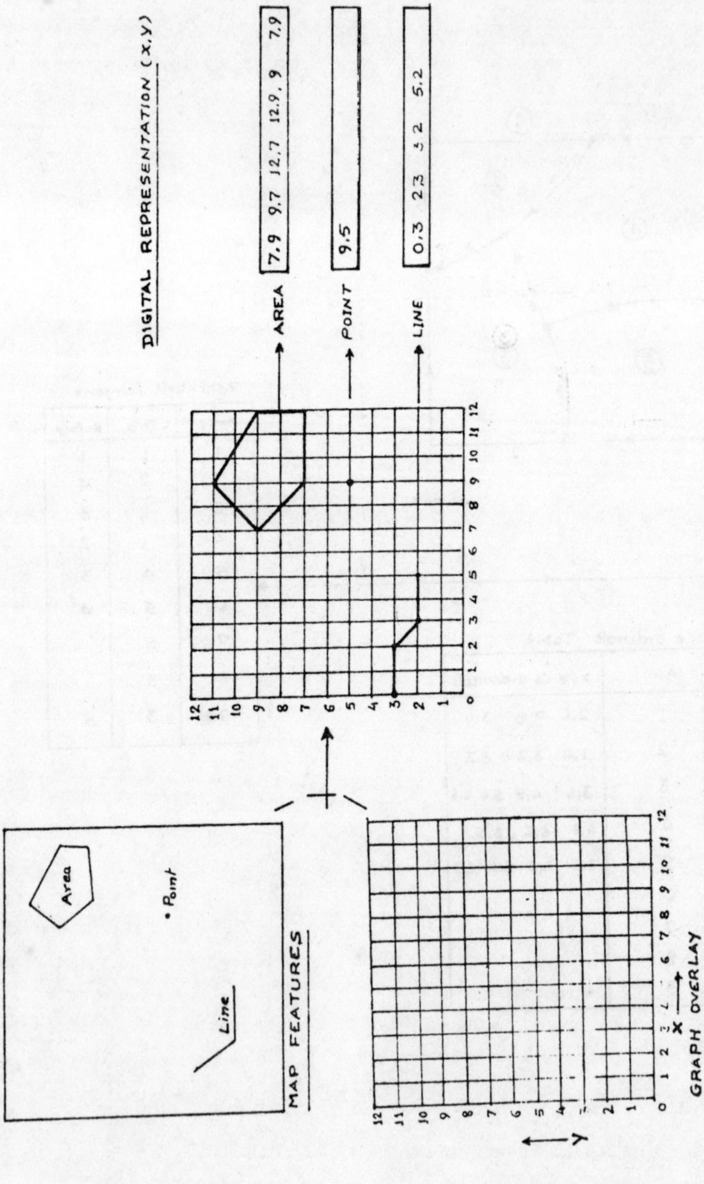

Fig. 23.6 : Digital Representation of Map Features.

Inaccurate space:—This type of error results due to over simplification. In other words, a feature digitised earlier, on scrutiny, may not tally the shape appearing in the map. In such case, the feature has to be carefully digitised again.

Missing data:—During the editing process, it may be possible that certain features shown in the original map have not been digitised. Such features are taken up for digitisation again.

Constructing Typology

When the map coverage is relatively free from errors pertaining to digitisation process, its typology is established. Typology refers to the details of connections between spatial objects such as polygons left and right of line segments or form or to nodes constituting an arc. Typological relationships have the following advantages:

— It enables efficient storage of data.
— It permits faster processing of data.
— It permits network overlay and other geographic analysis.

Building typology for each map coverage results in creation of a table in which left and right polygons to each arc are stored. If two polygons share a common arc, adjust to each other. Establishing typological relationship is a crucial step in geographic analysis.

Adding Attribute Information

Digitisation, as we have discussed, deals with input of graphic data. In a map, apart from the spatial data, there are many such non-graphical or textual data, which provides more characteristics, qualities and relationship to the graphic features. The examples of non-graphic data are name of the road, its width, name of the place. In a Geographic Information System (GIS) non-graphic data are correlated with the graphic data through common identifiers (Fig. 23.7). The graphic table contains co-ordinates for each map feature and the attribute table stores textual information corresponding to the graphic feature. The feature number acts as the common in both the tables and it functions as the link between graphic data and non-graphic data. After having established this link, queering of map features can be done with the help of attribute non-graphic data.

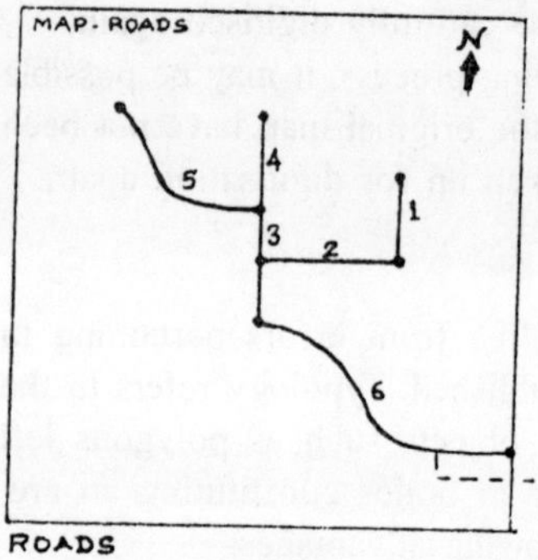

COORDINATES TABLE (Graphic)

Feature ID	Coordinates	
1	7.3	9.4
2	4.3	7.3
3	4.3	4.4
4	4.4	4.5
5	3.5	4.4
6	6.2	8.1

Attribute Table (Non-Graphic)

Feature ID	Width	Surface	Name
1	20	Concrete	street 1
2	10	Tar	street 2
3	5	Unpaved	street 3
4	20	Concrete	street 4
5	10	Tar	street 5
6	20	Concrete	street 6

Fig. 23.7 : Linkage of Graphic and Non-graphic Data through a Common Identifier Feature (ID).

Co-ordinates Transformation

Location points on the surface of the earth are referred in terms of latitudes and longitudes, which are angles measured from the centre of the earth. Such use of longitude and latitude does not provide accurate measurement of area or distance. It only indicates about the locating reference. In order to make the map coverage useful for spatial analyses like area and distance measurement, the earth's spherical surface has to be projected into a two-dimension plane surface where angle and area measurements are constant across the two dimensions. Using a map projection system does this. There are many map projection systems having their own distinctive characteristic effects on shape, area, distance etc.

The topographic map sheets of Survey of India in the scale of 1: 50,000 are in polyconic projection. Many GIS software packages have in-built transformation capabilities.

4. Geographic Analysis

Geographic analysis of data comprises the following components:

— Retrieval/Querying
— Spatial modelling.
— Manipulation technique.
— Data translator.

Retrieval/Querying

Data retrieval is the process of reviewing and presenting the information stored in the computer. For an example, retrieval involves asking for extent of all forest areas with 10 to 40 per cent crown density or areas with teak forests. Such information appears on the monitor. This will display only those forest pockets from an existing stock map or density map or forest type map.

Spatial Modelling

Spatial modelling comprises distance and area calculations overlaying, corridor analysis, proximity studies and network analysis. Distance and area calculation is a database querying. Distance between two points along a linear feature can be measured by making a sample query to that effect. Placing the cursor within the polygon and making the necessary query may retrieve areas of individual polygons in a thematic layer.

Overlaying is categorized into vector overlay and raster overlay and vector raster overlay.

(a) Vector overlay

Point and line on polygon overlays involve the overlay of point data and line data on vector polygon layer. Polygon on polygon overlays may involve overlaying soil texture, slope, and forest cover polygons to determine erosion susceptibility of an area.

(b) Raster overlay

This is mainly polygon overlay where the polygon are gridded, i.e. each pixel has its coordinates and values. By overlaying these, new data-sheet is created. This is used for measuring area and conducting multiple attribute modelling exercise.

(c) Vector-raster overlay

A village vector boundary may be overlaid on a raster surface such as forest area or agricultural area in order to determine the area.

Corridor or Proximity Analysis

This analysis is also called buffer zone analysis. This aims at defining areas of influence. This can be done around a point, line or around or interior of a polygon. For example, if a developmental plan requires the data regarding the number of villages situated within a zone of 5 kilometre away from the boundary of a National Park, then the GIS will easily demarcate the buffer areas around the park and show the number of villages.

Transformation Model

This is to create future prediction models. For an example, predicted loss of soil at a particular place may be estimated as a multiplicative function of many factors for which data is stored in the database. Based on the estimation, a final thematic layer is developed showing those locations where estimated erosion may pose serious problems to forest conservation.

Network Analysis

This is one of the spatial analysis and utility programme, which is developed by using different models created earlier.

Manipulation Techniques

Manipulation techniques are critical for qualitative designing of final output. There are different manipulation techniques. They are enumerated below:

1. *Map projection transformation*:—It establishes the mathematical relationship between the surface of the earth and the map.
2. *Rubber sheeting*:—This is a process of adjusting map features to enable the digital map to fit into a designated base.
3. *Map merging*:—This is a process of merging a portion of a digital map into an already available map layer. This technique is commonly used to update the existing data layers.
4. *Map generalization*:—Data are entered from a map having a certain scale. However, sometime it is required to produce a small-scale map on a data layer. In such case, maps are generalized or smothered in order to produce maps of a definite scale. For this, techniques like thinning dropline, edge matching, scale charge, projection change, coordinate rotation and translation are used.
5. *Raster/vector conversion*:—Raster polygons are converted to vector polygons by giving the grid cell size and directly cross hatched to produce maps. The *vice versa* is also possible, for feature editing and subtracting.
6. *Data translated*:—This is an important model, which is useful for converting the present data format into designed format. This has relevance in such situation where the required digital data such as satellite data is available in different formats.

III. An Organizational Context of GIS

The GIS functions effectively when it is placed in an appropriate organizational context (Fig. 23.8). In order to use GIS properly and effectively, apart from impending necessary investments in hardware and software, it is also necessary to engage personnel and managers to use the new technology in the proper organizational context. Considered in this perspective, the following are aspects which assumes importance:

GIS in Decision Support System

In order to implement GIS in an organization in its decision support system, the organization concerned should have sufficient amount of relevant data. GIS system is a potent tool, which helps the decision support system. As we know, that decision-making is a process of

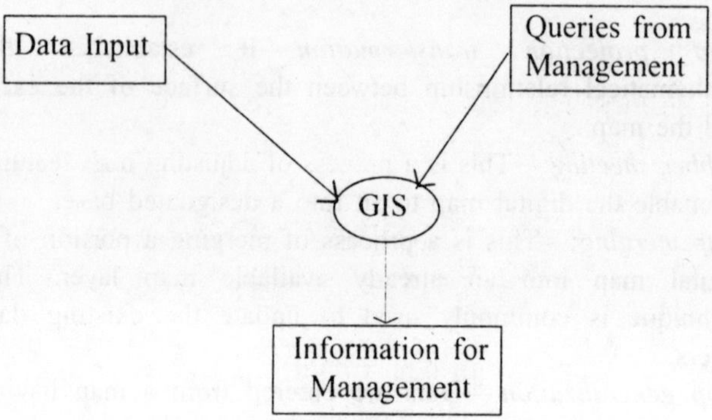

Fig. 23.8 : Organizational aspects of GIS.

selecting a course of action from an array of alternatives. Therefore, decision support system helps managers to react quickly to changing needs. Every organization has certain goals and objective, i.e. in forestry organization, one of the goals can be to know about the history of silvicultural treatments or know the extent of dense forests in the division. The GIS model should be designed in such a manner that the objective of storing data and querying desired information from the system is achieved. Therefore, GIS is an effective tool for management planning.

Data Reliability

Effectiveness of GIS depends upon the database available with the organization. The data has to be precise; so that they can be used in the GIS. In the forestry application, data is usually available in the form of maps and tables found in files and reports in different formats. These data are not compatible to the system. Therefore, there is a need to make these data compatible to GIS.

Responsibility

The responsibility of implementing GIS rests on the people concerned, who have to work together, bearing the following points in mind.

User Committee

A user committee has to be formed consisting of foresters and computer professionals who decide about the format of data to be used. The data should be concerned about which attributes are to be recorded, which are to be stored and which ones are important for a particular situation. The user committee, among other things, generally decides, whether a pilot project or a main programme is to be taken up for implementation or not.

Pilot Project

If the user committee decides about a pilot project to be taken, then it is necessary to analyse the possible problems and constraints one is supposed to face while handling GIS.

Database

To start working on the pilot project, one should have sufficient amount of relevant field data. Generally, such data are collected from primary and secondary sources. Such data collected is then fed into the GIS and accordingly oriented and used to meet the data need and to satisfy the analysis requirements.

Installation

Installation is an important aspect pertaining to GIS. However, before installing the system, one should decide about the type of hardware required. With the fast changing technology, one has to keep updating his information about the latest options with respect to hardware available. The hardware should be installed after careful consideration of all such requirements.

Training

Training is an important component in GIS. When the GIS is installed, a training programme for use of specific GIS modules is necessary. The duration of the training programme depends on the type of GIS application software. Nevertheless, the update systems supplied with GIS are quite complicated and need an extensive training for their use.

System Management

For operational working of a GIS, one needs an effective data management system. Data management necessitates proper training to be imparted to persons handling data. All data regarding inventory and other valuable information is generally handled by the data management system. Another important part of system management is handling upgrades. It can be either a hardware upgrade or a software upgrade to keep pace with the changing technology. Then comes data storage. Depending upon the quality and quantity of data to be handled, we go for different storage devices and depending on the importance and relevance of data, different backups are created.

Therefore, considered the whole gamut of issues related to GIS, it is thus obvious that the technology will be most effectively used when a proper organizational context has been established for implementing it.

Application of GIS

The GIS applications are multi-disciplinary and multipurpose. They provide a support for spatial data analysis in natural resource and socio-economic aspects. Mapping provides spatial framework on which GIS application depends. This system with its inherent capability of map overlaying, reclassification, proximity analysis and other mathematical operation, helps to carry out criteria based analysis.

The application areas of GIS are numerous. The following are some of the fields, which make use of this technology:

1. Automatic cartography.
2. Landscape designing.
3. Various types of mapping and map updating.
4. Land-use planning and management.
5. Environmental analysis, monitoring and prediction.
6. Natural resource mapping and management.
7. Hazard management.
8. Land and forest management.
9. Infrastructure planning.
10. Selection of highways routes, pipelines and supply lines.
11. Survey and engineering
12. Wildlife management.
13. Mineral exploration.
14. Geodemographic analysis.

GIS technology has wider applicability in analysing problems and formulating prescription for management plan. The following are some of application area. Few of most important advantages of GIS in the field of forest planning and management are explicated below:

1. GIS is helpful in formulating felling schemes in such a manner

that the prescribed yield is uniformly removed. For this, the growing stock available in each compartment is assessed from the inventory results. Then, it will constitute one attribute layer in the GIS. The compartment boundaries are digitized as another layer of information. Now marking and felling rules are defined objectively. The GIS is asked to assess the extent of felling and wood production in each compartments by superimposing compartment boundaries layer over the growing stock layer and applying marking and felling decisions.

2. Suitability maps for plantation, rehabilitation of degraded forest, wildlife management can be prepared with the help of GIS.

3. A variety of critical environmental data like slope, soil, drainage, erodibility can be analysed and processed easily. For this, each of these parameters are treated as separate layers. Combining these different layers with inventory results can help to evolve growth models.

Forest management necessitates improvement of site conditions and utilization pattern of the exiting forest resources. Such existing forest areas can be mapped with the help of GIS along with the database and potentiality of that area. The forest types, density, growing stock and other features like plantation can be represented by showing different colours. Suppose, we want to have a look at the teak plantation sites in the division or in different ranges. By simply giving a command, we get all such plantation areas on the map in a particular colour and scale. Other quantitative properties about plantation like survival percentage, age of the plantation, growth data, etc. can be retrieved from the data base any time we want. The development and changes made from time to time are fed into the database or plotted from the graphics. Therefore, the history of all treatments and changes made in the area can be kept in the GIS for future planning.

4. *Planning the Multipurpose Uses of Forests*: GIS helps in understanding the resource position, and their potentiality with spatial distribution. Apart from this, with the aid of this tool, multiple use of forest resources can be planned. As for example, protection zones like stream, river, banks, or bio-diversity conservation zone, wildlife habitat zone can be easily identified for special management.

5. *Producing Maps from GIS*: GIS can be used for producing maps. Such maps can be used for planning, management and monitoring of forest resources. Such maps can also be utilized to illustrate and visualize the effect of various proposals related to

diversion of forest land to non-forestry works.

6. *Soil Survey and Soil Resource Inventory*: GIS can be used for soil survey applications and for preparing soil resource inventory for a forest division or a region.

7. *Developing and Managing Watershed Projects*: The watershed projects can be very effectively managed by using GIS. The projects require an integrated effort to check the higher rates of run-off. Control of soil erosion facilitates a good vegetation cover on the base areas of the hills and as a result it helps to improve the condition of the people living in the vicinity of such areas. The imaginary and field survey data of a particular watershed area may be amalgamated to create a comprehensive database. Now, analyzing the problems in different sites, one can identify high priority areas. The denuded and high slope areas are treated by undertaking appropriate measures. The efforts are monitored at regular interval and database expanded to see the impacts of the works carried out.

The area conducive for minimizing erosion in a watershed can be well demarcated by over lapping the maps by different signs and digits. The population density and pressure on the adjoining areas of habitation can be quantified and database of the population living in that area be created covering all aspects of their social profile. The watershed area can be divided into segments for management purposes using GIS.

8. *Assessment of Land for Timber Production* : Assessment of land for timber production is one of the application areas of GIS. To determine the sites suitable for timber production, rating can be developed for soils and land from maps using their site indices. For each forest working circle or felling series all attributes (*e.g.* type, density, site quality, soil depth, etc.) should be compiled after collecting data from the field. The GIS allows to interact with data by asking such questions as the exact location of a particular type of a particular species etc. This query provides the answer in the form of both map and tabular data for each such area. In this way, it helps in taking proper decision in the course of planning, execution and monitoring.

9. *Forest Protection* : GIS helps in protecting forests. We produce map-using GIS. Such maps helps in identifying areas beset with various problems. The areas, which are prone to damages, are marked and planning for its control is carried out. The incidence of illegal activities can be recorded in the information system and the efforts

to control these activities can be monitored in regular interval.

10. *Wildlife Management* : Wildlife management requires multidimensional efforts to maintain a proper habitat with least disturbance to the fauna, properly balanced flora and fauna, and proper prey predator ratio. Forest sites, suitable for particular species or community can be easily selected using the GIS. Based on this analysis, management decision regarding developing proper habitat for any species can be taken. In order to accomplish it, it is necessary that all requirements of a species are fed into the GIS and superimposed upon one another along with other requirements. The results will be obtained in the form of well-defined and demarcated areas with required buffering.

Planning a GIS Project

A GIS project plan is a comprehensive guide that specifies the general methods as well as the detailed procedures for solving a problem. This helps in progressively classifying and integrating the detailed definitions and procedures with the overall objective of the plan. A project plan is evolved basing on the objective of the project. However, before beginning any GIS project, the following aspects require serious attention:

1. The objective of the project.
2. Capabilities of the GIS available packages.
3. The budget consideration and cost effectiveness of the exercise.
4. The staffing position.

The GIS project comprises the following aspects:

1. Formulating the problem and stating relationship.
2. Developing the plan and preparing for data collection.
3. Data collection.
4. Analysing the data.
5. Arriving at conclusion and reporting results.

These aspects of GIS plan can be arranged in four major steps. These steps are:

1. Determining the objective of the plan.
2. Building the database.

3. Performing the analysis.
4. Presenting the results of the analysis.

Determining the Objectives of the Plan

The design and execution of GIS plan is determined by the objectives of the main projects for which it is being taken up. Therefore, it is necessary to identify the objectives of the plan in line with the main project.

Building the Database

- All the data need for the project has to be specified and their sources have to be identified. Data collection is one of the important phases of GIS project.
- After having identified, the source of the data, they need to be identified in terms of geographic features and their attributes.
- Now the features are to be organized into layers of data. The layering of data are generally organized thematically by what they represent, such as contours, stream, river, villages etc. and may also be separated in terms of feature types such as point line or polygon.
- All the features need to be registered in terms of their geographic location. Different layers are registered to a co-ordinate system. The projection system and the scale of the project input and output maps is to be determined at the outset.

Data automation

After having collected the data, now there is a need to enter the GIS system. The process of digitization achieves this.

Performing the Analysis

Analysis of data begins after building a good comprehensive database. The process of retrieving and querying, provides a powerful tool for data analysis. This helps in analying the data against the backdrop of a problem and finding out an array of solution to the problem and its concomitant implication.

Presenting the Results

After comprehensive analysis of different options available to solve a problem, a set of results are obtained.

24
Participatory Rural Appraisal (PRA)

Introduction

Data collection is a vital part of planning exercise. The pace of collection of data, among other things, primarily depends on the methodology adopted. There are many methodologies adopted to collect data. However, such methodologies can be classified into the following two categories:

1. Conventional Methodology.
2. Participatory Methodology.

Conventional Methodology

The conventional survey method is a commonly used method for data collection. It involves surveying of selected areas by employing different tools like filling of questionnaires, interviewing or direct observation and recording. Such survey can be conducted by resorting to sample survey or complete enumeration or census.

— Complete enumeration or census involves enumeration of all subjects. Population census, live stock census are conducted for political, and developmental planning.
— Sample survey is another conventional methodology. This type of survey can be small, medium or large and it covers household / individuals / firms / other units. It can cover a single subject or multiple subjects. The data collection may be done by the following means.
— Through filling up of a questionnaire.

This is the most commonly used method for collecting socio-economic data. This involves preparing a written questionnaire for respondents. A questionnaire is defined as a group of questions designed to elicit information on a subject or a sequence of events, from a set of respondents.

— Through measurement, *e.g.* for estimated measurement of forest stand/crop.
— Through interviews.
— Through observation of the phenomenon under study.
— Through combination of these methods.

Participatory Methodology

Of late, there has been a paradigm shift from "development from above" to that "from below". It is increasingly felt by the policy planners that unless and until the local communities are involved in the developmental programme, no such efforts will succeed. Besides, it has been an accepted fact that indigenous knowledge system constitutes an alternative model for sustainable development of resources. Considered in this perspective, participatory methodology involves interacting with local communities, understanding them and learning from them on a continuous basis. Again, it is a way of learning from and with local community to investigate, analyse and evaluate constraints and opportunities and make pragmatic decision regarding developmental plans and projects. In other words, it is a method of generating a variety of data, identifying and mobilizing effective groups and evoking their participation in analysis, planning, decision making, execution, monitoring and evaluation.

Participatory Rural Appraisal (PRA)

Participatory Rural Appraisal (PRA) is a methodology for interacting with villagers, understanding them and learning from them. It involves a set of principles, a process of communication and a menu of methods for seeking villager's participation in putting forward their points of view about any issue and enabling them to do their own analysis with a view to make use of such learning. It initiates a participatory process and sustains it. Its principles and the menu of methods help in organizing participation (Neela Mukherjee, 1993).

PRA is a method of collecting variety of data, identifying and organizing intended groups and evoking their participation and initiating ways in which the said groups can interact and participate in different management processes like decision making, project design, implementation, monitoring and evaluation. It provides alternative frameworks for data collection, analysis and interpretation. Because of its participatory in nature, it is widely used to focus attention on people, their livelihoods and their inter-relationships with socio-economic and environmental factors.

Participatory Rural Appraisal (PRA) is not only a means of learning from the people but also a method to investigate, analyse, assess and evaluate prevailing constraints and opportunities in a rural society. It helps in taking timely decision about developmental projects in every sector. Now-a-days, this methodology is widely used for planning, executing, monitoring and evaluating different programmes in the field of rural development and environment management.

Objectives of PRA

Participatory Rural Appraisal (PRA) is conducted with various objectives in mind. Some of the objectives are given below:

1. To involve people by learning about their perceptions, experience and capabilities.
2. To collect data for preparing schemes/plans/projects/ programmes.
3. To understand the impacts of on-going schemes, plan, projects and activities.
4. To cross-check data collected from other sources.

The Kinds of PRA

The Participatory Rural Appraisal (PRA) methodology can be categorized into different kinds. This type of classification is based on the objective of participatory rural appraisal. They are as follows:

1. Explorative.
2. Topical.
3. Deductive.
4. Research, training and statistics.
5. Planning and implementation.
6. Monitoring and evaluation.

1. *Explorative PRA*

It is a PRA exercise conducted to understand broad patterns and trends prevailing in the rural society. This exercise provides inklings of socio-economic profile of the rural life. It throws open various burning issues afflicting the society at that particular point. Such issues need to be addressed forthwith. This again constitutes a basis for further in-depth investigation of any other complex and vital issue.

2. *Topical PRA*

When a PRA exercise is conducted to gather more insights and to collect more information on any pre-selected subject, then it is called a topical PRA. Dependence of people of a village on nearby forest can be studied by undertaking a topical PRA.

3. *Deductive PRA*

Here, the PRA exercise is undertaken for a subject or theme, which cannot be probed directly, because of the sensitive nature of the subject. Subjects like communal tension, corruption, police excesses can be taken up for undertaking deductive PRA. Different facets of village life covering wide range of socio-economic, socio-cultural, socio-political, and socio-environmental aspects can be taken for study under this method.

4. *PRA for research, training and statistics*

PRA exercise aims at collecting data and building a sound database. Such exercise can also be carried out to conduct research on any topical subjects. The exercise can also be designed to impart training to different group of professionals in the field of natural resource management and rural development.

5. *PRA for planning and implementation*

Here the PRA can be conducted for designing specific projects/ schemes/plans. This tool can also be applied to find out constraints and problems inherent in the implementation process of any project. Designing of all stages of a project cycle can be based on the information and experience gained by PRA. This helps to make the project or plan more people centred. As a consequence, it helps improve sustainability of the project or plan.

6. *PRA far Monitoring and Evaluation*

PRA is conducted to monitor and evaluate projects/plans/programme. With the identification of problems and constraints which come on the way to achieving the objective of the project, solution to the problem can be found out and as a result the performance of the project/plan can be improved considerably.

Participatory Appraisal of Natural Resources (PANR)

Participatory Appraisal of Natural Resources (PANR) is an appraisal conducted with the help of the people who live in proximity with such natural resources and have a close relationship with them. It is the local people who analyse, assess and evaluate use of such natural resources and take appropriate decision with regard to management of natural resources in the context of knowledge/experiences gained over the ages.

Participatory Appraisal of Natural Resources (PANR) is a way of learning from local community about conservation and management of natural resources. It is based on the method adopted in Participating Rural Appraisal (PRA). PRA has been described as a growing family of approaches and methods to enable local (rural or urban) people to express, enhance share and analyze their knowledge of life and conditions, to plan and act (Chambers, 1994).

PANR is a method for interacting with local people and learning from them about the natural resources and their related issues concerning to the people. PANR eventually involves participatory processes, which encompass a set of principles, attitudes and behaviour, congenial to learning and participation. Apart from this, it carries out participatory methods to enable local people to participate in knowledge building exercises, investigate and analyse their problems, evaluate opportunities and constraints and take informed decision regarding natural resource management. In spite of all such analogy in approach and methodology, PANR differs from PRA in the following two counts.

1. PANR is a participatory appraisal of natural resources. Therefore, it is specific to conservation and management of natural resources. Whereas, PRA is a methodology primarily focused on community issues of general kind.
2. PANR focuses on local natural resources from the viewpoint of local community and highlights the critical relationship of local community with the natural resources.

Participatory Methods

The process of participation necessitates application of certain methods. Such methods are the techniques, which help in seeking involvement of local people, enabling to analyse the issues and to take proper decision. In such an exercises, the role of facilitator is of immense importance. It is because of the fact that the facilitator, with his skill and expertise, facilitates involvement and participation of people in the exercise. The facilitator helps in conducting the exercise in the following way:

— In facilitating groups/communities without raising expectations.
— In identifying groups/key informants.
— In conducting semi-structured interviews.
— In conducting visual participatory methods.
— In initiating interacting participatory process.

There are different types of methods involved in the participatory approach. Some of the methods, which have relevance to field situation with respect to forest resources management, are explained below:

1. Rapport building.
2. Identification of key informants.
3. Semi-structured interviews.
4. Time line.
5. Participatory mapping.
6. Identification of groups.

   (i) Asking key informants.
   (ii) Participatory social mapping.
   (iii) Wealth ranking.
   (iv) Venn diagramming.
   (v) Social matrix method.

7. Venn or "Chappati" diagram.
8. Joint walk.
9. Vector scoring.
10. Matrix scoring.
11. Participatory seasonality analysis.
12. Trend analysis.
13. Flow diagram.
14. Source diagram.

Detail features and process of each method are discussed below:

1. *Rapport building*

Establishing an effective line of communication is essential for participation. A visit to a village with an objective of building communication link sets off the process of rapport building. In the initial stage, ordinary conversation with the people about the village life breaks the ground for establishing a dialogue. Sometime, it becomes necessary to identify a suitable person in the village to start a wide-ranging discussion, which gradually spills over to bigger cross-section of the people. Rapport building is a time consuming exercise and it needs considerable skill on the part of investigator. However, once a relationship is established then it becomes easy to apply other techniques of participatory enquiry. There are many methods by which rapport building can be achieved. One such observed method is "do it yourself or DIY". This means doing the task or activities like cleaning, harvesting, digging etc. in the same manner as the village community does in their day to day life. Another method "lend-a-helping hand" (LAHH) is also in vogue. Any help to community, in the hours of need and distress, stimulates the people to establish a communication with the investigator. Any help to an old man or a child or a patient can create a compatible atmosphere for developing an effective rapport with the community.

2. *Identification of key informants*

Power of observation, analysis, knowledge and wisdom differs from person to person. Therefore, it is necessary to identify those people who are endowed with such qualities. These people shall constitute as key informants. Identification of such key informants requires meeting a cross-section of people and then selecting such people who has an inherent capacity of knowledge, observation, analysis and impassionate expression, while taking part in a participatory method. Such informants help with crucial information regarding natural resources, while taking part in a participatory method of consultation. They also help in clarifying aberration and in eliminating errors and omissions. However, over-dependence on such key informants kills the enthusiasm of other participants. Therefore, it is imperative to adopt a trade-off while handling such key informants and other community members.

3. *Semi-structured interview*

Semi-structured interview contains flexible and open questions and not pre-determined questions. Therefore, such open question leads to an array of responses on which further impromptu question can be prepared and asked to the participants. In this method, the interviewer should have a checklist of issues on which instant questions can be asked. This helps the interviewer in conducting himself in proper way. Apart from this, it is always desirable that a multi-disciplinary team should conduct the participatory enquiry. This helps in approaching the problem or issues from different angles and perspectives. During the course of the interview people are interrogated with the use and help of words like 'why', 'how', 'when', 'where', 'who' and 'what' in order to elicit better information about any activity or phenomenon. Another important point, is jotting downs of the responses of the people

4. *Time Line*

It is a powerful participatory method used for understanding relationship of local communities with the natural resources. In case of degradation of forest resources, it is necessary to know the historical profile of such resources in different point of time. Apart from this, this tool helps in understanding aspects like relationship of people with the resources, the impact of resource degradation on the livelihood of village people, community efforts, if any, to conserve such resources, copping strategies of people in the situation of resource constraints, perception of local people towards different tree product, changes in wage rate, price line and food patterns.

Changes about resource status can be discussed with the people. Chronology of such changes or events can be constructed by the community members. This process helps in understanding the activity or phenomenon in a historical perspective. To initiate this learning process, one has to interact with the elderly person who symbolizes a repository of historical events having a relevance to the said natural resources. Such elderly person helps in tracing the historical patterns of changes. However, interaction with one or two such persons fails to establish a viable chain of communication. One has to rely on a number of elderly people to construct a line of events having focus on time as well as continuity. A time line constructed in this fashion is presented in Fig 24.1.

Year	Social profile	Agricultural Wage	Price	Role of forests and coping mechanism
Before independence < 1947	• 25 to 35 households • land ownership • Medium Land holders • Education. Only one person-literate. • Untouchability. • Barter system	2 annas per day or equivalent quantity of rice.	Rice: 4 paise per seer Pulses : 12 paise per seer	• Adequate CPR and forest resources. • Dependence on forest product during Magha to Ashadh (January to May/June.)
After independance till Emergency 1948 to 1975.	• 65 to 70 households • Land ownership • Non-agricultural occupation • Medium Land holders. • Education. 8 persons-literate. • 6 to 7 households migrated to towns.	Rs. 4 to 8 per day	Rs. 2 per kilo of rice and Rs. 5 per kilo of pulse.	• Pressure on CPR and Forest Resources increase. • Collection of food and non-food items from forests. • Marketing of forest produce especially NWFP • Degradation of forest adjoining to village.
After emergency to till today 1975 to 1995	• 160-190 households • Cattle population increased • Marginal Landholders increased • About 225 people literate	Rs. 8 to 20 per day.	Rice : Rs. 3 to 6 per kg. Pulse Rs. 9 to 18 per kg.	• Use of CPR and forest beyond carrying capacity. • Shrinkage of CPR • Employment increased becasue of plantation. • Increased use of cowdung as cooking fuel. • On-farm tree cultivation increased.

Fig. 24.1 : Timeline

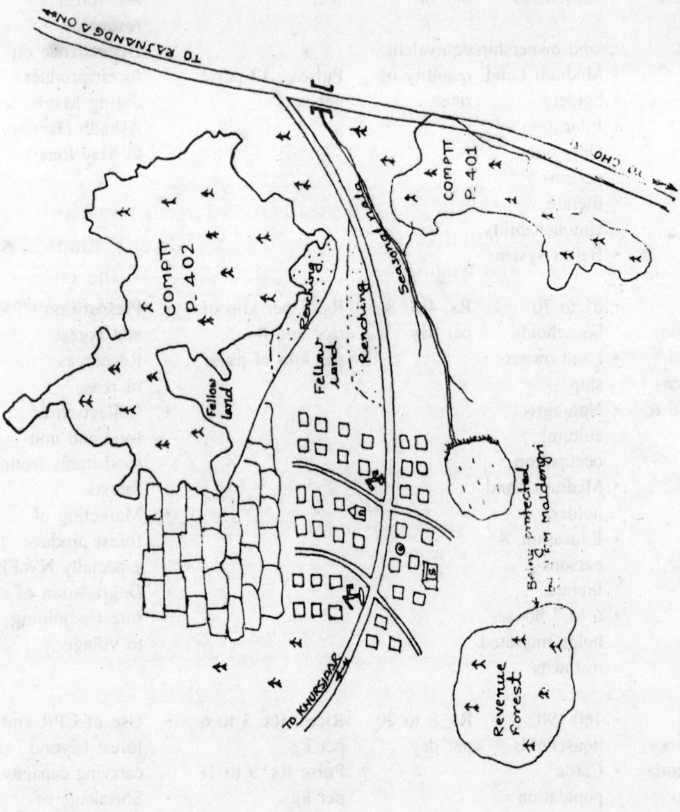

Fig. 24.2 : Participatory Reources map of V.F.C. Maldongri.

5. *Participatory Mapping*

Participatory mapping is a technique, which comprises spatial drawing of any one area. This is drawn generally on a piece of paper to show and explain their locality, physical attributes and events. The objective of such a method is to use the mapping for further in-depth probing of any issues. Maps prepared under this exercise generally shows households, farms, water bodies, forest, road and institutions. Sometime, events like literacy and epidemic incidences are shown on the map. Maps can also be prepared on the basis of separate themes like deforestation, environmental degradations, crop failure and epidemic incidences etc.

Participatory maps are generally drawn by key informants in the presence of other participants who help and qualify the mapping exercise. An intensive discussion among the participants, in the course of mapping, helps in adding new information and eliminating errors and omissions. Such participatory mapping exercise can be applied to identify:

1. Grazing areas
2. Forest areas prone to illicit felling.
2. Areas suitable for plantation and silvipasture.
3. Areas of forest used for collection of minor forest produce and fuel-wood.

6. *Identification of groups*

Communities are always socially heterogeneous in nature. Many different groups or sub-groups are existent in a community. Therefore, it is important to identify such distinguishable groups or sub-groups from the point of use of natural resources. There are many techniques to conduct the identification process in a village community. They are as follows:

1. *Asking key informants* : Key informants can be asked about the existing groups in a community. The views of key informants can also be elicited and such information is collected from groupings. Then this can be checked by other sources.

2. *Participatory social mapping* : Participatory social mapping is another method of identifying the socio-economic groups existing in a village community.

3. *Wealth ranking* : Wealth ranking is another tool for identifying groups. Grandin's (1988) method of card sorting is a method of

wealth ranking where a list of households is prepared with each household having a separate number along with the name of head of that household. Each household's name and number is transferred on a separate piece of paper and these papers are ranked by key informants according to their criteria. In the process, a set of criteria for wealth ranking emerges where each household has an allotted rank. This method helps in comprehending the socio-economic disparities amongst households and evolving local criteria for classification of wealth (Fig. 24.3).

4. *Vann diagramming* : Vann diagramming is another method of identifying groups in a community. This method is used to indicate the existence of different groups and their related importance. It helps in understanding the dynamics of economic and social relationship in the community as well as within groups.

5. *Social matrix* : Social matrix is another technique for identifying group. In a grid like set up, different criteria for different groups or sub-groups can be determined by eliciting information from key informants. This provides a broad idea about the existence of different groups in a community. The set of inferences can be put to a round of cross checking by other participants.

6. *Vann or "Chappati" diagram* : This is a very effective visual method. It helps in indicating mutual relationships and role of individuals, organization, institution or groups in decision making. Circles of different size representing different individuals/institution/organization/groups display their importance in decision making. Such diagrams are generally drawn on paper. The size of the paper is cut on a circular fashion and its distance generally indicates the degree of relationship (Fig. 24.4).

7. *Joint Walk* : This is another important method of learning and understanding specific issue in a community. Under this exercise, a joint walk with the participants is conducted to observe different aspects and issues related to development of natural resources. For example, while conducting a PRA, a joint walk to the degraded forest should be undertaken to understand the causes and extent of deforestation and the probable action for rehabilitation. This method has immense significance and relevance, so far forest resource planning and management is concerned.

8. *Vector scoring* : This is another visual method. It comprises scoring of criteria items for assessing their relative importance. This helps in prioritization of problems or issues. The selection of criteria

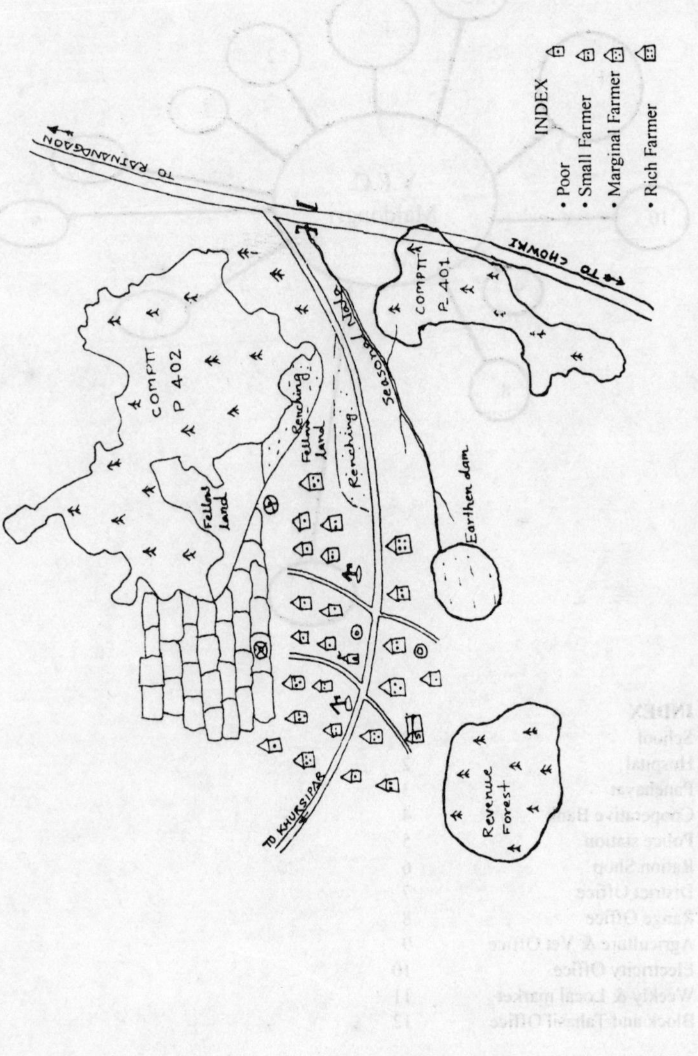

Fig. 24.3 : Participatory Wealth Grouping of V.F.C. Maldongri.

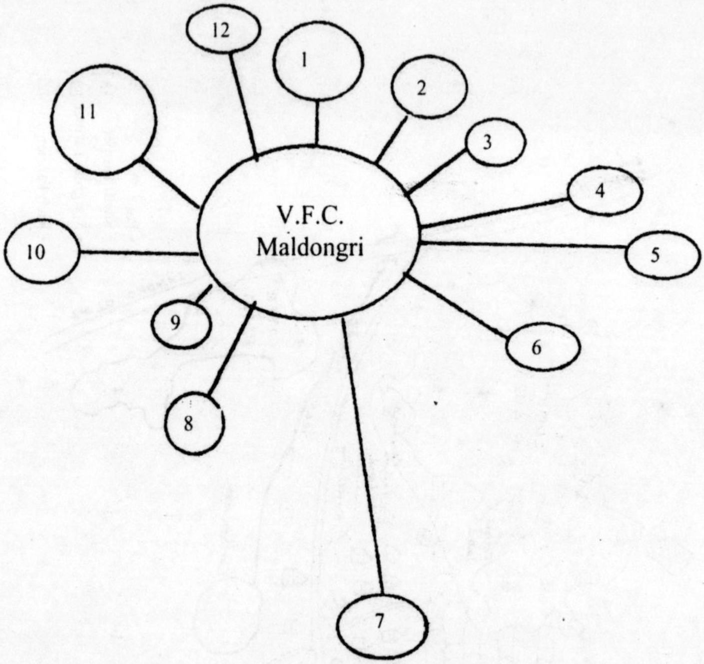

INDEX

School	1
Hospital	2
Panchayat	3
Cooperative Bank	4
Police station	5
Ration Shop	6
District Office	7
Range Office	8
Agriculture & Vet Office	9
Electricity Office	10
Weekly & Local market	11
Block and Tahasil Office	12

Fig. 24.4 : "Venn or 'Chapati' diagram" of V.F.C. Maldongri.

or items is a daunting task. It requires a continuous interaction with the participants. Once the list of criteria is determined, now, scoring process starts. Scoring or weightage is given by groups to different criteria with the help of indigenous symbols like seeds, stones etc. or putting points. Groups are required to place highest score on an item or criterion, which they consider most significant. On this basis, scoring can be done. Individuals can also do scoring, but in such case, again group scoring should be resorted to for a final assessment (Fig. 24.5).

S. No.	Criteri	Scoring by Local People
1.	Greenery	● ● ● ● ● ● ●
2.	Benefit to Farm	● ● ● ● ● ●
3.	Availability of Grass	● ● ●
4.	Availablity of Wood	● ● ● ● ●
5.	Effect on Soil Erosion	● ● ●
6.	Grazing of Live Stock	● ●
7.	Effect on Climate	● ● ● ●
8.	Availability of M.F.P	● ● ●
9.	Aesthetic Benefit	Not scored

Fig.24.5 : Vector Scoring : Benefit from Joint Forest Management

Note : Fixed scoring out of 10 seeds higher score indicates higher benefit.
Scoring done by V.F.C. Maldongri village (Rajnandgaon Dist.)

9. *Matrix Scoring* : This is another method of appraisal. It involves scoring of a range of criteria against a host of comparable items. The process begins with generating a set of items on which scoring is to take place and then listing of relevant set of criteria. Scoring takes places in a grid like set-up with criteria in the column and items on the rows, or the *vice versa*. This method is used to assess the choice of participants with respect to any issues. For an example, suitability of any tree species as a source of fuelwood or timber can be determined by using this matrix scoring method. Matrix scoring of fuel wood species is presented in Fig. 24.6 and utility scoring of minor forest produce is presented in Fig. 24.7.

Species Criteria	Landia	Palas	Saja	Dhawara	Karra	Awnla	Tendu	Wish Variety
Proximity	8	6	6	3	10	1	1	7
More Ash	8	10	6	4	10	5	4	4
Charcoal	6	1	8	10	1	4	10	10
Burn Longer	4	2	8	10	4	6	10	8
Easy to Light	10	10	6	4	10	8	4	6
Easy Availibility	10	6	4	2	8	4	2	4

Fig. 24.6 : Matrix Scoring on Fuel Wood Species.

Utility/Name of forest produce	Amla	Behra	Harra	Tendu	Mahua	Kusum	Bel
Fuel Wood	••	•	••	•••	•	–	•
Construction Purpose	•	•• •	•• ••	•• • ••	–	••	—
Fodder	••• ••	—	—	•	—	••	•• •
For Consumption Value	••• •••	—	—	•• ••	••• ••• •••	•• ••	•
Medicinal Value	••• ••• •••	•• •• ••	•• •• ••	—	—	—	••• •••
Commercial Value	•• ••	•• •	•• •	•••• •• ••••	••• •••	• •	•• •
As Fruit	•••• •••• ••	—	—	••• •••	••• •	—	••

Fig. 24.7 : Utility Scoring of M.F.P to Villagers of Maldongri (Rajnandgaon district)

Matrix scoring method is an important tool for decision making with regard to many issues pertaining to natural resource utilization. It provides a broad picture of preferences of group or individuals about the use of different forest products. It helps in prioritization of issues, which consequently leads to formulation of action plans.

10. *Participatory seasonality analysis :* This is another tool to understand seasonal dependency of local people on natural resources and degree and nature of relationship of local communities with the forests. The method involves asking people to enumerate and compare their activities, like livelihood patterns, food patterns, disease incidences, and rainfall pattern. These activities or events are generally attributed to occurrences of seasons. Based on such information elicited, a seasonal calendar is developed by using indigenous materials (Fig. 24.8). The process of documentation begins with a discussion with the participants about the seasons and activities undertaken in that period. Generally, local participants construct their own seasonal calendar to explain seasonal variations with respect to different activities or events and illustrate the nature and effect of such variation on the life of the villagers. Different seasons with accompanying activities can be depicted on separate piece of paper or represented through indigenous symbols, preferable on the ground. This marks the completion of visual chart of seasonality. However, this chart is again scrutinized by conducting semi-structured interviews for greater details and better clarity of information. This "Participatory Seasonality Analysis" technique has wide applicability in forest resource planning and management. It can be applied in the following situations:

Month	Labour Calendar of V.F.C. Member
Baisakh (April-May) Jeth (May-June)	Collecting tendu leave, mahua fruit achar fruit.
Asar (June-July) Sravan (July-Aug.)	Preparation of agriculture field and Sowing, planting paddy
Bhadra (Aug-Sept.) Kunwar (Sept.-Oct.)	Agriculture work and local labour at field.
Kartik (Oct-Nov.) Aghrayan (Nov-Dec.)	Harvestry of paddy and other crop and labour at field.

Poos (Dec.-Jan.) Magh Jan.-Feb.)	Labour at other's field and outside the village.
Phalgun (Feb.-March) Chait (March-April)	Collecting mahua fruit and achar fruit.

Fig. 24.8 : Seasonal Calendar.

1. To determine seasonal dependence of local community on. Major and Minor Forest Produces.
2. To prepare food calendar especially with respect to wild food.
3. To determine the coping mechanisms of local communities with respect to shortage in fuel wood, grass and fodder availability.

11. *Trend Analysis* : In this method, the local people describe the trends of use of natural resources. They also delineate other variables that affect the use. Based on their description, a historical perspective of the event or phenomenon is created. Extent of forest resources, in terms of stand density, species composition, bio-diversity can also be visualized and chronicled (Fig. 24.9).

Year → Horizontal
Score ↓ Vertical

1970	1980	1990	1998
• • • • • • • • •	• • • • • • • •	• • • • •	• • •
• • • • • • •	• • • • •	• •	• •
• • • • • •	• • • • •	•	•
• • • • •	• • •		
• • • •	• •		
• • •	•		
• •			
•			

Fig. 24.9 : Trend Analysis of Tree Population.

12. *Flow Chart:* In this participatory method, the local people are involved to describe and explain a cause-effect relationship. The description about factors and their linkage can be verbal or visual. Each problem can be taken up for appraisal separately. After putting the problems in a flow chart, possible solution will be thought of and tested against the responses and perception of local people (Fig. 24.10).

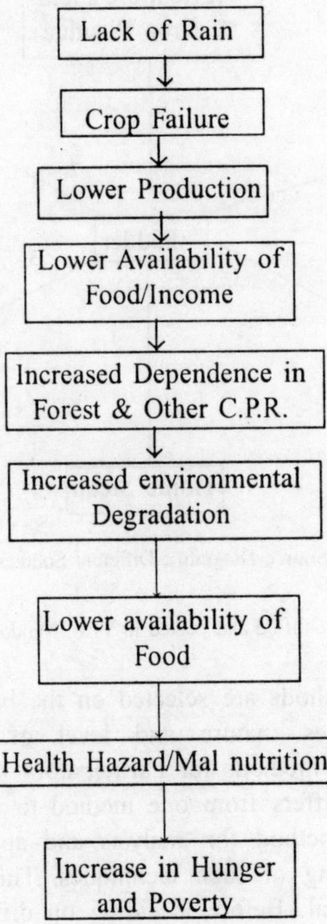

Fig 24.10 : Flow Chart : Result of Forest Degradation

Note : As stated by villagers of V.F.C. Maldongri (Rajnandgaon).

13. *Source diagramming:* This is another technique of participatory appraisal. It is used to identify sources of any activities, issues or events or phenomenon. In natural resource planning, the sources of fuelwood, fodder, grass, small wood can be identified by resorting to this tool. Apart from this, the dependability, and future position of such sources can be gauged from such analysis (Fig. 24.11).

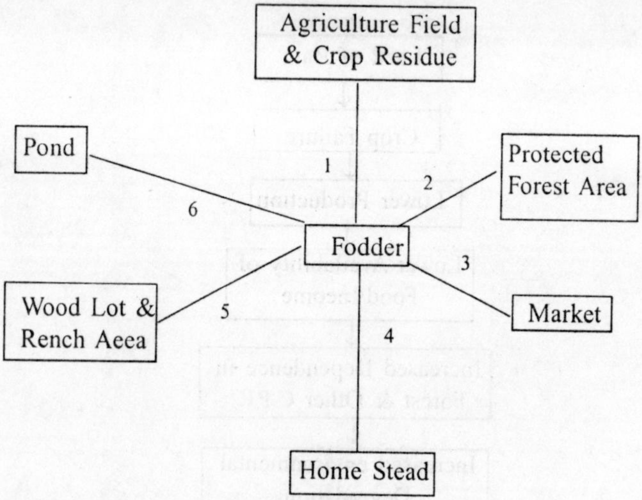

Fig. 24.11 : Source Diagram : Different Sources of Fodder.

Note : Sources of fodder identified and ranked in V.F.C. Maldongri (Dist : Rajnandgaon)

Participatory methods are selected on the basis of such factors like, objective, issues, nature and level of participation. The effectiveness of such methods for Participatory Appraisal of Natural Resources (PANR) differs from one method to another. Availability of more than one method for analysis and appraisal of an issue necessitates sequencing of such techniques. This provides synergy effect on the appraisal. Before, deciding on different methods and their sequence of use, it is necessary to comprehend the theme, nature of enquiry required, level of expression of the community, level of analysis and potentiality of local informant's knowledge. This helps in deciding the sequence of different methods to be adopted in the field.

In a case where the relationship of a community with the nearby forests is required to be explored, the method of vector scoring can help to prioritize problems of accessibility to forest resources. This can be followed by a well planned and semi-structured questionnaire based interview and mapping, which can be used to investigate further into the subject or theme. Participatory mapping is a useful tool to provide a spatial view of an issue. This can further be elaborated by using resource map indicating positioning of different natural resources such as trees, wildlife, pasture areas, degraded areas, etc. Specific location of such resource can be identified on the basis of joint walks carried out with the involvement of people. After identifying problems, with respect to their spatial position and intensity, one can think of prioritization by using vector-scoring method. Solution suggested by local people to such problems can be scored or enumerated by using matrix-scoring method. Based on the results of such scoring, appropriate actions can be planned. Therefore, conducting a series of such participatory techniques in a sequential basis helps not only optimum participation of local people but also unbiased identification and analysis of problems. As a consequence the entire exercise helps to find out solution in consonance with hopes and aspirations of people. It also helps in managing the natural resources on a sustainable basis.

25
Micro Planning

Introduction

Planning is an important function of forest management. It has evolved into different stages over the time. The usual planning being practised in the national development is called macro planning or board level planning. Over the years, it has been felt that success in any project or scheme requires detailed local level planning. This type of planning is called micro planning. Micro planning is planning with great detail. Micro plans are specific to a village or a group of villages.

With the adoption of sustainable development as the goal of present developmental planning, the relevance of micro planning has increased manifold. It is now an established fact that, before undertaking any planning for sustainable development of natural resources, the socio economic and environmental systems have to be studied and assessed in the backdrop of local variations. Such a study can be conducted only with the participation of people. Based on such participatory method of study, a micro plan for development would be developed. In other words, involvement of people is the integral part of the entire micro-planning process. Sustainable development of the local natural resources necessitates full participation of local people in the process of planning, implementation and monitoring and evaluation. The role of people's participation in planning stems out form the following premises:

— A local user inheriting traditional wisdom is fully conversant with the problems and potentialities of the natural resources.
— Actual users must be made aware about the drawback of the current practices and the positive relationship between environment and development.
— Local people should be made aware that they need to know the actual status of their resources to take conscious and

rational decision.

— The village body, as the local institution should chalk out their own development plans.

— Decision-making and programme implementation in a democratic set up is an integral part of the planning process.

Methods of Planning

The Participatory Rural Appraisal (PRA) methodology is employed to undertake planning. This method satisfies the requirement of micro-planning. It is because of the fact that under this method, planning for the resource development is done with active participation of the local community. This method can be supplemented with other methods like questionnaire type. Details of PRA methodology have been elucidated in separate chapter.

In the field of forestry and wildlife conservation, micro planning is conducted for two types of development activities. One is to prepare micro plan for joint forest management areas. This is also called as Joint Forest Management plan. The other is to prepare micro plan for eco-development. The essentials of micro planning for two different programmes are elucidated below.

The Micro Plan for Joint Forest Management

The objective of the micro plan is to produce a simple workable document to propose peoples' need based programme consistent with available resources and within the framework of national forest policy. To formulate the programme, we need to have relevant data on human and cattle population, resource status of the area, socio-economic profile, pattern of resource use, assessment of needs and aspiration of people, response of people toward different programme, etc. To generate such data, and prepare a sustainable work plan, there is a need for conducting micro planning. During this exercise, field data are collected. As already said, for collection of field data Participatory Rural Appraisal (PRA) method is adopted.

In order to facilitate the collection of field data relating to each of village forest committee by the field staff, the following proformae are in vogue.

Proforma I: Deals with basic information.
Proforma II : Deals with demographic information.
Proforma III: Deals with resource data information.

Proforma IV: Deals with need assessment information.

Preparation of micro plan involves two major steps. They are as follows:

1. Demarcation of joint forest management plan areas.
2. Preparation of joint forest management plan (Micro Plan).

The extent of area to be set apart for each village forest committee has to be decided on practical consideration like the ability of the committee to provide care and protection, extent of usufructs available to satisfy the expectation of the members. After making necessary inventory of the available resources and requirements of members, areas have to be allocated to the village forest committees. The areas thus set apart should be clearly demarcated on the ground. Depending on the local situations the demarcation could be as per the existing village boundaries or natural features like water courses, ridges of the hills roads, etc.

Joint Forest Management Plan or the micro plan for the area allotted to the committee has to be evolved by the local staff by resorting to the PRA method. The micro plan is a working scheme for the area selected. This is prepared after carrying out micro planning exercises. These exercises focus on traditional forest products available in the area, the resource kept at the disposal of the committee and identifies the measures necessary to increase the productivity from natural regeneration and by taking up enrichment planting. The choice of the species to be planted has to be finalised in consultation with the committee members.

Procedure to be Adopted

The sequence of activities for conducting micro-plan exercise is as follows:

(i) Demarcate the area on the ground.
(ii) Study the existing soil and condition of the vegetation of the area.
(iii) Prepare a treatment map for the area.
(iv) Divide the area into ten equal or equi-productive areas to the extent possible with the objective of managing the area for derivation of usufructory benefits and employment generation

on a sustained basis.

(v) Listing out various operations to be carried out, year-wise, in the area along with the financial input requirements in consultation with the members of the committee.

The operations that could be listed out should include the following:

- Demarcation of boundaries.
- Protection measures to be taken against the biotic interference like fire, grazing, illicit felling and transportation.
- Soil and moisture conservation works to be taken in the area.
- Silvicultural operations to be taken.
- Regeneration operations.
- Fodder development.
- Improving water resources.
- Improvement of existing roads in the forest areas.

The micro plan is prepared by the range officer and it is approved by the divisional forest officer after having inspected the areas and consulted the committee members on the management prescriptions.

The Micro Plan for eco-development

Eco-development is a process of continued action. It has been defined as a "site specific package of measures, developed through people's participation, with the object of promoting sustainable use of land and other resources, as well as on farm and off farm income generating activities which are not deleterious to PA values" (Pawar, 1992). It has also been defined by other writer as "limited rural development, designed with the participation of local people, for the purpose of reconciling genuine human needs with the specific aims of PA management" (Rodgeri 1992). The objective of eco-development is better conservation of natural resources as well as better quality of life for the local people. This is possible by reducing the dependency of local communities on diminishing forest resources in one hand and promoting more efficient use of the resources and providing meaningful alternatives to local communities in the other. This constitutes the essence of eco-dovelopment.

Micro Plan Format for Eco-development

Micro-planning is a more detailed planning exercise. Micro plans are specific to particular villages. A micro plan should be written in a plain, simple, straightforward style and be readily understood by people who have no technical training. These should be written in the local language, as far as possible. The descriptions should be kept as brief as possible (to keep the document handy) as the village level micro plans will not be a widely circulated. The micro-plan structure given below is based on the above logic. The users may modify this to suit their needs. Village level micro plans will be full-fledge documents bound in separate volumes and in adequate numbers so that each concerned party may have a copy.

CHAPTER 1

Description of the Village

This chapter constitutes an introduction to the village under the following sub-heads:

(a) *Location, Access and Communications*: (Also discuss the terrain, names of district, block, tehsil etc.)

(b) *Population and Social Structure*: Here describe the caste and tribe structure of various villages, give number of families, and population size of each group, briefly describe the salient features of the life-styles of each group, specially focusing on those aspects which define their relationship with forests and natural resources. Mention main economic activities of each group.

(c) *Infrastructure and Facilities*: Here discuss various facilities such as roads, school, hospital, veterinary facilities, irrigation, electricity, flour mill, rice sheller etc., year or period of development, current status, etc.

CHAPTER 2

Occupations, Forest Dependencies and Conflicts

1. *General economic conditions*

In this section, discuss the overall economy of the village, rich and poor groups (castes/tribes) people below and above poverty line etc.

2. *Occupations*

Describe forest dependent and non-dependent occupations. Forest dependent occupations should be described in detail especially highlighting the extent of dependence, seasonality, subsistence or commercial, domestic use or for trading, legality, sustainability and the impact on the PA/forest resources and management (*e.g.* degradation of habitat). Discuss whether the occupations are traditional or non-traditional.

Non-Forest-Dependent Occupations

Discuss about the following aspects:

Agriculture: Number of houses, sizes of holdings (small, medium and big farmers), castes/tribes, irrigation, extent of crop and crop rotations, improved seeds, fertilisers, pesticides, improved implements, extension facilities, marketing of produce etc. Mention shifting cultivation and encroachments on forest land, if any, only briefly as these are to be discussed in detail in the section on conflicts.

Business and Trade: (Only those other than forest produce).

Labour: (Agricultural, forest, other government departments, other seasonal).

Service: (Government or private but medium to long-term only). Handlooms and Handicrafts/Horticulture/Sericulture/Apiculture/ Poultry/Piggery/Others.

Discuss history, families involved, castes/tribes, small scale or large scale, income, technical, financial and marketing support from Government/NGOs, sources of raw materials etc.

3. *Forest-Dependent Occupations and Domestic Consumption of Forest Produce*

Here discuss all the occupations based on raw materials drawn from the forests. Give as much detail as possible, especially, the number of families involved, various castes/tribes involved, whether full-time or part-time, whether for domestic consumption or sale or both, (if both, estimate quantities for each) estimated incomes, seasonality, legal aspect, government regulations and the efficiency of implementation, government arrangements to meet the demand, if any, impact on the forest/wildlife and the impact on park-people relationship etc.

Estimated production/consumption in measurable units (quantification) be given especially highlighting the estimated carrying capacity and over-exploitation, if any. Parts of the forest (compartment number) affected be clearly determined. Classify affected areas into high, medium or low impact category. Use sketch maps to show areas affected by each activity. These activities/occupations can be:

Livestock and Grazing; Fuelwood collection; Timber & Small wood (for housing and agriculture); Thatching materials; Thorns and other fencing materials.

Collection of MFPs (discuss each item separately, *e.g.* Tendu (Kendu) leaf, Mahua, Amla, Chironji, Sal seed, Harra, Bahera, Mushrooms, medicinal plants and so on. These will vary from site to site.

Handicrafts and Cottage Industries: (*e.g.* bamboo baskets and other handicrafts, leaf plate making, beedi-making, wooden toys and carpentry, broom-making, forest-based sericulture and apiculture, etc.

Others: Mining, fishing, etc.

Man-Wildlife Conflicts : In this section, discuss all aspects of the problems caused by wildlife for local people, mainly, wild species involved, number of people or area affected, history of the problem, extent of damage or loss, government policy towards these problems, effect of government policies, seasonality, patterns etc. Show highly impacted, and mildly impacted village/areas with the help of sketch maps. The discussion would generally include the following items but may vary locally.

Crop-Raiding; Cattle Lifting; Death/Injury to Humans; Damage to Other Property, e.g. by elephants.

4. *Offences and Illegal Activities*

In this section, discuss the extent and seriousness of illegal activities, kinds of activities, reasons for each, areas affected, number of cases registered in the recent past (say 5 years), effectiveness of protection arrangements, attitude of the local people towards offenses and extent of cooperation in detecting case and apprehending offenders. These activities can be:

Illicit Felling (timber and bamboo); Illicit Grazing; Fires; Illegal Collection of MFPs; Poaching; Fishing; Shifting Cultivation and Encroachments; Others

5. *Summary of Park-People Relations*

In this section, summarise the overall relationship of the park (and park staff) with the local communities, whether people look upon the park as a resource or as an imposition, any agitation for or against the PA, any violent conflicts with the local people.

CHAPTER 3

Proposed Eco-development

1. *Goal and Objectives:* A plan should have a long-term goal and one or more problem-specific objectives. Objectives should be clear and well-defined (*e.g.* to reduce total herd size by 25 per cent and increase available fodder by 40 per cent over a period of five years). Progress towards the attainment of objectives (through the adopted strategies) will be monitored.

2. *Target Families and Prioritisation*: Discuss how beneficiaries of various proposals will be selected and prioritisation would be done (criteria: heavy forest dependence and poverty).

3. *Process*: Describe the basic approach of people's participation, institutional framework (Forest protection committees, Eco-development committees, user groups etc.), integration of the programmes from various departments, cost sharing between the people and other agencies, approach to the methods of fund raising, awareness and education etc.

4. *Problem Ranking:* Here discuss the priority assigned to each problem (both of the people as well as of the PA).

5. *Summary of Proposals:* Here give a gist of various activities proposed in the subsequent chapters. First, discuss the general approach, relation this section with the proceeding chapters and the use a table showing the titles, objective, physical and financial targets, and sources of funding and agency responsible for implementation.

CHAPTER 4, 5, 6,....

In these chapters discuss various programmes and activities individually in all aspects. There can be no common structural frame work for all possible proposals. However, scheme should clearly indicate the objective, target people, description of various components, annual physical and financial targets, sources of funding, responsibilities of the implementing agency, responsibilities of the people and user groups, expected benefits to the forests and PA, etc. Activities with common or related objectives should be described together under a common chapter.

CHAPTER ....

Monitoring and Evaluation

In this chapter, monitoring criteria as well as arrangements should be described. Monitoring is to be seen in two distinct categories. (1), monitoring of the achievement of annual physical and financial targets mentioned in the plan, and (2), monitoring of impact of eco-development on the PA and the surrounding forests. For the latter, monitoring parameters such as improvement in water table, in wells or period of flow in streams and the condition of vegetation, status of regeneration in sample plots will need to be identified and determined. Changes in life styles and extraction patterns of natural resources such as reduction in the use of fuel wood, number of headloads coming out of forests, changes in grazing pressure, frequency of offenses etc. can also be monitored. For scientific monitoring, benchmark studies will have to be done before the plan implementation starts. Keep the criteria and procedures as simple as possible and keep proper records. An outside agency, which will be responsible for coordinating monitoring, should be identified. People's role in monitoring and evaluation should be clearly mentioned. There should be a provision of at least two evaluations, one mid-term and another terminal. This will help in providing opportunity to evaluate the results of monitoring.

CHAPTER ....

Miscellaneous

In this chapter describe the process and timing of preparation of annual plans, list important documents and maps, show how common funds and bank accounts, etc. will be operated and managed. Any other relevant aspects not discussed elsewhere may be discussed.

26
Eco-Development

Introduction

Local people who live near forests are largely dependent on natural resources like, fuel, small timber, fodder, and other minor forest produces for their livelihood. Over the years, the pressure on the forest resources has increased manifold. Consequently, wide-spread depletion of forest has set in. Given the present pace of forest depletion, the local communities may not be able to continue to have the present level of forest resource use in the coming few decades. The reasons for this phenomenon are many. Over the year, Governments have initiated many rural development programmes, but such programmes never focused on the causes of depletion of natural resource. Therefore, depletion of precious forest resources accelerated unabatedly and it consequently reduced the capacity of the rural environments to sustain growing populations. This created a situation when it was imperative to adopt an approach to sustainable resource management with people's participation. Therefore, Government of India and various State Governments adopted community based sustainable forest management programmes. Such programmes address to the conservation need of the forest on the principle of sustainability and people's participation in resource management and community requirement. When such programmes are adopted ih and near Protected Areas (PAs), they are called as eco-development.

National Forest Policy, 1988, states: "forest should not be looked upon as a source of revenue" but as a "national asset to be protected and enhanced for the well-being of the people and nation. The people will be actively involved in programmes of protection, conservation and management of the forests". This policy directive is now being implemented by states under the name of Joint Forest Management (JFM). In the year 1990, Central Government launched a centrally sponsored scheme called "Eco-development Around National Park and Sanctuaries" in the Eighth Plan.

Eco-development is a process of continued action. It has been defined as a "site specific package of measures, developed through peoples participation, with the object of promoting sustainable use of land and other resources, as well as on farm and off farm income generative activities which are not deleterious to PA values" (Pawar, 1992). It has also been defined by other writer as "limited rural development, designed with the participation of local people, for the purpose of reconciling genuine human needs with the specific aims of PA management" (Rodgeri, 1992). The objective is better conservation as well as better quality of life for the local people. This is possible by reducing the dependency of local communities on diminishing forest resources in one hand and promoting more efficient use of the resources and providing meaningful alternatives to local communities in the other. This constitutes the essence of eco-dovelopment.

Planning for Eco-development

Planning for eco-development is a part of the general management planning process for the Protected Areas (PA). Therefore, eco-development plan is a sub-plan within the management plan. The objective of planning is to produce an eco-development plan and its constituent village level micro-plan. However, the planning process pertaining to eco-development generally consists of the following steps (Pabla *et al.* 1995):

1. Reconnaissance and Familiarisation.
2. Socio-economic and Resource Use Survey.
3. Identification of impact zone.
4. Prioritisation and selection of target villages.
5. Writing preliminary eco-development plan.
6. Discussion over preliminary plan.
7. Participatory micro-planning.
8. Writing Eco-development Plan and Micro-Plans.
9. Submission and approval.

Reconnaissance and Familiarisation

Reconnaissance exercise aims at obtaining the first impression of the area, people, resources, demand, shortages, location of villages *vis-a-vis* the PA pressure points, etc. This is essentially broad survey of the situation, both on the ground as well as in official records. This involves field visits, studying maps, reports, gazetteers, working plans, relevant literature and holding informal discussion with the people and officials. This stage constitutes a part of the trust-building stage or activities carried out to gain credibility of the people. Apart from this, this programme of interaction facilitates familiarisation. Acquaintances made during reconnaissance survey can be very helpful in the subsequent stages of eco-development planning. By the end of this stage, the planning officer acquires a good idea as to which villages are suitable for undertaking eco-development work and what kind of detailed surveys is required to accomplish the planning. During this stage the planning officer also interacts intimately with the management staff and elicits their perception of the situation.

Socio-economic and Resource-Use Survey

In the previous stage, a tentative identification of villages which are thought to be either impacting the PA or being impacted by the creation of the PA. is undertaken. Now a detailed survey, using a suitable questionnaire, should be done. Most of the socio-economic information is generally available in the office of the Block Development Officer (BDO) or some other similar agency. The questionnaire should be designed to fill the gaps in the information so available, and updating it if necessary. The main objective of this survey is to finalise the limits of the impact zone and to establish how various villages are linked to the PA and its management. In each village, the focus of the exercise will be on the determination of the kinds and extent of dependencies of the villagers on the park or the buffer and the location of the impact of each village. The questionnaire can be filled either by the local forest guard or specially deputed staff in interaction with the village elders. A door to door survey is neither necessary nor practical.

Identification of Impact Zone

Based on the information collected through the first two steps, an impact zone will be determined and defined. Impact will be defined in two ways. One is to define the impact of the human activities on the park resources, and the second is to define the impact of the park on the lives and economies of the surrounding villages. Impact can be classified as low, medium and high. Accordingly, impact maps should be prepared. The villages located inside will have the full impact on the PA area, whereas the villages situated on the boundary will have partial impact on the PA and rest on outside. The impact on the park will mainly depend upon the vicinity of a village to the park boundary. The degree of impact on the park can be determined from various indicators, such as the incidence of illegal grazing or tree cutting, high rate of offences and violent conflicts with the parkstaff and the extent and degree of degradation of the habitat. Similarly, the impact of the park on the communities will be in terms of cattle lifting, crop damage, etc.

Prioritisation and Selection of Target Villages

After having prepared the impact maps, identification of villages for eco-development inputs is to be carried. Not all the villages can be treated simultaneously because there is always paucity of the field staff, funds and infrastructural facilities. Apart from this not every village needs equal attention. Therefore, prioritisation of target villages is required. Target villages are those villages that would be given inputs during the proposed plan period. The principal criteria for the selection of a village will depend on its impact on the park and *vice versa*. Certain other factors such as the distribution of workload among various administrative units, and staff members will also have to be kept in mind. The total number of villages selected will depend upon an assessment of resources likely to be available and the administrative capability, the existing village organisation to undertake work.

Writing Preliminary Eco-development Plan

A preliminary eco-development plan will be prepared and presented for discussion with various prospective partners including departmental officials, other development agencies, NGOs and public institutions. The document will be prepared in the same format as the final plan document and would present the current socio-economic and resource status. Chapter 5 should be developed up to the selection of target villages, strategies and the general process to be followed. The strategies should be evolved after conducting a sample micro planning exercise in few villages, representing various degrees of impact of the park as well as socio-economic conditions. Detailed PRA should be undertaken in these villages to arrive at a common perception of the situation and to decide on workable solutions. Possible solutions and strategies should be discussed with the line agencies before incorporating them in the plan. Based on the conclusions drawn from these sample micro plans and discussions with other agencies, general prescription and projections for the entire area should be developed and discussed in the preliminary plan. An outline of the remaining chapters of the plan should be elucidated, broadly indicating the overall approach and likely conclusions. Apart from this, the possible role and responsibilities of various agencies should be discussed and highlighted.

Discussion on Preliminary Plan

This discussion on preliminary plan is always in the shape of meetings/workshops involving peoples institutions, NGOs, district development departments, funding agencies and departmental officials. Specific comments of these agencies should be sought on their likely role and responsibilities during the currency of the plan, and an attempt should be made at this stage to finalise future mechanisms for effective co-ordination. Outcome of this interaction should determine further development of the plan.

Participatory Micro planning

This constitutes the village level planning stage. It commences after broad agreements between various participant agencies have been arrived at. This stage is the core of the entire planning exercise. Here the collection of information and data becomes more formal and structured and ideas start taking a concrete shape. It is now a common practice to adopt Participatory Rural Appraisal (PRA) techniques for undertaking eco-development planning. PRA is a set of semi-structured methodology designed to facilitate a productive interaction with rural communities. Various exercises such as resource mapping, transact walk, Venn diagram, timeline, livelihood analysis etc. are aimed at eliciting people's perceptions about their lives, resources, development agencies, etc. The observations and perceptions emerged out of such PRA exercise can be used for any rural development planning work.

The participatory planning process is a complex process involving PRA, collecting information of various government plans, establishing contact and rapport with various agencies (both government and non-government), and setting up co-ordinating mechanisms at village, range and division/district levels. It also involves preparing comprehensive plans for various components of eco-development such as joint forest management, watershed management, agricultural development, irrigation, cottage industries and handicrafts, apiculture, sericulture, fisheries, alternative energy and energy conservation programmes, etc. However, before taking up detailed PRA and planning work, the planning team has to strive to gain complete trust of the concerned people. Trust building involves such actions/gestures on the part of the planning team, which promotes mutual confidence between the PA staff and the people. Trust building must start with the PA staff as a clear understanding of the approach necessary for implementing it.

Writing Eco-development Plan and Micro Plans

The micro planning exercise culminates with preparation of a micro plan. Each village will have a separate micro plan. Description of micro plan is given in separate chapter. As far as possible, the micro plan should be written in the local language so that the villagers are able to understand it easily. All the micro plans will be incorporated into the PA level 'Eco-development Plan'. This document will be prepared by further developing the preliminary plan. The physical and financial targets from all the micro plans will be compiled into suitable tables under Chapter 5 on 'Proposed Eco-development'. The strategies already proposed will be further examined and improved and more details will be added. Deficiency in data, information and description should be removed by using secondary source. The plan copy should be prepared in sufficient number so that it can be as widely distributed as necessary.

Submission and Approval

Micro plans are generally approved by the village level organisations. The consolidated 'Eco-development Plan' is approved by such agencies, which have a role to perform during implementation. Since the micro plans are developed with the full participation of the concerned people, a separate approval may not be necessary. Similarly, as the preliminary plan will have been discussed and approved by various organisations, it can be presumed that the final plan would also get their approval. However, a formal approval of the entire local, district and state level organisations, both government and non-government will be essential to formalise their participation. The plan document, in draft form, may be circulated to the key agencies to solicit their final comments and approval. The plan should finally be discussed in a meeting/workshop in which various views could be elicited and incorporated if found acceptable. After obtaining various approvals from the local and regional level, the plan document can submit to the government for a formal adoption.

The Plan Format

The eco-development plan would normally have two components: a combined, consolidated 'Eco-development Plan' document and Micro plans for all the target villages. A structure for a typical eco-development plan and micro plan is proposed hereafter. The format (Pabla *et al.,* 1995) proposed here would have to be adjusted to the local requirements. Photographs, sketches, graphs, charts and other visual aids should be used as and when necessary.

CHAPTER 1

Description of the PA and its Surrounds

This chapter forms an introduction to the PA and provides information about the following:

1. Location.
2. Access and Communications.
3. Core/Buffer area description, Administrative control.
4. History, Establishment and Legal Status.
5. Climate.
6. Drainage and Topography.
7. Rocks and Soils.
8. *Vegetation* - Discuss forest types and overall diversity, rare plants.
9. *Fauna* - Discuss total diversity and highlight the occurrence of rare and endangered species.

CHAPTER 2

Impact Zone, Human Settlements and Social Structure

1. *Impact Zone:* It includes all the surrounding villages that have an interaction with the PA. Either they impact the PA (illicit grazing, tree cutting, fires, MFP collection, etc.) or their lives have been impacted by the creation of the PA. Discuss, how the zone has been identified, whether through actual analysis, or by drawing an arbitrary

circle (2km, 5km or 10km radius) around the park. Describe, whether all of it, falls within the core/buffer complex or it spills out. Use a sketch map to illustrate.

2. *Human Settlements:* Use a table to show the names and population of the villages. Clearly, mention which villages are to be relocated from the core area. Discuss, which villages have the most impact on the park and *vice versa* and why.

3. *Social Structure:* Here describe the social structure of various villages like caste and tribe. Provide number of families, and population size of each group; briefly describe the salient features of the life styles of each group, specially focusing on those aspects which define their relationship with forests and natural resources. Mention main economic activities of each group.

CHAPTER 3

Occupations, Forest Dependencies and Conflicts

1. *General Economic Conditions:* In this section, discuss the overall economy of the village; elucidate rich and poor groups, castes/ tribes, people below and above poverty line.

2. *Occupations:* Describe forest dependent and non-dependent occupations. Forest dependent occupations are described in detail, especially highlighting the extent of dependence, seasonality, subsistence or commercial, domestic use or for trading, legality, sustainability and the impact on the PA/forest resources and management (*e.g.* degradation of habitat). Discuss, whether these occupations are traditional or non-traditional.

Non-Forest Dependent Occupations

Agriculture: Number of houses, sizes and distribution of holdings (small, medium and big farmers), castes/tribes, irrigation (methods and extent), crops and crop rotations, improved seeds, fertilisers, pesticides, improved implements, extension facilities, marketing of produce, etc. Mention shifting cultivation and encroachments on forest land, if any, only briefly as these are to be discussed in detail in the section on conflicts.

Business and Trade: Only those, which are not trading forest produce.

Labour: Agricultural, forest, other government departments, others.

Service: Government or private (medium to long-term only).

Handlooms and Handicrafts, Horticulture, Sericulture, Apiculture, Poultry, Piggery, Others: Discuss history, families involved, castes/tribes, small scale or large scale, income, technical, financial and marketing support from government/NGOs sources.

Forest-Dependent Occupations and Domestic consumption of forest produce: Here discuss all the occupations based on raw materials drawn from the forests. Give as much detail as possible, especially, the number of families involved in each village, various castes/tribes involved, whether full-time or part-time, whether for domestic consumption or sale or both, estimated incomes, seasonality, legal aspect, government regulations and the efficiency of implementation, government arrangements, if any, to meet the demand, impact on the forest/wildlife and the impact on park-people relationship etc. Estimated production/consumption in concrete units be given especially highlighting the sustainability of such practices and the extent of over-exploitation on the forest resources. Use sketch maps to show areas impacted by each activity. These activities/occupations can be:

Livestock and Grazing, Fuel wood collection, Timber and Small wood removal, Removal of thatching materials, thorns and other fencing materials; Collection of Minor Forest Produce (MFPs). Discuss each item separately, *e.g.* Tendu (Kendu) leaf, Mahua, Amla, Chironji, Sal seed, Harra, Bahera, Mushrooms, medicinal plants and so on. These will vary from site to site.

Handicrafts and cottage industries: Bamboo baskets and other handicrafts, leaf-plate making, beedi-making, wooden toys and carpentry, broom making, forest-based sericulture and apiculture etc.

Others:- Mining, fishing, etc.

Man-Wildlife Conflicts: In this section, discuss all aspects of the problems caused by wildlife for local people, mainly number of people or area affected, history of the problem, extent of damage or loss, government policy towards these problems, effect of government policies, seasonality, patterns, etc. Show highly impacted, and mildly impacted villages/areas with the help of sketch maps. The discussion would generally include the following items but may vary locally.

Crop Raiding/Cattle Lifting/Death/Injury to humans; Damage to other property, *e.g.* by elephants.

Offences and Illegal activities: In this section discuss the extent and seriousness of illegal activities, kinds of activities, reasons for each, area affected, number of cases registered in the recent past (say 5 years), effectiveness of protection measures, attitude of the local people towards offences and extent of co-operation in detecting cases and apprehending offenders. These activities can be:

Illicit Felling: For example timber and bamboo; Illicit Grazing; Fires; Illegal collection of MFPs; Poaching; Fishing; Shifting cultivation and Encroachments; Others.

Summary of Park-People Relations: In this section, summarise the overall relationship of the park and its park staff with the local communities. Explore whether people look upon the park as a resource or as an imposition. Also, describe about any agitation, if undertaken, for or against the PA, or any violent conflicts with the people that have taken place in recent past.

CHAPTER 4

Current Eco-development Activities

1. *History:* In this section, discuss when, how and in what shape the concept of eco-development came into the area, mention any programmes which can be seen as forerunners to eco-development.

2. *Current Eco-development Plan:* Discuss if the area has an existing plan or not. If yes, the name of the author, year of writing, proposed plan period, activities proposed, sources of budget (whether departmental or extra departmental), etc. may be mentioned. Special mention may be made whether the plan was developed through a dialogue with the people, NGOs and other departments or not. The status of its implementation and any other salient features should be mentioned.

3. *On-going Eco-development Activities:* Mention the activity, year of implementation, size, villages, objective, impact, name of the implementing agency, source of funding and involvement of NGOs.

4. *Constraints:* In this section, discuss whether the PA management has adequate staff, mobility, budgets, communications, etc. to effectively carry on an intensified eco-development programme. External problems and constraints, if any, should also be discussed.

CHAPTER 5

Proposed Eco-development

1. *Target Villages:* As it will be difficult to start ecodevelopment in all the villages simultaneously, a selection of target villages will have to be made. Here, mention the names of the villages selected for priority inputs and the reasons for their selection. The selection criteria can be more than one. Although the major criteria will be made on the basis of administrative and financial reasons.

2. *Goal and Objectives:* A plan should have a long term goal and one or more problem-specific objectives. Objectives should be clear and well defined (e.g. to reduce total herd size by 25 per cent and increase available fodder by 40 per cent over a period of five years). Progress towards the objectives (through the adopted strategies) will be monitored.

3. *Strategies:* A strategy is a programme evolved for achieving one or more objectives. Each strategy consists of one or more activities directed towards achievement of one or more objectives. These will be described under two categories, namely, General process and Sectoral strategies.

General Process

(a) *Participatory Process:* In this section, describe how the people participated in problem identification process, and selection of various strategies and schemes. Also, explain how this process is to be continued during implementation and monitored. The process of institutionalisation of participation by various agencies including the local community, NGOs and government agencies need to be explained.

(b) *Integration of Rural Development Programmes:* Discuss the ways and means of integrating different activities of various departments.

(c) *Institutional Framework:* Discuss various committees and user groups existing and proposed.

(d) *Fund-Raising and Cost Sharing:* Discuss the various modes of funding, i.e. forest departments and other government departments, departments, private donations, sharing by villagers either in cash or physical labour.

(e) Training and Orientation of Staff.

(f) Awareness, Education and Interpretation.

Sectoral Strategies and Summary of Proposals

In the plan, the sectoral strategies will be outlined concisely explaining how the objectives will be achieved. Different strategies and sub-strategies may be applied to deal with a single complex problem. The package of strategies and sub-strategies may incorporate a hierarchy of eco-development measures. For example, the strategy for controlling over-grazing may have one or more of the following sub-strategies or activities:

(a) Rotational grazing,
(b) Pasture development,
(c) Promoting fodder crops in agriculture,
(d) Reducing livestock populations,
(e) Promoting stall feeding,
(f) Improving livestock breeds, and
(g) Promoting dairy, etc.

The proposals to be discussed here are to be based on village level micro-plans. Provide an outline of the scheme, total target and time span, village-wise and year-wise targets, expected output, financial requirements, sources of funding and fund sharing, responsibilities of various partners, etc. Each proposal should be discussed as a distinct sub-chapter.

(4) *Financial Forecast:* Here annual financial targets for all the schemes (activities) and sources of funding should be tabulated. This section will consist of a number of tables, *e.g.* village-wise annual requirement of funds, activity-wise annual requirement of funds, agency-wise annual contribution, year-wise total requirement of funds.

CHAPTER 6

Monitoring and Evaluation

In this chapter, monitoring criteria as well as arrangements should be described. Monitoring is to be seen in two distinct categories. (1) monitoring of the achievement of the annual physical and financial targets mentioned in the plan, and (2) monitoring of the impact of eco-development on the PA and the surrounding forests. For the latter, monitoring parameter such as improvement in water table in wells or period of flow in streams, changes in silt load and the conditions of vegetation, status of regeneration in sample plots will need to be identified and determined. Changes in life styles and extraction patterns of natural resources such as reduction in the use of fuel wood, number of head loads coming out of forests can also be monitored. For scientific monitoring, benchmark studies will have to be done before the plan implementation begins. Keep the criteria and procedures as simple as possible and keep proper records. An outside agency, which will be responsible for co-ordinating monitoring should be named. At least two evaluations, one mid-term and one terminal should be provided to evaluate the results of monitoring.

CHAPTER 7

Miscellaneous

In this section, detail account of responsibilities of various participating agencies and interest groups should be given and the implementation process should be discussed. A list of various references, important notifications and government orders, agreements between various agencies should be listed and appended. A list of various maps prepared should also be included.

27

Technique for Optimizing Forest Land Use: Developing Inventory of Forest Soil Resources for a Forest Division

Introduction

Forest, one of the most important naturally renewable resources plays a significant role not only in socio-economic development but also in maintaining ecological balance of the nation. Therefore, it is imperative to evolve appropriate strategies and practices for sustainable management of forest resources. In order to achieve this avowed objective, it is necessary to develop, *inter alia*, inventory of forest soil resources based on modern technologies. Considered in this backdrop, development of soil resource map for almost all the states by National Bureau of Soil Survey and Land Use Planning, Nagpur is a significant step towards this end. Now the question is how to develop an inventory of forest soil on the basis of the aforesaid soil map. The chapter, recapitulating the attributes, and uses, presents a methodology to prepare a divisional soil resource map and inventory of soil resources in the context of a forest division.

Soils are vital natural resources. Sustainable use of soil resources determines the life-supporting system and socio-economic development of any country. Deterioration of forest soil is a matter of great concern to the country. Degradation of forest soil resources, *inter alia*, is a cause as well as an effect of deforestation. Presently, forest soil resources are under different degrees of degradation and are getting further deteriorated at an alarming rate. Rapid degradation of soil resources jeopardizes the ecological security of the country. Therefore, there is a great need to maintain the forest soil health and quality. This can be best possible by developing strategies at micro level to reverse the processes of degradation and to ameliorate the soil quality.

Making effective use of available soil and water resources shall improve the forest cover and productivity in India. This requires a systematic and scientific appraisal of soils, which forms an important base for planning sustainable forest development. Soil resource maps are prepared during working plan exercises. But, such maps are sketchy and are based on maps prepared by state soil survey and other organization after having adopted inferior technology and method. Apart from this, no uniform technology and sound criterion were used to generate soil resource map on a scale, which can be used for optimizing forest land use.

In view of the aforesaid problem, the National Bureau of Soil Survey and Land Use Planning (an unit under Indian Council of Agricultural Research), Nagpur, in co-operation with soil survey staff of the state department of agriculture, took up soil resource mapping of the state of M.P. on the scale of 1:250,000 during the period from 1989 to 1994. It adopted a three-tier approach in mapping, *viz.* satellite image interpretation, field surveys and lab analysis, and cartography and printing in order to generate soil resource map. Till today, the organization has published soil resource maps for the states like West Bengal, Pondichery, Gujarat, Haryana, Punjab, Tamil Nadu, Karnataka, Kerala, Orissa, Bihar, Rajasthan, Meghalaya, Maharashtra, Arunachal Pradesh, Manipur, Himachal Prasdesh, Madhya Pradesh, Sikkim, Andaman & Nicobar Islands, Jammu & Kashmir and Assam. For procuring the bureau's publications, the readers are advised to contact "The Documentation Officer, NBSS&LUP, Amravati Road, Nagpur- 440 010."

The soil resource maps prepared by NBSS&LUP indicates the kinds, extent and distribution of soils on the landscape. It specifies different characteristics of soil of different sites. It unfolds the erosion status of the soil. Therefore, the soil resource map can be used for a variety of purposes. Few of such uses are illustrated below:

- To generate forest soil resource map indicating the kinds, extent and distribution of different soils in a forest division.
- To delineate problematic, degraded, eroded soils for their amelioration.
- To generate several thematic maps as per the requirement of forest management practices.
- To promote forest research on mapped and defined soils for generating soil-based technology and its analogous transfer to other areas having comparable soil site characteristics.
- To generate database for storage and retrieval as and when

desired for future interpretation and use.

- To predict forest productivity based on soil site suitability models.
- To monitor the soil health and to forewarn the emerging constraints.
- To prepare perspective land use plan for the area in order to optimize land use for sustainable forest management.

Methodology for Preparing Soil Resource Map

The project of soil resource mapping was undertaken by National Bureau of Soil Survey and Land Use Planning (An organization under Indian Council of Agricultural Research, Nagpur, in collaboration with state departments of agriculture. It adopted a three-tier approach, comprising image interpretation, field surveys and laboratory investigation and cartography. The LANDSAT imageries on 1:250,000 scale was interpreted after developing a detailed legend for generating a landform analysis map of the area. The physiographic delineations were then transferred into the Survey of India topographical sheet (1:250,000 scale).

The soil samples were collected horizon strip-wise and analysed for various physical and chemical characteristics, such as soil texture, Ph, organic carbon, calcium carbonate, CEC and base saturation, etc. by following the standard procedures.

The states have been divided into different major landforms, which are again divided into regions. In order to elucidate the matter further the example of M.P. has been taken. Different regions in the state of Madhya Pradesh are presented in Table 27.1. Such regions constitute a group of homogeneous mapping units and are elucidated with their descriptions, soil use, soil constraints, soil potentialities and the strategy required to minimize the constraints. Based on landform analysis, field surveys laboratory investigations and field reviews, the soil resource map of the state was prepared on 1:250,000 scale comprising 64 sheets of the state. The same was printed and published on 1:500,000 scale (9 sheets) without any generalization. Soil family association with dominant phases formed the soil mapping unit.

The state of Madhya Pradesh has been divided into three major physiographic regions. Depending upon elevation, slope and ruggedness of terrain, these regions are: Northern Decan plateau, Central Highlands and Eastern plateau. The state is mostly drained by Narmada, Son, Chambal, Betwa, Ken, Sind, Kali Sind, Parvati,

Dhasan, Mahanadi, Indravati, Sabari, etc. The geological formations are diverse in origin. It comprises the archaean, metamorphic to sub-recent and recent alluvial deposits, while the remaining part encompasses a wide variety of rock formations like sedimentary, crystalline and Deccan trap and basic flows. The state has a wide variations in climate ranging from semi-arid (dry moist, dry and moist), sub-humid to humid tropical. The mean annual rainfall varies from 500 to 3500 mm covering 20 to 80 per cent of the mean annual potential evapo-transpiration (PE). It receives rainfall in the monsoon period from June to September.

The mapping units have been assembled after having adopted a three-tier approach. The upper tier shows diagnostic properties of the dominant soil and its phases. It includes order, sub-order, great group, sub-group, soil depth, particle size of family control section, surface texture, slope, erosion class and other phases of the dominant soil family, such as salinity, flood hazards etc. The middle tier gives landforms attributes and the lower tier gives the mapping unit (polygon) number.

The soil map shows the soil family association number (polygon numbers), such as 001, 002, 003.... , 851. These polygon numbers are typical of each mapped unit. The mapped units are organized in a systematic manner in the legend given on all the nine sheets. The legend depicts major characteristics of each polygon number (mapped soils). The soil maps of different states have been published by the said organization. So far the soil resource maps are concerned the soil map of Madhya Pradesh has been published by NBSS&LUP on 1:500,000 scale in nine sheets. The Index map of Madhya Pradesh, a part of soil map provides the location of districts in the states. In order to know the soils of an area, first locate with the help of index map, the respective sheet out of 1 to 9 in which the area of one's interest falls. After locating the area on the relevant sheet, see the soil map unit number(s) appearing in that area. The details of each soil map unit are given in the legend. The soil map unit is an association of soil families. The legend provides brief description of the dominant and associated (sub-dominant) soil families within a particular map unit, highlighting the soil depth, drainage, texture, slope, erosion, salinity, etc. The taxonomy of the soils, as per USDA System of Soil Classification, and the extent of the area covered by each mapped unit along with its percentage of the total area of the state are given in separate columns.

Procedure for Preparing Division Specific Soil Resource Map

First decide the study area. The study area can be a range or forest division or a district. Suppose the study area decided is Rajnandgaon forest division situated in the Chhatisgarh belt of Madhya Pradesh. Now find out the category of landform/region that falls in the study area. The said study area falls in the land-form categories of 19,20,22, and 23. For better illustration, the land-form categories of state of M.P. are presented in Table 27.1. The mapping units of the study area, i.e. Rajnandagaon division are also given in Fig. 27.1 Each mapping unit is described in terms of attributes such as surface form, parent material, soil depth class, particle sizes class, minerology class, calcareous class, soil temperature regulation class, reaction (ph) class, drainage class, ground water depth, compact layer, (type, depth and extent), surface texture class, slope class, erosion class, salinity or sodicity class and extent, surface stoniness, flooding, soil taxonomy, organic carbon, calcium carbonate, base saturation, CEC. The study area encompasses 41 number of polygons or map units.

Table 27.1 : Districts and mapping units in different landforms of M.P.

S.No.	Name of the major landforms or Region	Districts	Mapping Units
1	2	3	4
1	North Deccan Plateau Satpura Range (Basalt)	Jhabua, Khargone, Hosangabad, Betul, Seoni, Jabalpur, Narsinghpur, Mandla, Shahdol.	01 to 88
2	North Deccan Plateau Satpura Range (Granite)	Betul, Chhindwara, Seoni, Balaghat, Durg	89 to 122
3	Northern Deccan Plateau Satpura Range (Sedimentary)	Khandwa, Hosangabad, Betul, Chhindwara, Narsinghpur	123 to 135
4	Northern Deccan Plateau Satpura Range (Laterite)	Chhindwara, Betul, Seoni	136 to 139
5	Central High land, Narmada Valley (Basalt)	Raipur, Raigarh, Ambicapur, Sidhi	140 to 166
6	Central High land, Narmada Velly (Alluvium)	Jabalpur, Raisen, Hosangabad, Narsinghpur, Khandwa, Dewas, Indore, Sehore, Khargone, Dhar, Jhabua,	167 to 199
7	Central High land, Vindhyan Ranges (Sedimontary)	Raisen, Bhopal, Sehore, Damoh, Sagar, Narsinghpur, Panna, Satna, Rewa, Sidhi	200 to 277
8	Central High land, Vindhyan Range (Alluvium)	Sagar, Damoh, Satna, Panna, Rewa	278 to 298

Contd.

Table 27.1:*Contd.*

1	2	3	4
9	Central High land, Vindhyan Range (Granite/ Basalt).	Vidisha, Raisen, Sagar, Chhatarpur	299 to 317
10	Central High land, Malwa Plateau (Basalt)	Dhar, Jhabua, Indore, Dewas, Sehore, Bhopal, Ratlam, Ujjain, Shajapur, Mandsaur, Rijgoch, Vidisha, Raisen, Sarar, Guna and Shivpuri	318 to 419
11	Central High land, Malwa Plateau (Granite Gneiss)	Jhabua	420 to 423
12	Central High land, Pathar and Bundelkhand upland (Sedimentary Vindhyan)	Chhatarpur, Sagar, Damoh, Panna, Satana, Rewa, Sidhi	424 to 483
13	Central High land, Pathar and Bundelkhand upland (Granite Genesis)	Chhatarpur, Tikamgarh, Panna	484 to 527
14	Central High land, Pathar and Bundelkhand (Alluvium)	Bhind, Morena, Gwalior, Shivpuri and Datia	528 to 574
15	Eastern Baghelkhand plateau (2 uactzite and schist)	Jabalpur, Shahdol	575 to 579
16	Eastern Baghelkhand plateau (Laterite)	Jabalpur, Shahdol, Satana	580 to 585
17	Eastern Baghelkhand plateau (Sedimentary Gondwana 2 Uactizite)	Sidhi, Shahdol, Ambicapur, Bilaspur, Raigarh, Mandala	586 to 649
18	Eastern Baghelkhand plateau (Basalt)	Shahdol	650 to 652
19	Eastern plateau, Mahanadi Basin (Sedimentary)	Balaghat, Rajnandgaon, Mandala, Bilaspur, Durg, Raipur, Raigarh	653 to 720
20	Eastern Dandakaranya plateau (Granite, Gneiss)	Raipur, Bastar, Durg, Rajnandgaon	721 to 777
21	Eastern Dandakaranya plateau (Sedimentary)	Bastar, Durg, Raipur	778 to 805
22	Eastern Dandakaranya plateau (Laterite)	Bastar, Durg, Rajnandgaon	806 to 819
23	Eastern Dandakaranya plateau (Basalt)	Bastar, Durg, Rajnandgaon and Raipur	820 to 825
24	Eastern Chhota Nagpur plateau (Granitic - Laterite)	Ambicapur and Raipur	826 to 843
25	Eastern Chhota Nagpur plateau (Sedimentary Dhaiwar)	Raipur, Raigarh, Ambicapur, Sidhi	844 to 845
26	Eastern Chhota Nagpur plateau (Basatic - Laterite)	Ambicapur	846 to 851

Source : NBSSAUP (1994)

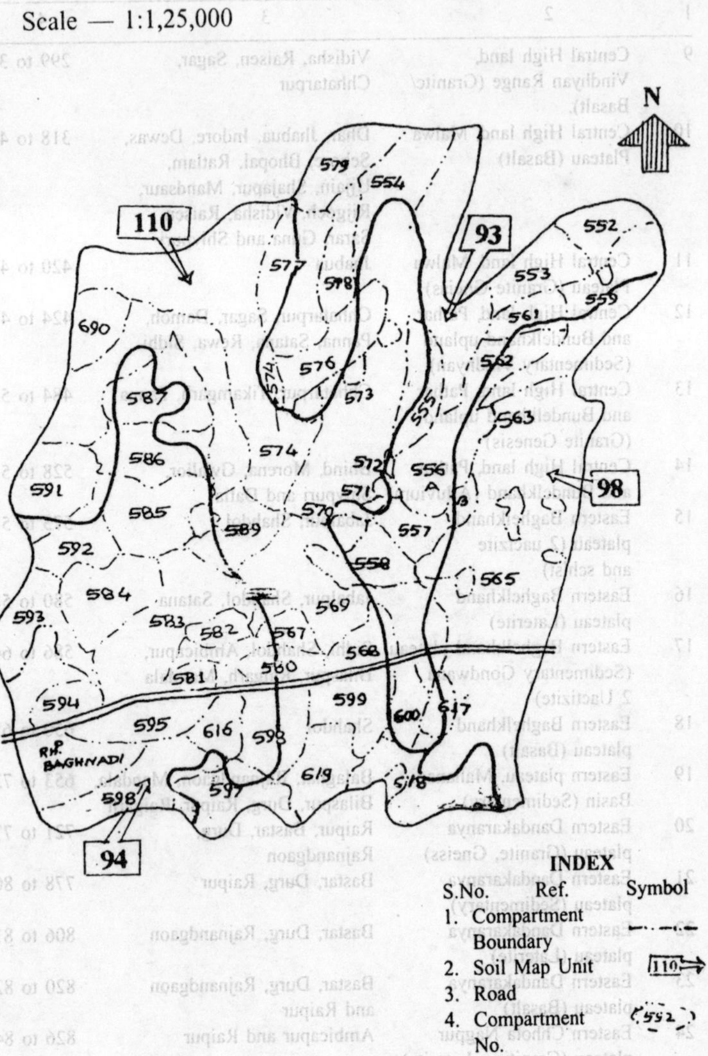

Fig. 27.1 : Map of Baghnadi Range of Rajnandgaon Forest Division (Chattisgarh) Showing : Soil Map Units.

Now the map units or the polygons which fall in the study area are identified. A soil resource map of the division is first prepared on the basis of soil resource map of 1:500,000 scale. This map contains the entire map units or polygons falling within the forest area of the division. This map is enlarged to 1:250,000 scale map. Now divisional map of 1:250,000 scale indicating compartment boundary and number is super-imposed on the soil map to find out the forest compartments falling in different polygon or mapping unit. The list containing different mapping units or polygons with their respective compartment of Rajnandgaon is presented in the format/ Table 27.2. Now, with the help of this table we can find out the corresponding map unit number or polygon number of the forest block or compartment of the division. The soil resource status of the compartment can be determined by referring to the soil characteristics of the respective map unit or polygon number. The attributes of map unit number or polygon number is presented in tabular form as shown in Table 27.3 and Table 27.4. The key to different legends is given in Table 27.5.

Table 27.2 : Forest Compartment and Soil Map Unit Index
Range : Gandai of Rajnandgaon Division

Compt. No.	Soil Map Unit No.
1	689
2	689, 707
3	689, 669
4	669, 689
5	669, 707
6	93, 669, 707
7	93, 669, 707
8	93, 669
9	93
10	93

Note:- Only 10 compartments have been taken on a sample basis.
Source: Working Plan Rajnandgaon Division, 1999.

Table 27.3 : Forest Soil Resorce Inventory (Short List)
Rajnandgaon Forest Division

S. No.	Map Unit No.	Description	Soil Classification
		(6)	(7)
1	71	Deep, Well drained Loamy soil on. Gently sloping plateau with mounds with moderate erosion. Associated with: Deep, Well drained, loamy soils on gently sloping with moderate erosion.	* Fine - loamy, mixed, hyperthermic, typic. Ustochrepts. * Fine, mixed, hyperthermic, Typic Haplustalfs.
2	93	Slightly deep, somewhat excessively drained, loamy soils on moderately steep, sloping hills and hill ranges with severe erosion, associated with: Shallow, well drained, loamy soils on moderately sloping with severe erosion and slightly stony.	* Fine - loamy, kaolinitic, isohy - perthermic, vertic Ustropepts.
3	94	Deep well drained, clayey soils on gently sloping hills and hill ranges with moderate erosion, associated with: Very shallow, somewhat excessively drained, loamy soils on moderately sloping with severe erosion and slightly stony.	* Fine - mixed, isohyperthermic, Vertic Ustropepts. * Loamy, kaolinitic, isohyperthermic, Lithic Ustorthents.
4	98	Deep, well drained, clayey soils on gently sloping hills with pediment with moderate erosion, associated with: Deep, well drained, loamy soils on gently sloping with severe erosion and slightly stony.	* Fine - loamy, kaolinitic, isohyperthermic, Typic Ustropepts. * Fine - loamy, kaolinitic, isohyperthermic, Typic Haplustolls.

Contd.

Table 27.3: *Contd.*

		(6)	(7)
5	110	Deep, moderately well drained, clayey soils on gently sloping beds (moderately dissected) with moderate erosion, associated with: Deep, moderately well drained, clayey soils on very gently sloping with moderate erosion.	* Fine, Mixed, isohyperthermic, Typic Haplusterts. * Fine, Mixed, isohyperthermic chromic Haplusterts.

Source: Working Plan, Rajnandgaon Forest Division, 1999.

Table 27.4 : Different Parameters and Their Codes

Surface Form

Code	*Type*
D	Dissected
H	Hummocky
L	Level
G	Gently sloping
U	Undulating
O	Rolling
S	Steep
R	Ridges
V	Valleys
T	Terraces
P	Plateau

Miscellaneous Land

Code	*Type*
MR	Extremely Rocky land
MS	Extremely Salty land
MZ	Extremely Sandy land

Parent Material

Code	*Type*
A	Alluvium
C	Colluvium
E	Aeolian
G	Granite
N	Gneiss
S	Schist
D	Sandstone
B	Basalt
L	Limestone
T	Glacial
M	Marine sediments
U	Undifferentiated

Particle Size Class

Code	*Class*
G	Fragmental
Z	Sandy-skeletal
K	Loamy-skeletal
P	Clayey-skeletal
S	Sandy
L	Loamy
R	Coarse-loamy
M	Fine-loamy
T	Coarse-silty
Y	Fine-silty
C	Clayey
F	Fine
V	Very fine

Mineralogy

Code	*Class*
c	Carbonatic
f	Ferratic
g	Gypsic
a	Micaceous
s	Silicaceous
k	Kaolinitic
m	Montmorillonitic
i	Illitic
x	Mixed

Calcareousness

Code	*Class*	$CaCO_3$(%)
0	Non-calcareous	Nil
1	Slight	<5
2	Moderate	5-15
3	Strong	>15

Soil Depth

Code	*Class*	*Depth* (cm)	*Map symbol*
0	Ext. shallow	(<10)	I
1	Very shallow	(10-25)	I
2	Shallow	(25-50)	I
3	Sli. deep	(50-75)	t
4	Mod. deep	(75-100)	t
5	Deep	(>100)	t

Soil Temperature Regime

Code	*Class*	*MAT(^{0}C)*
F	Frigid	<8
M	Mesic	8-15
T	Thermic	15-22
H	Hyperthermic	22-28
I	Isothermic	<5°C
P	Isohyperthermic	difference between MSST & MWST

Soil Reaction (pH)

Code	*Class*	*(pH)*
1	Strongly acidic	<4.5
2	Mod. acidic	4.5-5.5
3	Slightly acidic	5.5-6.5
4	Neutral	6.5-7.5
5	Slightly alkaline	7.5-8.5
6	Mod. alkaline	8.5-9.5
7	Strongly alkaline	>9.5

Slope

Code	*Class*	*Slope (%)*
a	Level to nearly level	(0-1%)
b	Very gently sloping	(1-3%)
c	Gently sloping	(3-8%)
d	Moderately sloping	(8-15%)
e	Moderately steep sloping	(15-30%)
f	Steeply sloping	(30-50%)
g	Very steeply sloping	(50%)

Soil Drainage

Code	*Class*
0	Extremely poor
1	Very poor
2	Poor
3	Imperfect
4	Moderately well
5	Well
6	Somewhat excessive
7	Excessive

Erosion

Code	*Class*
1	Slight
2	Moderate
3	Severe
4	Very severe

Salinity

Code	*Class*	*Ece (dSm^{-1})*
S1	Slight	2-4
S2	Moderate	4-8
S3	Mod. strong	8-15
S4	Strong	15-25
S5	Severe	25-50
S6	Very severe	50

Groundwater Depth

Code	*Class*	*Depth (m)*
1	Shallow	(<1)
2	Mod. shallow	(1-2)
3	Mod. deep	(2-5)
4	Deep	(<5)

Sodicity

Code	*Class*	*ESP*
N1	Slight	<5
N2	Moderate	5-15
N3	Strong	15

Compact Layer

Code	*Class*
D	Duric
P	Placic
R	Rock
Y	Clayey
F	Petroferric
G	Petrogypsic

Surface Stoniness

Code	*Class*	*Coverage*
t1	Slight	<15%
t2	Moderate	15-40%
t3	Stong	> 40%

C	Petrocalcic		**Flooding**	
A	Compacted (man's action)		*Code*	*Class*
			1	Slight
Surface Texture			2	Moderate
Code	*Class*		3	Severe
s	Sandy		4	Very severe
l	Loamy			
c	Clayey			

Source : NBSSAUP (1994)

Advantages of Forest Soil Resource Map

The following are the advantages of Forest Soil Resource Map:

1. *For collection of basic soil resource information*

The soil resource status of the compartment can be determined by referring to the soil characteristic table. This information can be stored in the compartment history during the working plan exercise.

2. *Selection of suitable species for afforestation*

While undertaking plantation, soil suitability for the species to be planted can be analyzed basing on these information.

3. *For forest management planning practices*

The better productive areas can be identified on the basis of soil resource characteristic and standing crop. Different management regimes can be evolved by forming different working circles and felling or treatment series within a working circle.

4. *For preparing database in GIS*

Different soil characteristic like, slope, soil depth, drainage, texture, etc. can constitute different layers. This, when analysed with other thematic layers, provides a comparative picture of forest crop and growth potentialities with relation to soil characteristic.

5. *For evolving strategy for soil and water conservation measure*

The soil erosion prone areas in the division can be identified with the help of soil characteristic table. Then, effective strategy can be evolved to minimize the soil erosion problem.

6. *For evaluating plantation*

Plantation can be evaluated with the help of soil characteristic table and reason for its success or failure can be determined. On the basis of evaluation corrective steps can be taken to improve the performance.

7. *For evolving different forest growth models*

Different forest growth models can be evolved after delineating equiproductive areas in a forest division.

Conclusion

Developing strategies to increase the forest cover, density and productivity is the key issue, which confronts the country. In order to address such burning issue, it is imperative that we have a systematic forest soil inventory of forest soil resources on 1:250,000 scale. Such maps shall provide information on the kind, distribution, and extent of different soils; their nature, characteristics, problems and potentials, so as to develop management strategies and practices not only to increase forest cover, density and productivity but also to maintain the ecological balance. In nutshell, developing an inventory of forest soil resources for a division on the basis of NBSS&LUP soil resource map and evolving management strategies and practices will go a long way in sustainable management of forest resources.

28

Monitoring and Evaluation: Principle and Practices

Introduction

Monitoring and evaluation are two important components of project cycle. Besides, they constitute important management functions. They play very important role in any plan or project or any programme. Forestry programmes are no exception to it. Monitoring of plan, project or programme are always in vogue in forest departments. While revising any forest management plan, terminal evaluation of the last working plan is conducted as a part of the revisionary process. Findings of such evaluation is generally presented in the working plan in the Chapter 'Results of last plan'. Likewise, on-going evaluation are also carried out in plantation, joint forestry management and other forestry programme. Therefore, monitoring and evaluation of forestry programmes have been built into the management systems in forest department since the introduction of scientific management in Indian forestry.

Evaluation means, "assessing the value of". Nevertheless, in planned works, evaluation is carried out during the implementation of the works or after they have been completed. The purpose of evaluation is to document the results of the present or past project or plan or programme. Results include output and impact or specific services which an activity is expected to produce from its inputs. Impact indicates the outcome of project or plan effects, which are broad and long-term in nature.

Monitoring can be broadly defined as frequent, largely routine wise collection and analysis of and reporting on information about the performance of the work in a programme or project, comparison of this with the programme or project plans, and connected discussion about and proposals for any corrective action (Dale Reidar, 1998). Thus, monitoring satisfies the purpose of gathering information required for implementation of current plan, and programme.

Monitoring covers all issues pertaining to inputs and outputs in relation to time. Evaluation is broadly defined as a more thorough examination, specified points in time, of programmes or projects, or parts of them, usually with emphasis on impacts and additionally commonly on efficiency, effectiveness, relevance, replicability and sustainability (Dale Reidar,1998).

Monitoring

Monitoring is continuous/periodic review undertaken by management, at every level of implementation, of an activity to ensure that input deliveries, work schedules, targeted outputs and other required actions are proceeding according to plan. Another way of defining monitoring is that it is a process of measuring, recording, collecting, processing and communicating information to assist project management decision making. However, precisely, it can be defined as an information system for management decision making.

A project's operation and performance are the aspects of concerns in monitoring with a view to keep track of the technical or economic "efficiency" of the project. In monitoring, the purpose is to ascertain whether project objectives are achieved or not. Therefore, the exercise of monitoring involves examination of the following points:

- Whether the various tasks are carried out according to schedule.
- Whether project impact is in accordance with project objectives.
- Whether project objectives/targets/execution need adjustments.

Thus monitoring is a management function. It begins with the initiation of the project and ends with the completion of the project.

Evaluation

Carlsson, Kohlin and Ekleem (1994) defined evaluation as "an activity" for finding out the value or results of something. Evaluation can be conducted either by personnel engaged in project, programme or plan or independent persons or organization. Evaluation means, "assessing the value of". One undertakes this common exercise in all spheres of life, in formal way or informal way. The objective of this exercise is to determine the effect of some action, or event or programme.

Evaluation, as a management function, is carried out in case of planned and organized development work. By, planned, we mean that there is a plan document or a report has been evolved after going through the process of planning. Considered in the same view, "organized" encompasses activities which are not only undertaken by some organization but also executed according to principles and regulations prescribed by the plan. Evaluation may be undertaken during the implementation of the plan or programme or after it has been completed.

In evaluation of purely development work, the focus remains on people. Whereas, in a work concerning to forest management practices, the focus shifts to the forests. The purpose of such evaluation is to document the results and utility of work for society or for the forests. Besides, it aims to explore how the results have been created.

Evaluation necessitates study of the events or effect in a broader perspective. That means, we should go wider covering factors internal to the development schemes or programmes as well as external (in its surrounding). In other words, evaluation should be comprehensive. For this, one has to use wide range of methods, ranging from hard statistical areas to purely qualitative assessments and personal interpretation. In practice, the correct choice of methods depends on the nature of work to be evaluated.

In day to day working, one comes across two terms frequently. They are (1) project, (2) programme. Project has been defined as a planned information for achieving one or more objectives, encompassing a set of interrelated activities, which are undertaken during a specified period of time, using limited human, financial and physical resources (Wickramayake, Ebel, 1993). A programme is a less clearly bounded entity. It can most easily be defined into a project, as a less specified and commonly more comprehensive, long-term intervention. Both the terms, however, are loosely used.

Forest management plan aims at improving the resources. Therefore, the evaluation encompasses an examination of effects of such management-plan or programme. Generally, a combination of direct and indirect examination of benefits is required in evaluation. In a plantation project, the direct effects in the forms of plant growth are evaluated. Whereas, in a programme of management function, the indirect effect of such programme is in the form of improvement in the crop density, composition, growing stock, or regeneration status.

Objectives of Evaluation

The following are the objectives of evaluation. They are :

- It records achievements.
- It measures progress.
- It improves monitoring.
- It identifies strengths and weaknesses.
- It helps to assess the effects and change.
- It provides base for cost-benefit analysis.
- It provides base for analysing replicating the successful programme.
- It improves the effectiveness of the programme.
- It facilitates better planning.

Features of Evaluation

Evaluation has the following basic features. They are:

- It is always with reference to stated criteria.
- It is always with reference to a point of time.
- It starts where progress reporting or monitoring exercises end.
- It adopts qualitative approach and defines relationship between policies and results.
- It is diagnostic in nature. It investigates and finds out factors, which are responsible for success or failure of the project or programme.
- It provides remedial measures.
- It focuses on broad and crucial aspects like policy issues, problem analysis, organizational pattern, administrative practices, implementation of technical measures, level of people's participation in the programmes, attitude of the people towards the programmes and impact of the programmes on the community or on the environment as a whole.

Benefits of evaluation

Evaluation helps in understanding and assessing the process and impacts of a programme and accountability of major agencies involved in accomplishment of the programme. Evaluation also helps in designing similar programmes. This understanding is then integrated into the organization and management framework of that programme. Such integration helps in improving the delivery system. The important benefits of evaluation are as follows:

— It records achievement.
— It explores the reason of failure/success of a programme and helps the management to learn while executing the project or plan.
— It measures progress.
— It improves monitoring process.
— It helps to identify strengths and weakness.
— It improves effectiveness of the programme.
— It determines the relevance, efficiency, effectiveness and impact of project/plan activities systematically and objectively.
— It strengthens management in future planning and decision, making.
— It provides a plethora of information about the past or on-going activities as a basis for modifying or redesigning future strategies.

Difference between Evaluation and Appraisal

Generally, it is a common place to see that people use evaluation and appraisal interchangingly. However, both these words are not synonymous. Both have different management connotation. Appraisal deals with economic viability, technical feasibility and social desirability of project that is under consideration. Therefore, appraisal is a part of planning process. Appraisal is carried out before the launching of the project. Whereas, evaluation is carried out after the launching of the project. Besides, it has completely different management functions. Therefore, it is not a part of planning, though it aids planning.

Monitoring and Evaluation

The following are the differences between monitoring and evaluation. These are:

— Monitoring is continuous. It starts and ends with a programme. Whereas evaluation is a short of exercise or an operation. Generally, evaluation is a carried out at the point of completion or during the currency of the programme.

— Monitoring is necessary to correct and improve the on-going programme. Whereas evaluation is generally used for future planning.
— Monitoring is generally done by implementing agency or staff. Whereas, evaluation is done by an outsider agency.
— Monitoring covers all aspects. It is also rapid. Whereas, evaluation aims at in-depth study and covers only a sample.
— Monitoring is basically a management function. It helps in correcting the delivery system. Whereas, evaluation is a learning process.
— Monitoring provides a picture indicating the trend of implementation. Whereas, evaluation is diagnostic and indicates about realization of project objectives and goals.

Monitoring and Progress Reporting

Monitoring and progress reporting are not synonymous. However, in most of the time these two management functions are misunderstood. But, they have also some similarities. These are:

1. Both are organized and consist of a systematic flow of data of different technical or administrative operation in progress.
2. Both can be quantitative and qualitative in their approach. But, the degree differs.

Apart from the similarities, there are many counts where monitoring differs from progress reporting. An account of differences between progress monitoring and monitoring is presented in Table 28.1.

Table 28.1 : Progress reporting and monitoring

S.No.	Progress reporting	Monitoring
1.	Regular/ ad hoc	Regular and continuous.
2.	Suitable for smaller projects.	Necessary for bigger and complex projects.
3.	Comprehensive. It covers all aspects.	Selective. It selects only four items which has been given priority.
4.	Required during implementation	Required for planner as well as controlling authority.
5.	It originates from the primary level.	It may originate at any level.
6.	Nature of reporting is passive; because it records only. It broadly responds.	It regulates the pace of development. It responds to the need of improvement.

Variables of Evaluation

As discussed earlier, there are four variables of evaluation. Besides, there are two subsidiary variables in evaluation. These variables are basis of evaluation. Any evaluator, entrusted with the tasks of evaluation should address to such variables. The six variables are as follows:

1. Efficiency.
2. Effectiveness.
3. Relevance.
4. Impacts.
5. Sustainability.
6. Replicability.

These variables are better understood with project matrix approach. Project matrix, also known as logical frameworks, is a tool for structuring the aims and components of a project. A sketch of project matrix is given in Table 28.2. The project matrix is divided into three columns. The first column and the bottom cell of the second column spell out the immediate and long-term objective, output, implementation tasks and inputs. Outputs are required for achieving the objectives. Task constitutes the work required to be done to produce the outputs. Inputs are those resources like capital, manpower, information, etc., required to perform these tasks. The three upper cells of the second column contain indicators of the respective outputs and objectives. The third column contains external factors, which encompasses conditions, events, policy or decision outside the sphere of the project or plan. These external factors are required to be fulfilled in order to ensure that the outputs lead to achievements of immediate objective. Because of the fact that these external factors are lying below the realm of the project or plan, uncertainty always looms large over the success of the project or plan.

While formulating plan or project, certain assumptions about factors outside the sphere of the project are made. Generally, such assumptions are not documented in the plan or project. Now, against the backdrop of this logical framework, the variables of evaluation are explained as under:

Table 28.2 : Basic Project matrix (Logical Framework)

Development Objective The impacts which the project is intended to generate for the intended beneficiaries.	*Indicators* Measures to verify to what extent the development objective is fulfilled with means of verification.	*External Factors* External conditions, events which are necessary for sustaining the impacts.
Immediate Objectives. The effects which the project is expected to create for the intended beneficiaries.	*Indicators.* Measures to verify to what extent the immediate objectives are fulfilled with means of verification.	*External Factors.* External conditions, events or decisions which are necessary for attaining the intended impacts.
Outputs The results of the work that the project management should be able to guarantee.	*Indicators* Measures to verify to what extent the outputs are produced with means of verification.	*External Factors* External conditions, events or decisions which are necessary for creating the expected effects.
Implementation Tasks The tasks that have to be undertaken by the project in order to produce the outputs.	*Inputs* The resources needed to undertake the work.	*External Factors* External conditions, events or decisions which are necessary for producing the outputs.

Source: Reidar Dale, 1998.

1. *Efficiency* : Efficiency may be defined as the amount of outputs created and their quality in relation to the resources invested. It indicates how the resources have been used productively. Thus, evaluation of output in relation to input is conducted. The evaluator has to use his own experience, knowledge and judgment about relation between the magnitude of inputs of different kinds and the amount of quality of outputs.

2. *Effectiveness* : This indicates the extent to which the planned output, immediate objective and development or long term objective are being produced or have been achieved. In practice, it is appropriate to focus mainly on the effects on the outcome side. Where the immediate objective or intended effects have not been precisely stated, the evaluator is required to specify the area.

3. *Relevance* : The issue here is to analyse the extent to which the programme or project has been addressed or the problems of high

priority has been addressed, mainly from the point of stake holders or beneficiaries.

4. *Impacts* : These are the long term, largely indirect consequences of the programme or project for the intended beneficiaries. The main impacts are expected to be positive. However, there may be negative impacts on the beneficiaries. These impacts should be analysed.

5. *Sustainability* : It aims at exploring the maintenance or augmentation of positive changes induced by the programme or project after the later has been terminated. Sustainability relates to all aspects of outcome. In more precise terms, sustainability means:

- — Maintenance of physical facilities produced (for instance, a road or a community centre).
- — Continued use of physical facilities (for instance, a road) or intangible qualities (for instance, knowledge).
- — Continued ability to plan and manage similar development work (by organizations, which have been incharge of the programme or project or any other organization).
- — Continued production of the kinds of outputs created.
- — Maintenance of impacts created.
- — Multiplication of effects and impacts.

6. *Replicability* : Replicability is another evaluation criterion. By that, we mean the feasibility of replicating the particular programme or project or parts of it in another context, i.e. at a later time, in other areas, or among the other groups of people.

Type of Evaluation

Evaluation is categorized into different kinds on the basis of focus, agency and stage of the project.

Focus:

Depending on how the information is to be used, there are two kinds of evaluation. They are:

1. *Formative evaluation* : It aims at improving the performance of the programme, which is evaluated. It is carried out at set intervals or according to needs during the course of implementation.

2. *Summative evaluation* : It is done after the programme has been completed or the programme has achieved a stable state of operation. The objective of such an exercise is to assess the worthiness

and appropriateness of design and management of the programme.

By agency:

Evaluation can be categorized on the basis of agencies to which the task has been entrusted. These categories are:

1. *Internal evaluation* : It is a progress monitoring or impact monitoring undertaken by the management itself, for example, on going or concurrent evaluation.

2. *External or independent evaluation* : It is an evaluation undertaken by some independent or external agencies or body. It is aimed at getting unbiased, objective and detailed assessment of the project performance.

By stages:

Evaluation can be categorized into three types on the basis of stages of the project or programme.

1. *On going evaluation* : On going evaluation is the most important exercise in the project management. It is done during the currency of the project. The exercise aims at assessing continuing relevance, present and future output, effectiveness and impact during implementation. Such an exercise culminates in recommending major corrections in the project. This may be done by internal or external agencies. This helps to determine the relationship between project inputs and outputs and the influence of external constraints on project performance. This is, in other words, a type of effect and impact monitoring exercise in long-term project or programme. This aims at adjusting or redefining policies, objectives, institutional arrangements and resources affecting the project during its operation.

2. *Terminal evaluation* : It is done at the end of or immediately after completion of project. It helps in making better planning in future.

3. *Ex-post Evaluation* : It is carried out after a reasonable time from the completion of the project or programme. It, like terminal evaluation, helps to draw lessons for future planning.

Methods of Evaluation

Evaluation needs methods. The methods may be divided into two categories:

1. Qualitative method.
2. Quantitative method.

Qualitative method involves information expressed in words whereas, quantitative method involves numerical data. Generally evaluation needs primary data. However, secondary data are also required to provide important background information.

1. *Qualitative method*

While analysing the impact, it is found that there is much information, which cannot be quantified or expressed, in numerical terms. Sometimes, there are certain information that can be quantified, but it does not serve the purpose. Therefore, evaluation of certain work is done through qualitative methods only. Sometime, the method symbolizes a combination of qualitative as well as quantitative methods.

The qualitative method is used in the following situations.

— When the changes are the results of a complex process.
— Where there is a value judgment involved.
— While dealing with organizational issues.

The qualitative method has the following advantages.

— It exposes different situations and compares them critically.
— It conducts sampling of units through personal judgment.
— It provides a flexible evaluation design.
— It provides opportunity to conduct enquiry through more than one method.
— It helps in recording data in flexible form.
— It provides enormous scope for participation of people in the exercise.

The following are some of the qualitative methods in vogue in government organization.

— Case studies.
— Questionnaires survey.
— Participatory rural appraisal.
— Rapid assessment.

Case study technique is one of the most appropriate techniques for community development programmes. It is highly suitable to study the impacts of programmes on the well-being of the communities. It helps to reveal the conditions for success. This technique is generally used to study organizational issues pertaining to execution of any project or programme. It is also suitable to use this method to evaluate the efficacy and performance of any institution and its works. A separate discussion about significance and methodology of case study method has been added to this chapter.

Questionnaire survey is another important method. It has the following advantages:

- It is flexible.
- It facilitates dialogues.
- It uses semi-structured questionnaire format in which many questions are open-ended.
- It can be used to study effectiveness and impacts of any programme.
- It can be used to study organizational issues.
- It takes relatively less time.

Rapid assessment is another quick and simple method. It is generally undertaken by a team of professionals belonging to different fields. It takes less time. Apart from this, it addresses to specific issues by a team of said professionals. The techniques which are employed under this method are:

1. Direct observation.
2. Informal communication with key informants.
3. Brief group discussion.

This method of rapid assessment is used for evaluating forestry works. Generally, it is advisable to resort to this method of evaluation by a team of technical staff with sufficient knowledge and adequate skills. Besides, this method can be used to educate and sensitize the team of evaluators before the real exercise of evaluation by any other methods starts. In this way, this method plays a supplementary role to the main field method.

Participatory Rural Appraisal (PRA) is another important method used in evaluation works. Over the years, this method has gained enormous importance. This method is extensively used to evaluate different community development programmes. In forestry, this method is used for evaluating village resource development programme or any other programmes where the focus is on community development. This method has been dealt with in separate chapter.

2. *Quantitative method :*

Quantitative methods require quantitative data. Such methods are used for assessing and measuring effects and impacts of the programme or project. The methods of collecting primary data in forest resource management sector are as follows:

1. Forest Resource Survey or sampling.
2. Questionnaire survey for socio-economic study.
3. Case studies.
4. Other technical or economic analysis.

In all the aforesaid methods, the data are recorded in nùmerical form (i.e. number of trees, diameter and height of trees in resource survey and income of different households, land use pattern, consumption pattern in socio-economic survey).

To sum-up, a qualitative method is the most appropriate one for evaluating community development works, whereas the quantitative approach is most appropriate for evaluation of scientific work with specific and physically measurable indicators.

Process of Evaluation

The process of evaluation encompasses many components. In the words of Reidar Dale (1998) "Evaluation are basically about describing, judging and explaining what has been done, how things have been done, what has been achieved, and commonly, what future prospects or options may exist. This means, that one must seek to link past status and event to present status and events, by specifying any difference between them and exploring how such differences have been created. In addition to it, if one is looking into the future, he has to ask whether observed changes are likely to continue and what factors are likely to influence the continuation of such changes. Commonly, the following models are in vogue in forestry management practices.

1. *First model* : This is a conventional method of evaluation. Here, base line information is collected before the initiation of the programme or project. Corresponding information is collected after the completion of the said project. These two separate exercises provide 'before' scenario and 'after' scenario. The study related to 'before' is called as base line and the studies conducted after the project is called follow-up studies. Under this method, quantitative information will be collected. Such information pertaining to pre and post-situation will be presented in relevant tables and differences between two values will be compared. Qualitative information can also be collected such as people's perception about forest management or adequacy of forest produces. Such information should be recorded in some standard form or should be ranked to enable comparison. Certain things can directly be observed and judged by the evaluator himself.

2. *Second Model* : Under this model, the base line information is recorded before the beginning of the project and corresponding information is collected after the completion. Now the evaluation begins with examination of processes and relation among internal as well as external factors in the course of the programme. Both the processes of work and change is analyzed in the exercise. While conducting the exercise, the evaluator has to resort to certain assumptions. Assumptions are opinion about the little known factors. This approach of evaluation helps to provide an understanding of what has happened and how it has happened. This method is complex and time consuming too.

3. *Third model* : This model applies to evaluation exercise in forestry as well as non-forestry works. The evaluator records the situation at the time of evaluation and simultaneously collects information with regard to the situation prevailing before the inception of the project. This is accomplished by undertaking household interview or talking to key informants or examining the untreated patch of forest. The questionnaire method of inquiry for socio-economic survey can also be resorted to. This model is well suited to socio-economic programmes. Here, the participatory method of exploring things is generally adopted.

Indicators

Indicators symbolize brief and concise expression about any phenomenon. They are used as an approximation or proxy for the phenomenon under study. They are expressed in qualitative as well as quantitative terms. They are used to assess the effects and impacts in the course of evaluation. They also facilitate comparison among different situations. An indicator is indirect statement or measure pertaining to a phenomenon, or event or development or change. Indicators may be direct or indirect. However, the use of indicators in evaluation has its own disadvantages. They cannot explain beyond a point. They provide only a pre-desired explanation about an event or phenomenon or change.

Monitoring and evaluation indicators

These are three broad categories of indicators. They are:

1. Operation and performance indicators:

(a) Input indicators, e.g. staff arrangement, recurrent expenditure.
(b) Operation indicator, e.g. measure of execution, timeliness.
(c) Performance indicator, It is done by comparing results with targets.
(d) Efficiency indicators, they relate inputs to performance.

2. Impact Indicators:

(a) Direct impact indicators/effect indicators, e.g. target V*s.* achievement.
(b) Indirect impact indicators; measures of all other effects.

3. Indicators of external conditions, prices, market conditions, climate, etc.

Organization of Evaluation

Evaluation is generally conducted in every sphere of forestry works. However, over the years it has gained importance in the following forestry practices:

— Plantation projects.
— Works done under joint forest management.
— Eco-development.
— Externally aided forestry project.
— Management practices undertaken in the last plan.

Evaluation is a systematic and planned exercise. Therefore, in order to accomplish it, there is a need for an organizational framework. In forest departments, evaluation is generally done by departmental agencies. Such evaluations are more formal in nature. However, evaluation can be conducted by external agencies also. Such evaluation provides independent and unbiased judgment about the effect and impact of the programme. Such an approach assumes significance because of the fact that it helps to improve the performance standard of the programme.

While defining the evaluation system for any project or programme, it is necessary to emphasize the following areas:

- Purposes of evaluation.
- Type of information required.
- Method of collection of information.
- How the findings of evaluation exercise will be used and what are its benefits.
- Organization of evaluation and their management.
- Cost of the exercise.

The Evaluation Report

Evaluation report is the end of evaluation process. It contains all the outcomes of evaluation study. The basic characters of the report are simplicity, reliability and cohesive presentation. The general principles on which the report should be based are as follows:

- Structuring of the report.
- Presentation of strength and weakness of different methods used and quality source of information.
- Organization of the reports containing.
- A summary.
- A main body—presenting the finding and arguments.
- Appendices—giving additional supportive information on specific and relevant topics.
- Clearly spelt recommendation.
- References to source of information.
- Representation of appropriate data in tabular, pictorial or graphical forms.

A general format for evaluation report has been suggested in

Table 28.3. This table can be used as a guide for writing such evaluation report. Improvement or change can be made, as the situation requires. It may so occur that on account of specific focus on some areas all the category of evaluation like relevance, efficiency, effectiveness, impacts, sustainability or replicability may not be relevant to evaluation. Therefore, the evaluator has to decide the variables, which are relevant to the said task.

Table 28.3 : Indicative format of comprehensive evaluation reports

- — Summary, emphasizing conclusions and recommendations.
- — Introduction. Back ground and rationale of the evaluation. Placing the evaluation in general context.
- — Brief description of the project or programme. Its rationale and objectives, components, organizational structure, and process of implementation.
- — Evaluation design and process. Focus of the evaluation, evaluation methodology and the process of information collection, analysis, and report writing.
- — Recommendation. Listed by major topic and in order of importance. Specifying decisions, which are recommended to be taken, and a timetable.
- — General conclusion. Lessons learned from the evaluation for the planning and implementation.
- — Appendices.

Case Study: Principles and Method

Case study method is a widely used field research technique in the field of sociology. Frederic Le Play introduced this method to the field of social investigation. Herbert Spenser used this technique to make a comparative study of cultures. Similarly, many sociologists, anthropologists, historians, etc. used this method to study different problems pertaining to their area of interests. Now-a-days, many management experts are applying this method to analyse several intractable management problems. In brief, case study method is widely used and accepted technique in every discipline of arts and management.

Case study method is one of the most important techniques used for qualitative analysis. It involves careful and complete observation of a social unit. Apart from this, it is a method of study that requires in-depth analysis. Commonly, it lays emphasis on comprehensive analysis of few activities or events and their dynamic interrelationship. The objective of case study method is to identify factors that account for the behaviour of any unit as an integrated totality. In other words, this method of study primarily aims at exploring the causative factors, which leads to occurrence of any phenomenon or activity or event.

According to H. Odum, "The case study method is a technique by which individual factors, weather it be an institution or just an episode in the life of an institution or an individual or a group is analysed in its relationship to any others in the group". Burgess, another sociologist used the words "the social microscope" to describe the case study method of analysis. Pauline V. Young, a sociologist describes case study as "a comprehensive study of a social unit, be that an unit, a person, a group, a social institution, or a district or a community."

To sum up, a case study method can be said to be a form of qualitative analysis wherein careful and complete observation of an individual or a situation or an institution is made. Based on such analysis, inferences are drawn and applied to such analogous situation.

Characteristics of Case Study Method

The important characteristics of the case study method are as under:

- The researcher takes one or more social unit(s) or event or a situation for comprehensive study.
- The selected subject or situation is studied to the extent of possible minute details.
- The approach in case study method is qualitative and not quantitative. Therefore, all aspects of life are investigated with more emphasis on qualitative interpretation.
- Mutual inter-relationship of casual factors responsible for the occurrence of the event or situation is studied.

Process involved in the Method

The following are some of the steps involved in the preparatory process for a case study:

- Identification of problem or event or phenomenon or situation.

- — Determination of the status of the phenomenon to be investigated.
- — Collection of data, examination and portrayal of historical account of history of the given phenomenon.
- — Diagnosis and identification of casual factors as a basis for remedial and developmental treatment.
- — Application of remedial measures (Case Work).
- — Follow-up programme to determine effectiveness of the remedial measures.

Advantages of Case Study Method

The following are some of the important advantages of this method.

1. It helps the investigator to understand the complete behaviour pattern of the phenomenon. In the words of Charter Horton Cooley, "case study deepens our perception and gives us a clear insight into life. It gets at behaviour directly and not by an indirect and abstract approach".

2. The investigator finds an exposure of the man's strivings, tensions, emotion and motivation that are responsible for a certain pattern of behaviour.

3. It helps the investigator to find out the historical perspective of the phenomenon, with clear depiction of its social, economic, physical and natural environment.

4. This method provides new knowledge to the existing concept and approach. This helps in suggesting improvement in the programme.

5. It provides full opportunity for understanding social dimension of the phenomenon.

6. Other method of study can also be harmonised with the case study method to add more qualitative information and perception into any multi-dimensional problem.

7. This method helps in improving the skill, expertise and analysing ability of the investigator.

8. This method helps in studying social changes in the society. On account of an in-depth study of different facts of a social phenomenon, the investigator understands the pace, trend and qualities of social change then and now. This helps in formulating suitable management practices.

Limitations

In spite of all advantages, the case study method suffers from few limitations. They are as follows:

- — In a society, phenomena appear similar apparently. But, in reality, it differs. So, on the basis of study of a phenomenon in a society, a general inference should not be drawn.
- — The data generated in case study method are not scientific and objective. The investigator is more interested in subjectivity of the phenomena.
- — Case study method is based on several assumptions, which may not be based on realism all the times.
- — Case study method is more prone to personal bias of the investigator.

Despite the aforesaid limitations, case study method has emerged as one of the powerful tools to study social problems. Most of the limitations can be corrected according to necessity to improve the method. This will increase the applicability of the method in many fields.

In the field of forestry, the case study method has wide applicability. As we know, forestry is an interdisciplinary subject. The field of sociology, economic, anthropology having relationship with forestry are the application areas for case study method. The following are some of the social problems, which are generally studied by employing case study method.

- — Community response to joint forest management.
- — Changing consumption patterns among people living near forest.
- — Adaptation of village people to fuel wood shortage situation.
- — People's response to work carried under world food programme in the forestry sector.
- — Dynamics of people's response in case of forestry fire incidences.
- — Effects of eco-development programme on the village life.

GUIDELINES FOR PREPARATION OF WORKING PLANS AND FELLING IN FORESTS

Guidelines for Preparation of Working Plan and Felling in Forests

Development of forests largely depends upon the quality of working plans and various prescriptions proposed to be undertaken in the particular forest division. The forests have achieved much importance for their productive, environmental and protective functions. Preparation/revision of working plans for proper management of forests is, therefore, of great importance. The set of guidelines proposed here is to ensure quality of working plans covering matters which, in recent times have been recognised as of great importance.

1. Need for Working Plans

In India 69 per cent of the forest area is covered by working plans. The balance 31 per cent have no working plans. State/Union Territory Governments should take steps for preparation of working plans for the areas yet uncovered a crash, time bound programme. All forest areas should be covered by working plans by the end of this decade. In case, preparation of regular working plans would sometime, preparation of working schemes, should be taken up immediately for the uncovered areas and be completed by 1985.

In the North-Eastern region many forests are managed by District Councils. Generally, these forests do not have working plans and are not properly managed. District Councils should be pursuaded to prepare working plans for such forest areas with the assistance of the Forest Department.

Wherever private forests exist, owners of such forests should be pursuaded to prepare working plans with the help of the Forest Department.

2. Objectives of Management

The working plan should clearly enunciate the objective of management. Such objective should be in tune with the present policy which emphasises conservation as the most important function of forests. Wherever production forestry is proposed, environmental and social considerations should prevail. Revenue should not be the prime criterion in the objectives of management and it should be clearly stated that meeting the local needs of forest produce should be the first charge on the disposal of such produce.

3. Working Plan Code

In many States/Union Territories (UTs), Working Plan Codes are available. These Codes give guidelines for preparation of working plans. In some States/UTs, however, Working Plan Codes are not available. Immediate steps should be taken by the States/UTs to prepare Working Plan Codes where they do not exist. In case of smaller States/UTs, it may be possible to adopt the Codes prevalent in adjoining Sates/UTs. For preparation or adoption of such Codes, Forest Research Institute may be contacted for guidance.

4. Preliminary Working Plan Reports

The preparation of working plan of a Forest Division/Unit should be preceded by a Preliminary Working Plan Report in two stages.

The First Preliminary Report, in case of revision of a working plans, should briefly review the results of the past management, justify the needs for its revision and make suggestions regarding modifications necessary to improve upon the prevailing management techniques.

The First Preliminary Working Plan Report should also be prepared for the first working plan of any Division/Unit, incorporating therein suggestions for management practices to be followed. The report should be prepared by the Conservator of Forests in charge of the Forest Division for which the report is to be prepared. The First Preliminary Working Plan Report should be approved by the Chief Conservator of Forests after discussions in which a representative of the Ministry of Agriculture should also be associated.

The Working Plan Officer, on the First Preliminary Working Plan Report should reconnoitre the entire area and submit outlines for the working plan, classifying the areas into various Working Circles and suggesting the silvicultural system to be followed in each of the Working Circles depending on the objectives of the plan. This Report, which will be the second preliminary working plan report, will be approved by the Chief Conservator of Forests in consultation with the representative of the Government of India and expert from disciplines like, Environment, Botanical Survey of India, whose assistance should be sought particularly for determination of protection areas and preservation of endangered species.

5. Contents of the Working Plan

The Working Plan Codes, wherever they exist, outline preparation of the plan in two parts, the first part deals with "Summary of facts on which the proposals are based" and the second part covers "Future management strategies with prescription of operations". It will be necessary to deal with the following issues as well, in detail, in the two parts of the working plan, in case these issues are not mentioned in the Working Plan Code:

(A) *Rights and Concessions :* Besides dealing with the existing rights and concession on forests, the tribal rights and practices should be properly highlighted together with mechanism proposed for regulating such rights and concessions.

(B) *Grazing and fodder :* Grazing is a serious problem in most of the forest areas. A detailed study should be made of this problem and clear prescription used as to how grazing need can be adjusted in relation to carrying capacity for forests. The scope for improvement of grazing potential through better management is also required to be brought out. Cultivation and propagation of fodder species including grasses in various forest areas to help meet forage needs especially during distress conditions should be included in the prescriptions.

(C) *Shifting cultivation :* In many forest areas, shifting cultivation is a tradition bound practice. A study of the local practice should be made and measures to control shifting cultivation, as far as practicable, should be proposed without upsetting the local economic situation and social customs. Alternatives to the existing practices should be worked out and recommended.

(D) *Encroachment :* Encroachment is fairly wide-spread in many forest areas. An assessment of the problem should be made and to what extent and how this can be controlled should be suggested.

6. Clear Felling of Natural Forests

Over the years, many silvicultural practices like Selection System, Uniform System, Coppice with Reserves, which are based on the concept of harvested trees through natural regeneration, had to be replaced progressively by clear felling with artificial regeneration system. The reasons are very many, important of them being anticipated better productivity and operational convenience. However, in view of evidence of continuing depletion of natural forests and need for preservation of all biological diversities which natural forests sustain, it would be necessary to review the existing and proposed management systems involving clear felling with artificial regeneration. As a matter of principle, clear felling of natural forests with artificial regeneration should be avoided as far as possible. However, if it has to be resorted to for compelling reason, *e.g.* lack of natural regeneration of desired species, poor condition of the standing crops and need for restoration of biological diversity of the original natural crop, a thorough study should first be made. Again, wherever clear-felling is resorted to, it should be over very restricted areas of not exceeding 10 hac. in the hills and 25 hac. on the plains and, in no case, good natural forests should be clear-felling in preference to poorly stocked natural forests.

7. Restriction of Fellings in the Hills

In view of the poor state of forests in the hills and mountains of the country and necessity of preservation of forests in those areas for soil and water conservation, State Governments should consider banning of all fellings above certain altitude, say 1000 metres, at least for some years. It will also be necessary to identify critical areas in the hills and mountains, in the catchments of river valleys, areas prone to landslips, erosion and geologically unstable formations which require protection from felling of forests and need immediate vigorous afforestation within the shortest possible time and a review of the conditions of such critical areas should be made every five years. Detailed prescriptions of management should be embodied in the working plans for these critical areas.

8. Afforestation of Barren and Degraded Areas

It is estimated that about 35 million hac. of forest lands are barren or in degraded conditions. Vigorous plantation programmes should be launched for quick re-afforestation of these barren and degraded areas. The Working Plans should prescribe a quick and time-bound programme of afforestation of such lands within the shortest possible time, if necessary, even suspending the practice of clearfelling and planning in areas where these are in vogue.

9. Protection Areas

The working plans should deal in detail the management practices relating to the protection areas like, Wildlife Sanctuaries, National Parks, Biosphere Reserves, Preservation Plots etc. Internationally, 10 per cent of the geographical areas of the country is being considered for protection in order to protect representative eco-systems in the region and preservation of germ-plasm of diverse nature. At present, in India about 2 per cent of the geographical areas are being treated as protection areas. Working plan should identify more protection areas and aim at reaching 4 per cent of the geographical area. Such areas should, in particular, include tropical/sub-tropical temperate, wet/moist evergreen/semi-evergreen forests, littoral forests, mangroves and alpine formations. Detailed management strategies and protection measures should be outlined in the working plans for the areas. In fact, it may be advisable to devote a separate chapter on Protection Areas.

Wherever separate management plans are being prepared for Wildlife Sanctuaries, National Parks, etc., a gist of prescription of such plans should be incorporated in the working plan of the territorial forest division.

10. Tribal Welfare

The Scheduled Tribe live mostly in the forest areas. Forests represent for them a whole way of life home, culture, worship, food and source of employment and income. Forestry occupies a central position in the tribal economy.

Forest development and Tribal development should reinforce each other and these should gain wide recognition in the working plans. Working plan prescriptions should aim at the welfare of the tribals who depend heavily on the forest for their livelihood ensuring simultaneously that the forests are conserved.

Tribals are dependent on several forest species like sal (*Shorea robusta*), mahua (*Bassia species*), kusum (*Schleichera trijuga*), tendu (*Diospyros species*) and several other species of food and medicinal value. All species vital to the tribal economy should be identified, listed and prescribed for preservation and perpetuation.

Tribals derive usually very low wages by working as forest labourers or by selling firewood, and as they do not share in the profits made from the forests, their wages always remain low as compared to the industrial labourers. The tribals should be a participating community in harvesting, collecting, processing of all forest produce especially minor forest produce like, oilseeds, tendu leaves, etc.

11. Monoculture

A review should be made to see whether monoculture practices should be dispensed with in view of its adverse impact on environment. It has also been alleged that such practices lead to less productivity than the natural forests which monoculture replaced. Culture of mixed species of indigenous origin should be preferred to monoculture, indigenous species should receive preference over exotics. Wherever exotics are planted for reasons of particular need or higher productivity, such exotics should preferably be intimately mixed with species of natural origin.

In artificial regeneration operations, choice of species should be made judiciously. The choice will not only depend on the site and the silvicultural needs of the species, it should also take into account the consideration of preference of the local population where they are the principal users and for uses to which species will ultimately be put.

12. Reduction of Yield Prescription

There is continuous evidence of gradual degradation of the forest stock due to demands on various types of forest produce, principally firewood, and due to illicit fellings which have assumed serious proportion in some areas. Such a degradation is obviously the result of removal of forest produce far in excess of that which is prescribed in the working plans. To build up the stocking of the forests it would be desirable to reduce the fellings by curtailing the fellings. It would be advisable to reduce the felling by 50 per cent of the yield, that is arrived at by normal calculations.

Whenever extra fellings outside working plan prescriptions are done, say for a development project, salvage operations of timber due to flood, snow-damage, landslides, etc. such fellings should be adjusted against the working plan yield during the following year.

13. Forest Area, Stock Maps and Growing Stock

At the time of preparation/revision of the working plans, the Working Plan Officer should make an accurate survey of the forest area of the division/unit. For this purpose, data available with the Survey of India and the resources division of the states should also be made use of.

Preparation of stock maps of compartments and enumeration of growing stock of natural and man-made forests together with an analysis of projected increment and productivity of the forest should be carried out. For preparation of forest area and density, the use of aerial photographs may also be made with advantage.

While making enumerations, the growing stock of important plant entities like herbs, shrubs and climbers should also be assessed and a list of endangered plant species should be prepared in consultation with Botanical. Survey of India. Protection of such species should be ensured by making specific prescriptions.

Although only two parts of the working plan have been suggested earlier, additional volumes on stock maps, enumeration of growing stock and a list of protection areas should be added to the working plan.

14. Forest Industry Catchment

In every State/UTs small, and large industries based on wood and other forest produce as raw material from forests, are functioning. Generally, their requirements are met from the natural forests, thus causing considerable pressure on these forests. Although the need of various industries on natural forests may not be completely eliminated due to specific requirements, there should be consistent endeavour to reduce the pressure of forest industries on natural forests.

The working plan should identify possible catchments for various forest industries in barren and degraded lands and assess the production potential for such lands after afforestation.

The plan should identify potential for forest-based industries in small and cottage industries sector in particular, and highlight their development for the benefit of weaker sections of the community, especially tribals. Indigenous processing of oilseeds and use of tendu leaves are two examples.

15. Forest Development Corporations

Areas leased to the Forest Development Corporations of the States/ Union Territories should also be brought under working plans, where no such plans exists. It is understood that for clear felling and regeneration without proper working plan. The plans for such areas should be prepared immediately and completed within the next 18 months.

16. Checking of Boundaries

In view of the fact that the encroachment of the forest areas is a serious problem, these should be assessed and enlisted at the time of field work relating to preparation/revision of the working plan.

17. Working Plan for Social Forestry

Working Plans for Social Forestry programme should be prepared. In Social Forestry programme, afforestation of barren land in possession of the authorities other than the Forest Department should also be included.

18. Working Plan Officer

As working plan is a very important document and its preparation requires considerable skill, it is recommended that a forest officer of at least 10 years service should be employed as Working Plan Officer.

19. Period of Working Plan

The need for frequent revision of working plans to conform to changed need of the situation, cannot be over-emphasised. Normally, the period of working plan should be 10 years but a review should be made after a period of 5 years and prescriptions modified if considered necessary.

20. Committee for Approval of the Working Plan

The working plan should be approved by the State Government after it has been examined by a Committee headed by the Secretary of the State/UT Forest Department. Experts from other disciplines like Environment, Botanical Survey of India, etc. besides Chief Conservator of Forests of the State/UT should be included in the committee. A representative of the Forestry Division of the Government of India should be associated in these discussions and his views given due consideration.

21. Working Plan Cell in the Government of India, Ministry of Agriculture

A Working Plan Cell has since been created in the Ministry of

Agriculture under the Inspector General of Forest. The primary functions of this Cell is to assist the State/UT Forest Department for preparation/revision of every working plan. This Cell should be closely associated at various stages of preparation revision of every working plan in the State/UT.

The Working Plan Cell should also be associated while considering any deviation from the prescription of the approved working plan. Deviation should be allowed only after obtaining the approval of the Inspector General of Forest.

No. 13-/39-F.P.

GOVERNMENT OF INDIA

Ministry of Environment and Forests

Department of Environment, Forests and Wildlife

Paryavaran Bhavan, CGO Complex, B-Block, Vth Floor

Lodi Road, New Delhi

Dated 11th June, 1989

To,

The Forest Secretaries

All States/UTs

Subject : Plantation and protection of species useful as minor forest produce to the tribals and other poor people.

Sir,

As you might be aware the National Forest Policy, 1988 suggests that special attention be paid to protection, regeneration and optimum collection of minor forest produce for providing gainful employment to the tribals.

2. Normally, the working plan prescriptions are expected to retain fruit bearing and other multiple utility species in forestry operations. It would, however, be desirable to advise again the Working Plan Officers the Conservator of Forests, Incharge of the Working Plans and other concerned officials to ensure that the species which are useful to the local inhabitant for domestic consumption or collection of minor forest produce should be scrupulously retained in silvicultural operations. Such species ought to be valued primarily for their utility to the local people and production of timber in such cases should be considered incidental to it.

3. Apart from protection of existing growth efforts would be needed for supplementing growing stock of minor forest produce. This could be done by undertaking plantations of such species in various afforestation works, especially in the degraded forests around habitation. An illustrative list of the species which may prove useful to tribal and other local people is enclosed.

Yours faithfully,

Sd/- K. M. Chadha

Joint Secretary to the Govt. of India.

Copy to :—

1. Principal Chief Conservator of Forests/Chief Conservator of Forest (all States/UTs).
2. Director of Social Forestry (all States/UTs).
3. Managing Directors, Forest Corporations (all States/UTs).
4. NWDB.

Sd/-
K. M. Chadha
Joint Secretary
to the Govt. of India

Species generally useful to tribal people

S.No.	Botanical Name	Local Name
1.	*Acacia nilotica*	Babul
2.	*Agave Sisilana*	Sisal
3.	*Annona Squamosa*	Sitafal
4.	*Artocarpus heterophyllus*	Kathal
5.	*Artocarpus lakoocha*	Salap (lakooch)
6.	*Azadirachta indica*	Neem
7.	*Bamboos spp.*	Bans
8.	*Bauhinia spp.*	Kachnar
9.	*Buchnania latifolia*	Achar
10.	*Bursera serrata*	Garuda Nimburu
11.	*Dalbergia lanceolata*	Chhota sisu
12.	*Dendrocalamus spp.*	bamboo
13.	*Diospyrous melanoxylon*	Tendu
14.	*Eagle marmelos*	Bel
15.	*Emblica officinalis*	Aonla

S.No.	Botanical Name	Local Name
16.	*F. glomerata*	Gular, Umar
17.	*Ficus bengalensis*	Bargad
18.	*Garuga pinnata*	Kasaramb
19.	*Grewia spp.*	Dhaman
20.	*Juglans regia*	Akhrot
21.	*Madhuca Indica*	Mahua
22.	*Mangifera Indica*	Am
23.	*Moringa oleifera*	Sanjna
24.	*Morus alba*	Shahtut
25.	*Phylienthus distichus*	Harphul
26.	*Pongamia pinnata*	Karanj
27.	*Schleichera trijuga*	Kusum
28.	*Ptorocarpus marsupium*	Bijasal
29.	*Shorea robusta*	Sal
30.	*Spondias pinnata*	Amra
31.	*Syzygium cumini*	Jamun
32.	*Tamarindus indica*	Imli
33.	*Terminalia arjuna*	Arjun
34.	*Terminalia balerica*	Baheda
35.	*Terminalia chebula*	Harida, Harra
36.	*Terminalia tomentosa*	Asan
37.	*Zizyphus rotundifolia*	Jhadi Beri
38.	*Zizyphus jujuba*	Ber

No. 3-1/86-FP
GOVERNMENT OF INDIA
Ministry of Environment and Forests
(Department of Environment, Forests and Wildlife)

Paryavaran Bhavan, CGO Complex,
Lodi Road, New Delhi - 110003
Dated the 7th December 1988.

RESOLUTION
National Forest Policy, 1988

1. Preamble

1.1 In Resolution No. 13/52-F, dated the 12th May 1952, the Government of India in the erstwhile Ministry of Food and Agriculture enunciated a Forest Policy to be followed in the management of State Forests in the country. However, over the years, forests in the country have suffered serious depletion. This is attributable to relentless pressures arising from ever-increasing demand for fuelwood, fodder and timber, inadequacy of protection measures; diversion of forest lands to non-forest uses without ensuring compensatory afforestation and essential environmental safeguards; and the tendency to look upon forests as revenue earning resource. The need to review the situation and to evolve, for the future, a new strategy of forest conservation has become imperative. Conservation includes preservation, maintenance, sustainable utilisation, restoration, and enhancement of the natural environment. It has thus become necessary to review and revise the National Forest Policy.

2. Basic Objectives

2.1 The basic objectives that should govern the National Forest Policy are the following:

- Maintenance of environmental stability through preservation and, where necessary, restoration of the ecological balance that has been adversely disturbed by serious depletion of the forests of the country.
- Conserving the natural heritage of the country by preserving the remaining natural forests with the vast variety of flora

and fauna, which represent the remarkable biological diversity and genetic resources of the country.

— Checking soil erosion and denudation in the catchment areas of rivers, lakes, reservoirs in the interest of soil and water conservation, for mitigating floods and droughts and for the retardation of siltation of reservoirs.
— Checking the extension of sand dunes in the desert areas of Rajasthan and along the coastal tracts.
— Increasing substantially the forest/tree cover in the country through massive afforestation and social forestry programmes, especially on all denuded, degraded and unproductive lands.
— Meeting the requirements of fuelwood, fodder, minor forest produce and small timber of the rural and tribal populations.
— Increasing the productivity of forests to meet essential national needs.
— Encouraging efficient utilization of forest produce and maximizing substitution of wood.
— Creating a massive people's movement with the involvement of women, for achieving these objectives and to minimize pressure on existing forests.

2.2 The principal aim of Forest Policy must be to ensure environmental stability and maintenance of ecological balance including atmospheric equilibrium, which are vital for sustenance of all lifeforms, human, animal and plant. The derivation of direct economic benefit must be subordinated to this principal aim.

3. Essentials of Forest Management

3.1 Existing forests and forest lands should be fully protested and their productivity improved. Forest and vegetal cover should be increased rapidly on hill slopes, in catchment areas of rivers, lakes and reservoirs and ocean shores and on semi-arid and desert tracts.

3.2 Diversion of good and productive agricultural lands to forestry should be discouraged in view of the need for increased food production.

3.3 For the conservation of total biological diversity, the network of national parks, sanctuaries, biosphere reserves and other protected areas should be strengthened and extended adequately.

3.4 Provision of sufficient fodder, fuel and pasture, especially in areas adjoining forest, is necessary in order to prevent depletion of forests beyond the sustainable limit. Since fuel wood continues to be the predominant source of energy in rural areas, the programme of afforestation should be intensified with special emphasis on augmenting fuel wood production to meet the requirement of the rural people.

3.5 Minor forest produce provides sustenance to tribal population and to other communities residing in and around the forests. Such produce should be protected, improved and their production enhanced with due regard to generation of employment and income.

4. Strategy

4.1 *Area under Forests*

The national goal should be to have a minimum of one-third of the total land area of the country underforest or tree cover. In the hills and in mountainous regions, the aim should be to maintain two-third of the area under such cover in order to prevent erosion and land degradation and to ensure the stability of the fragile eco-system.

4.2 *Afforestation, Social Forestry & Farm Forestry*

4.2.1 A massive need-based and time-bound programme of afforestation and tree planting, with particular emphasis on fuel wood and fodder development, on all degraded and denuded lands in the country, whether forest or non-forest land, is a national imperative.

4.2.2 It is necessary to encourage the planting of trees along side of road, railway lines, rivers and streams and canals, and on other unutilised lands under State/corporate, institutional or private ownership. Green belts should be raised in urban/industrial areas as well as in arid tracts. Such a programme will help to check erosion and desertification as well as improve the micro-climate.

4.2.3 Village and community lands, including those on foreshores and environs of tanks, not required for other productive uses, should be taken up for the development of tree crops and fodder resources. Technical assistance and other inputs necessary for initiating such programmes should be provided by the Government. The revenues generated through such programmes should belong to the panchayats where the lands are vested in them, in all other cases, such revenues should be shared with the local communities in order to provide an incentive to them. The vesting, in individuals, particularly from the weaker sections (such as landless labour, small and marginal farmers, scheduled castes, tribals, women) of certain ownership rights over trees, could be considered, subject to appropriate regulations. Beneficiaries would be entitled to usufruct and would in turn be responsible for their security and maintenance.

4.2.4 Land laws should be so modified wherever necessary so as to facilitate and motivate individuals and institutions to undertake tree-farming and grow fodder plants, grasses and legumes on their own land. Wherever possible, degraded lands should be made available for this purpose either on lease or on the bases of a tree-patta scheme. Such leasing of the land should be subject to the land grant rules and land ceiling laws. Steps necessary to encourage them to do so must be taken. Appropriate regulations should govern the felling of trees on private holding.

4.3 *Management of State Forest*

4.3.1 Schemes and projects which interfere with forests that clothe steep slopes, catchment of river, lakes, and reservoirs, geologically unstable terrain and such other ecologically sensitive areas should be severely restricted. Tropical rain/moist forests, particularly in areas like Arunachal Pradesh, Kerala, Andaman & Nicobar Islands, should be totally safeguarded.

4.3.2 No forest should be permitted to be worked without the Government having approved the management plan, which should be in a prescribed format and in keeping with the National Forest Policy. The Central Government should issue necessary guidelines to the State Government in this regard and monitor compliance.

4.3.3 In order to meet the growing needs for essential goods and services, which the forests provide. It is necessary to enhance forest cover and productivity of the forests through the application of scientific and technical inputs. Production forestry programmes, while aiming at enhancing the forest cover in the country, the increasing gap between demand and supply of fuelwood. No such programme, however, should entail clear-felling of adequately stocked natural forests. Nor should exotic species be introduced, through public or private sources, unless long-term scientific trials undertaken by specialists in ecology, forestry and agriculture have established that they are suitable and have no adverse impact on native vegetation and environment.

4.3.4 *Rights and Concessions*

4.3.4.1 The rights and concessions, including grazing, should always remain related to the carrying capacity of forests. The capacity itself should be optimized by increased investment, silvicultural research and development of the area. Stall-feeding of cattle should be encouraged. The requirements of the community, which cannot be met by the rights and concession so determined, should be met by development of social forestry outside the reserved forests.

4.3.4.2 The holders of customary rights and concessions in forest areas should be motivated to identify themselves with the protection and development of forests from which they derive benefits. The rights and concessions from forests should primarily be for the *bona fide* use of the communities living within and around forest areas, specially the tribals.

4.3.4.3 The life of tribals and other poor living within and near forests revolves around forests. The rights and concessions enjoyed by them should be fully protected. Their domestic requirements of fuel wood, fodder, minor forest produce and construction timber should be the first charge on forest produce. These and substitute materials should be made available through conveniently located depots at reasonable prices.

4.3.4.4 Similar consideration should be given to scheduled castes and other poor living near forests. However, the area, which such consideration should cover, would be determined by the carrying capacity of the forests.

4.3.4.5 Wood is in short supply. The long-term solution for meeting the existing gap lies in increasing the productivity of forests but to relieve the existing pressure on forest for the demands of railway sleepers construction industry (particularly in the public sector), furniture and panelling, mine-pit props, paper and paper board, etc. substitution of wood needs to be taken recourse to. Similarly, on the front of domestic energy, fuel wood needs to be substituted as far as practicable with alternate sources like bio-gas, LPG, solar energy. Fuel-efficient "Chulhas" as a measure of conservation of fuel wood need to be popularized in rural areas.

4.4 *Diversion of Forest Lands for Non-forest Purposes*

4.4.1 Forest land or land with tree cover should not be treated merely as a resource readily available to be utilized for various projects and programmes, but as a national asset which requires to be properly safeguarded for providing sustained benefits to the entire coummunity. Diversion of forestland for any non-forest purpose should be subject to the most careful examinations by specialists from the standpoint of social and environmental costs and benefits. Construction of dams and reservoirs, mining and industrial development and expansion of agriculture should be consistent with the needs for conservation of trees and forests. Projects, which involve such diversion, should at least provide in their investment budget, funds for regeneration/compensatory afforestation.

4.4.2 Beneficiaries who are allowed mining and quarrying in forest land and in land covered by trees should be required to repair and re-vegetate the area in accordance with established forestry practices. No mining lease should be granted to any party, private or public, without a proper mine management plan appraised from the environmental angle and enforced by adequate machinery.

4.5 *Wildlife Conservation*

Forest Management should take special care of the need of wildlife conservation, and forest management plans should include prescriptions for this purpose. It is specially essential to provide for "corridors" linking the protected areas in order to maintain genetic continuity between artificially separated sub-sections of migrant wildlife.

4.6 *Tribal People and Forests*

Having regard to the symbiotic relationship between the tribal people and forests, a primary task of all agencies responsible for forest management, including the forest development corporations should be to associate the tribal people closely in the protection, regeneration and development of forests as well as to provide gainful employment to people living in and around the forest. While special attention to the following:

- One of the major causes for degradation of forest is illegal cutting and removal by contractors and their labour. In order to put an end to this practice, contractors should be replaced by institutions such as tribal cooperatives, labour cooperatives, government corporations, etc. as early as possible;
- Protection, regeneration and optimum collection of minor

forest produce along with institutional arrangements for the marketing of such produce;

— Development of forest villages on par with revenue villages;

— Family oriented schemes for improving the status of the tribal beneficiaries; and

— Undertaking integrated area development programmes to meet the needs of the tribal economy in any around the forest areas, including the provision of alternative sources of domestic energy on a subsidized basis, to reduce pressure on the existing forest areas.

4.7 *Shifting Cultivation*

Shifting cultivation is affecting the environment and productivity of land adversely. Alternative avenues of income, suitably harmonized with the right landuse practices, should be devised to discourage shifting cultivation. Efforts should be made to contain such cultivation within the area already affected, by propagating improved agricultural practices. Area already damaged by such cultivation should be rehabilitated through social forestry and energy plantations.

4.8 *Damage to Forests from Encroachments, Fires and Grazing*

4.8.1 Encroachment on forest lands has been on the increase. This trend has been arrested and effective action taken to prevent its continuance. There should be no regularization of existing encroachments.

4.8.2. The incidence of forest in the country is high. Standing trees and fodder are destroyed on a large scale and natural regeneration annihilates by such fires. Special precautions should be taken during the fire season. Improved and modern management practices should be adopted to deal with forest fires.

4.8.3 Grazing in forest areas should be regulated with the involvement of the community. Special conservation areas, young plantation and regeneration areas should be fully protected. Grazing and browsing in forest areas need to be controlled. An adequate grazing fees should be levied to discourage people in forest areas from maintaining large herds of non-essential livestock.

4.9 *Forest Based Industries*

The main considerations governing the establishment of forest-based industries and supply of raw material to them should be as follows:

- As far as possible, a forest-based industry should raise the raw material needed for meeting its own requirements, preferably by establishment of a direct relationship between the factory and the individuals who can grow the raw material by supporting the individuals with inputs including credit, constant technical advice and finally harvesting and transport services.
- No forest-based enterprise, expect that at the village or cottage level, should be permitted in the future unless it has been first cleared after a careful scrutiny with regard to assured availability of raw material. In any case, the fuel, fodder and timber requirement of the local population should not be sacrificed for this purpose.
- Forest based industries must not only provide employment to local people on priority but also involve them fully in raising trees and raw material.
- Natural forest serves as a gene pool resource and help to maintaining ecological balance. Such forests will not, therefore, be made available to industries for undertaking plantation and for any other activities.
- Farmers, particularly small and marginal farmers, would be encouraged to grow, on marginal/degraded lands available with them, wood species required for industries. These may also be grown along with fuel and fodder species on community lands not required for pasture purposes, and by Forest department/corporation on degrade forests, not earmarked for natural regeneration.
- The practice of supply of forest produce to industry at concessional prices should cease. Industry should be encouraged to use alternative raw material. Import of wood and wood products should be liberalised.
- The above considerations will, however, be subject to the current policy relating to land ceiling and land-laws.

4.10 *Forest Extension*

Forest conservation programme cannot succeed without the willing support and co-operation of the people. It is essential, to inculcate in the people, a direct interest in forests, their development and conservation, and to make them conscious of the value of trees, wildlife and nature in general. This can be achieved through the involvement of educational institutions, right from the primary stage. Farmers, and interested people should be provided opportunities through institutions like Krishi Vigyan Kendras, Trainers Training Centres to learn agri-silvicultural and silvicultural techniques to ensure optimum use of their land and water resources. Short-term extension courses and lectures should be organized in order to educate farmers. For this purpose, it is essential that suitable programmes are propagated through mass media, audio-visual aids and the extension machinery.

4.11 *Forestry Education*

Forestry should be recognised both as a scientific discipline as well as a profession. Agriculture universities and institutions dedicated to the development of forestry education should formulate curricula and course for imparting academic education and promoting post-graduate research and professional excellence, keeping in view the manpower needs of the country. Academic and professional qualifications in forestry should be kept in view for recruitment to the Indian Forest Service an the State Forest Service. Specialised and orientation course for developing better management skills by in-service training need to be encouraged, taking into account the latest development in forestry and related disciplines.

4.12 *Forestry Research*

With the increasing recognition of the importance of forests for environment health, energy and employment, emphasis must be laid on scientific forestry research, necessitating adequate strengthening of the research base as well as new priorities for action. Some broad priority areas of research and development needing special attention are:

(i) Increasing of productivity of wood and other forest produce per unit of area per unit time by the application of modern scientific and technological methods.

(ii) Revegetation of barren/marginal/waste/mined lands and watershed areas.

(iii) Effective conservation and management of existing forest resources (Mainly natural forest eco-systems).

(iv) Research related to social forestry for rural/tribal development.

(v) Development of substitutes to replace wood products.

(vi) Research related to wildlife and management of national parks and sanctuaries.

4.13 *Personnel Management*

Government policies in personal management for professional foresters and forest scientists should aim at enhancing their professional competence status and attracting and retaining qualified and motivated personal, keeping in view particularly the arduous nature of duties they have to perform, often in remote and in hospitable places.

4.14 *Forest survey and Data Base*

Inadequacy of data regarding forest resources is a matter of concern because this creates a false sense of complacency. Priority needs to be accorded to completing the survey of forest resources in the country on scientific lines and to updating information. For this purpose, periodical collection, collation and publication of reliable data on relevant aspects of forest management needs to be improve with recourse to modern technology and equipment.

4.15 *Legal Support and Infrastructure Development*

Appropriate legislation should be undertaken, supported by adequate infrastructure, at the Centre and State levels in order to implement the Policy effective.

4.16 *Financial Support for Forestry*

The objective of this revised Policy cannot be achieved without the investment of financial and other resources on a substantial scale. Such investment is indeed fully justified considering the contribution of forest in maintaining essential ecological processes and life-support system and in preserving genetic diversity. Forests should not be looked upon as source of revenue. Forests are a renewable natural resource. They are a national asset to be protected and enhanced for the well being of the people and the Nation.

(K.P. Geethakrishnan)
Secretary to the
Government of India

Bibliography

Alexander, S. Mather (1990); *Global Forest Resources*; International Book Publishers and Distributors, Dehradun.

Anon, (1953),. Glossary of Technical Terms, British Commonwealth Forest Terminology (BCCFT) (modif); Empire Forestry Research Association, England.

Anon, (1961), 100 Year of Forestry, Volume 1 & 2, Forest Research Institte, Dehradun.

Anon (1966), Indian Forest and Forest Products Terminology, Part-I Forests Indian Forest Record, FRI, Dehradun.

Anon, (1983), Abridged Glossary of Technical Terms, Forest Research Institute, Dehradun.

Anon, (1983), Guidelines for Preparation of Working Plan and felling in forests. Ministry of Environment and Forests. Govt of India.

Anon, (1985), Report on Forest Resources of : (A), Balaghat, Mandla and Seoni, (1985), Rajnandgaon and Durg, Forest Survey of India, Dehradun.

Anon, (1988), National Forest Policy - 1988, Ministry of Environment and Forests, Govt. of India.

Anon, (1995), India 1995 : Publication Division; Ministry of Information and Broadcasting, Govt. of India.

Anon; India 1995; Publication Division, Ministry of Information and Broadcasting, Govt. of India.

Anon; India 1995; Publication Division, Ministry of Information and Broadcasting, Govt. of India.

Anon (1996), Volume equations for forests of India, Nepal and Bhutan, Forest Survey of India, Dehradun.

Aronoff S. (1989), *Geographical Information System : A Management Perspective*. WDL Publications, Ottawa, Canada.

Asthana, B.N.; (1990), *Elements of Statistics*, (Part I & II), Chaitanya Publication Home, Allahabad.

Bebarta, Kailash Ch., (1999) *Teak: Ecology, Silviculture, Management and Profitability* : International Book Publishers and Distributors; Dehradun.

Bebarta, Kailash, Ch. (1999), Working Plan for Rajnandgaon Forest

Division. M.P. Forest Department. Bhopal.

Benerjee, S.K. ; Misra, T.K. ; Singh, R.B., Lal, R.B.; (1996), *Forest Vegetation and Soil*; Indian Council of Forestry and Education, Dehradun.

Brandis, D. (1906), *Indian Trees* (fifth edition) (1971) Mahendra Pal Singh, 23-A New Connaught Place, Dehradun.

Burrough P.A., (1986). *Principles of Geographical System for Land Resources Assessment*. Clarendon, Oxford.

Burton, Gene; Thakur, Manab, (1995). *Management Today: Principles and Practices*. Tata McGraw-Hill Publication Company Ltd., New Delhi.

Carlsson, Jerkier; Kohlin, Gunnar and Ekbom, Anders. 1994. The Political Economy of Evaluation : International Aid Agencies and Effectiveness of Aid; New York : St Martin's Press.

Chaturvedi, A.N., and Khanna, L.S. (1982) *Forest Mensuration*. International Book Dirstributors, Dehradun.

Chambers, Robort, 1994. "Participatory Rural Appraisal. (PRA) Analysis of Experience". *World Development*, vol. 22. No 9, pp 1253-1268.

Champion J.L. and Reiss, M.J.; (1995); *Ecology: Principles and Application*; Cambridge University Press, U.K.

Champion Sir Harry. G. and Seth. S.K. ; (1968) GOI. Publication.

Champion, G. Harry and Seth, S.K.; General Silviculture for India. (1968). Forest Research Institute, Dehradun.

Connor, E.F. and E.D. McCoy. (1979). The Statistics and Biology of the species - Area Relationshiop. *American Nationalists*. Volume 13. 791-833.

Cowen, D.J. (1988). GIS versus CAD versus DBMS : What are the differences ? Photogrametric Engineering and Remote Sensing 54:155 1-4.

D'Arcy. W.E., (1910); Forest Working Plan, Govt. Press.

Dale, Earnst., (1976). Organization. American Management Association, New York.

Dale, Reider; (1998), *Evaluation Framework for Development Programme*; Sage Publication, New Delhi.

Department of Environment (DOE), (1987). Handling of Geographical Information. HMSO. London.

Dueker, K, J. (1979). Land Resoruce Information system : A review of fifteen years experience. *Geo-processing* 3 105-28.

Dunster, J.A. (1988), Land use Planning in Canada : An Overview of the Forestry Aspect. *Land Use Policy*. 6:83-93.

Dwivedi, A.P., (1980); *Forestry in India*, Jugal Kishore and Company, Dehradun.

Dwivedi, A.P., (1998); *Forests: The Non-Wood Resources*, International Book Distributors, Dehradun.

Elhance, D.N.; Elhence, Veena and Aggarwal, B.M.; (1996), *Fundamentals of Statistics*; Kitab Mahal, Allahabad.

FAO (1963) World Forest Inventory. FAO. Rome.

FAO (1971), Planning a Forest Inventory.

FAO. (1991), Sustainable Management of Tropical Forests. Anex F. In rep. Committee on Forest Development in the Tropics. Tenth Sessions, 10-13, December, 1991, Rome.

FAO, (1993). The challenge of sustainable forest management : What future for the World's forest? Rome.

F.A.O. (1994) : Reading in Sustainable Forest Management; FAO Forestry Paper No. 122.

FAO, (1996) Wood : World Trends and Prospects. *Unasylva*, 20; 1-135.

Forest Research Institute, Dehradun (1972), GOI, 100 years of Indian Forestry (Volume I &II).

Forest Research Institute, Dehradun (1972), GOI, *Indian Forest Utilization,* Volume II.

Forest Research Institute; (1983) *Abridged Glossary of Technical Terms*; GOI Publication.

Forest Survey of India (1987) State of Forest Report. Ministry of Environment and Forests. Dehradun.

Forest Survey of India, (1995) State of Forest Report - 1995. Ministry of Environment and Forest, Dehradun.

Forest Survey of India, (1999) State of Forest Report - 1999. Ministry of Environment and Forest, Dehradun.

Forest Survey of India, Dehradun (1980-1987); Various Reports on Field Inventories and Vegetation Mapping.

Govt of India (GOI) (1984) Conservation of Biological Diversity in India : An Appraisal. Ministry of Environment and Forests. pp. 13.

Grandin, Barbara. (1988). Wealth ranking in small holders' communities : A field manual. Intermediate Technology Publications, London.

Griffin, Ricky, (1990). *Management,* 3rd Edition. Houghton Mifflin Company, Boston.

Gupta, Tirath; and Guleri, Amar. (1982). Potential Production of Minor Forest produce-An overview. *Indian Forester*. Vol 94(1).

Hamilton, A.C., (1984). *Deforestation in Uganda*. Oxford University Press. Nairobi.

Indian Council of Forestry and Education, Dehradun, (1995), Forestry Statistics in India, Indian Council of Forestry and Education, Dehradun, (1996), *Soil Vegetation and Studies*.

International Tropical Timber Organization (ITTO) (1998) Criteria and Indicators for Sustainable Forest Management of Natural Tropical Forests. Yokahama. Japan.

IUCN *et al*. (1990). Conserving the world's Biodiversity. IUCN/WRI/CI/WWF-US/World Bank.

Jacob, M. (1988). The tropical rain forest. p. 12; 256-258. Berlin-Heidelbury. Germany.

Janjen, D.H. (1975). *Ecologoy of Plants in the Tropics*. Arnold Press. London.

Johnstone, D.R.; Grayson, A.J.; Bradlay, R.T. ; (1967) Faber and Faber Ltd., Nieml xvii.

Khanna, L.S. and Ram Prakash (1979). *Principles and Practice of Silviculture*. Khanna Bandhu. Dehradun.

Koontz, Harold and Weihrich, Heinz; (1998), *Essentials of Management*, Tata McGraw-Hill Publication Company Ltd., New Delhi.

Kotwal, P.C.; and Benerjee, Sujoy; (1998); *Bio-diversity Conservation in Managed Forests and Protected Areas*; Agro-botanical, Bikaner, Rajasthan.

Lal, J.B., (1992), *Forest Ecology*, Natraj Publishers, Dehradun.

Lal, J.B.; (1989); *India's Forests*: *Myth and Reality*; Natraj Publication, Dehradun.

Lanley, J.P. (1982). Tropical Forest Resources, FAO Forestry Paper - 30, Rome.

Mackinon John; Cathey, Child, G. and Thorseli, J. (1986). Managing Protected Areas in the Tropics. IUCN.

Margalef, Raman. (1968). *Perspectives in Ecological Theory*. University of Chicago Press, Chicago, 112pp.

Mather, Alexender S. (1990). *Global Forest Resources*, Belhaven Press. 25 Floral Street, London.

Mathur, V.P., (1981), *Forest Management*; Jugal Kishore and Company, Dehradun.

Ministry of Environment and Forest (MOEF) (1994). Consevation of

Biological Diversity in India. An Approach. Govt of India. New Delhi.

Ministry of Environment and Forests, (MOEF) (1992). Govt. of India: National Conservation Strategy and Policy Statement on Environment and Development (NCSPED), New Delhi.

Misra, R. (1968). Ecology Work Book, Oxford. New Delhi, IBH Publishing Co.

Mukerjee, Amitava & Agnihotri, V.K.; (1993); *Environment and Development*, Concept Publishing Company, New Delhi.

Mukherjee, Neela; (1993). *Participatory Rural Appraisal. Methodology and Application*. Concept Publishing Company, New Delhi.

Mukherjee, Neela; (1997), *Participatory Appraisal of Natural Resources*, Concept Publishing Company, New Delhi.

Mukherjee, Neela; (1997); *Participatory Appraisal of Natural Resources*; Concept Publishing Company, New Delhi.

Myers, N. (1988). "Threatened Biota : Hot-spots' in Tropical Forests." *The Environmentalists*, Volume 8 : 187-208.

Myers, N. (1990). 'The Biodiversity Challenge : Expanded Hot-spots analysis'. *The Environmentalists*, Volume, 10, 243-256.

NAtional Bureau of Soil Survey and Land use Planning (NBSSAUP). (1994) Soil Resource Mapping of M.P. Indian Council of Agricultural Research, Nagpur.

National Commission on Agriculture (NCA) (1976) Ministry of Agriculture and Irrigation, Government of India (1976) : Report on Agriculture (Part ix - Forestry), Governnment of India, New Delhi.

National Remote Sensing Agency. (1984), Landsat Imagery. Department of Space Government of India, Hyderabad.

Noss, R.F. (1990). 'Indicators for Monitoring Biodiversity : A Hierarchical Approach.' *Conservation Biology* vol. 4 : 355-356.

Odum, Eugene P. (1971), *Fundamentals of Ecology*. (Third edition), W.B. Saunders Company USA.

Odum, Eugene, P., (1971), *Fundamentals of Ecology*; Natraj Publication, Dehradun.

Odum, Howard T; Cantlon, J.E.; Kornicker, L.S. (1960). An organizatonal hierarchy postulate for the interpretation of species individuals distribution, species entropy and ecosystem·evolution and the meaning of a species variety index. *Ecology*, 41:395-399.

Osting, H.J. (1958). The study of Plan Communities, W.H. Freeman & and Co, San Fransisco.

Ozemoy, V.M.; Smith, D.R. and Sicharman A. (1981). Evaluating computerized geographic information systems using decision analysis. *Interfaces* 11:92-8.

Pabla, H.S.; Pandey, Sanjeeva; Badola, Ruchi. (1995). Guidelines for Ecodevelopment Planning. Wildlife Institute of India.

Panwar, H.S. (1992). Ecodevelopment; An Integrated Approach to sustainable development for people and Protected Areas in India. Paper presented in the iv world congress on National Parks and PRotected Areas, Carcus, Venezuela.

Parker, H.D. (1988). The Unique qualities of a Geographical Information System : A Commentary : Photogrametric Engineering and Remote sensing 54 (ii) : 1547-49, Part I, Empire Forestry Association, London.

Phillips, E.A. (1959) *Methods of Vegetational Study*. Henry Holt & Co. Inc.

Prakash, Ram and Khanna, L.S.; (1979), *Theory and Practice of Silvicultural Systems*, Periodical Expert Book Agency, Delhi.

Raunikiar, C. (1934). The Life Forms of Plants and Statistical Geography. Clarendon Press. Oxford.

Reid, V. Walter; McNeelay, Jeffrey, A : Tunstall, Daniel, B; Bryant, A Dirk; Winograd, Manuel. (1993) *Biodiversity Indicators for Policy Makers*. WRI, New York.

Rodgers, W.A. and Pawar, H.S. (1990). Planning wildlife in Protected area Networks in India. Vol. I&II. Wildlife Institute of India, Dehradun.

Roy, P.N. (1993). A study on the optimum size of sample plots for forest inventory. *Indian Forester*. !993, pp. 1-10.

Sagreiya, K.P. (1967) *Forest and Forestry*. National Book Trust. India, New Delhi.

Sawarkar, V.B. (1995) A Manual for Planning Woldlife Management in protected area nad managed forests. Wildlife Institute of India.

Shagriya, K.P.; (1967), *Forest and Forestry*; National Book Trust, New Delhi.

Shannon, C.E., and Weaver, W. (1963). The Mathematics Theory of Communication. University of Illinois Press, Urbana, 117 pp.

Simberloff, D. (1986). 'Are we in the verge of a Mass Extinction', ed. D.K. Elliot, pp. 165-180, New York, NY : John Wiley and Sons.

Simpson, E.H. (1934). Measurement of Diversity. *Nature*, 163-688.

Smith, T.R., Menon S., Star J.L. Estes J.E. (1987). Requirement and principles for the implementation and constructions of large-scale geographical information systems, *International Journal of Geographical Information Systems* 1; 13-31.

Trevor, C.G. and E.A. Smithies. (1923). *Practical Forest Management*. Govt. Press, U.P.

Troup, R.S. (1921), *The Silviculture of Indian Trees*. Govt Press.

UNECE/FAO. (1985). The Forest Resources of ECE Region, ECE/FAO, Geneva.

WCED (World Commission on Environment and Development - The Brundtland Commission) (1987). Our Common Future. Report of the WCED. Oxford. Oxford University Press.

Wickramanauake, Ebel. (1994). How to check the feasibility of projects. Bangkok : Asian Institute of Technology.

Wittaker, R.H. (1965). Dominance and Diversity in land Plant communities, *Science*, 147:250-260.

WRI. (1986). World resources 1986. Basic Books, New York.

Index